ORGANIC CHEMISTRY

A FIRST UNIVERSITY COURSE IN TWELVE PROGRAMS

by

F. W. EASTWOOD
M.Sc., D.Phil., A.R.A.C.I.
Reader in Chemistry, Monash University

J. M. SWAN
D.Sc., Ph.D., F.A.A., F.R.A.C.I.
Professor of Organic Chemistry, Monash University

JEAN B. YOUATT
M.Sc., Ph.D.
Senior Lecturer in Chemistry, Monash University

THIRD EDITION

Programs 1 to 6

CAMBRIDGE
AT THE UNIVERSITY PRESS
1972

Published by
The Syndics of the Cambridge University Press
Bentley House, 200 Euston Road, London, N.W.1.
American Branch: 32 East 57th Street, New York, N.Y. 10022.

Published in Australia and New Zealand by
Science Press, Marrickville, N.S.W.

ISBN 0 521 07951 9

PRINTED IN AUSTRALIA
BY
MACARTHUR PRESS, SYDNEY

ORGANIC CHEMISTRY

A FIRST UNIVERSITY COURSE
IN TWELVE PROGRAMS

Programs 1 to 6

PREFACE

Seven years experience has convinced us that programmed learning is a most valuable aid in teaching organic chemistry. Obvious benefits are derived from the step by step account and the frequent testing of comprehension. Less apparent, but important, advantages are that programs aid concentration and give the student a definite objective in private study. It is for the last reason that we have deliberately issued this book as a set of twelve programs, many of which can be completed in an evening's study. The method is not intended to replace a lecture course or the study of conventional texts, but can be used as a supplement to them.

We wish to thank Dr G. R. Inglis of St Andrew's University who allowed us to use his programs in 1965. These convinced us of the value of programmed instruction, and in 1966 we produced a set of programs to cover a large part of our first year syllabus. These were used not only by first year students, but also by second year students for revision. As a result of this testing, the programs were extensively rewritten and published as the first edition of this book. We acknowledge with pleasure many helpful comments received from our colleagues in this department and from Dr P. G. Simpson of the University of Sydney.

The programs cover a typical first-year university course in Organic Chemistry, and a knowledge of chemistry up to this level is assumed. Programs 1 and 2 should be completed first, after which programs 3-11 can be attempted in any order. In our own teaching we are using Program 11 at an early stage. Program 12 draws heavily on the earlier programs. This new edition gives a brief introduction to the sequence rules for naming stereoisomers. Other minor changes have been made throughout the book. In discussing theoretical ideas we have tried to avoid a dogmatic approach by indicating the existence of alternative hypotheses.

We thank all those who have contributed to the production of this third edition, and especially Mr B. A. Baxter, Mr A. Boden, and Mr Grahaeme King.

Chemistry Department
Monash University
Clayton, Victoria,
Australia.
March, 1972

F.W.E.
J.M.S.
J.B.Y.

HOW TO USE THE PROGRAMS

A teaching program is divided into "frames". Each frame provides some new information or consolidates material given earlier, and asks one or more questions designed to test understanding. Place the cover card over the page, and then uncover one frame at a time by moving the card down to the next horizontal line. Write your answers on a separate sheet of paper so that the program can be used more than once. Check your written answer against that given below the line, and proceed to the next frame.

The program is designed to allow you to work at your own pace, and the special Test Frames allow pause for revision. You do not have to complete the program at one sitting, and you may refer back to earlier material if necessary. Understanding and memorising the contents of a program will usually require more than one reading. Between readings, the Test Frames can be studied as a guide to your progress in mastering the subject.

At first sight you may be surprised to find half the pages in this book upside down. We have arranged the material in this way so that as you turn a page, you have to cover only one immediately following page. It has been found that where it is possible to see the material immediately ahead as when working on a left hand page with the program continuing on the right hand page and remaining uncovered, one unconsciously absorbs the answers and the testing aspect of programmed learning is lost.

Many teaching programs break up the subject matter into a large number of small steps, and the same idea may be presented in several different ways. In these programs most frames include sufficient discussion to allow searching questions to be put and the reiteration of factual material is minimised. In this way we have endeavoured to provide intellectual stimulus and maintain interest.

Some examples of "flip cards" are included with the programs. Each card has interrelated facts or ideas on front and back. The cards are examined at random and an attempt is made to recall the material given on the other side. As in programmed learning the answer is immediately checked by turning the card over. Students will find that preparation and use of their own set of flip cards is a valuable aid to memorising.

CONTENTS

Programs 1 to 6 are on the right-hand pages when reading the book from this end and Programs 7 to 12 when reading from the other end.

Program 1: NAMING ORGANIC COMPOUNDS

CONTENTS

NAMING ORGANIC COMPOUNDS

Program 1

In most organic compounds, carbon atoms are found bonded to each other and to hydrogen atoms and, in addition, other elements such as oxygen, nitrogen, sulphur, and halogens may be present. There are very many organic compounds and it is essential to have systematic ways of naming them. It has proved convenient to relate the name of the compound to the type of functional group which the molecule contains. The aim of this program is to introduce a systematic way of naming molecules which has been agreed to by the International Union of Pure and Applied Chemistry (IUPAC). This system provides a means whereby chemists can describe chemical structures without ambiguity. Existing side by side with these systematic names are other semi-systematic names and trivial names, the latter being used initially to identify compounds of unknown structure. It may seem undesirable that the older trivial names persist but to abandon their use completely would render much of the literature incomprehensible. It is therefore necessary to learn two names for many common chemicals—a systematic name and a trivial name.

ALIPHATIC HYDROCARBONS—ALKANES, ALKENES, AND ALKYNES

ALKANES

1. As the name implies, hydrocarbons contain, in addition to carbon, the element. . . . . (1).

---

(1) hydrogen

---

2. If all the bonds in the molecule are single bonds, the compound is said to be saturated and the arrangement of the bonds about the carbon atom is tetrahedral. The simplest hydrocarbon, CH_4, is called methane and its structure can be drawn. . . . . (1).

---

(1)

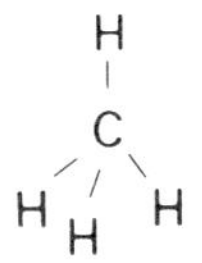

INDEX

3. As the number of carbon atoms increases we form an homologous series with the general formula C_nH_{2n+2}. These compounds are known as *alkanes* (or *paraffins*). Write the formulae and draw the structures of the alkanes which have two carbon atoms (*ethane*) . . . . (1) and three carbon atoms (*propane*) . . . . (2).

---

(1) ethane is C_2H_6

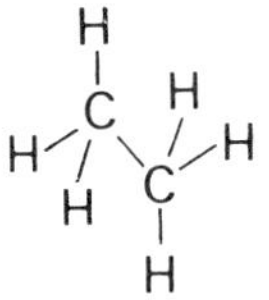

which is abbreviated to CH_3CH_3

(2) propane is C_3H_8

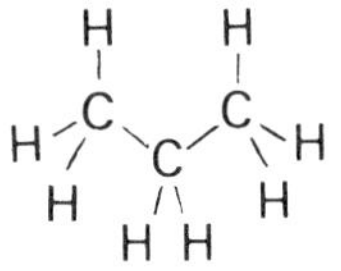

which is abbreviated to $CH_3CH_2CH_3$

---

4. Only one arrangement of the atoms is possible for ethane and propane but for alkanes with four or more carbon atoms, alternative arrangements are possible. Compounds having the same molecular formula but different molecular structure are called *isomers*. These isomers are distinct compounds having their own individual chemical and physical properties. Write the formulae and draw the structures of the alkanes with four carbon atoms . . . . (1) and five carbon atoms . . . . (2). Abbreviated structures may be used.

---

(1) C_4H_{10} $CH_3CH_2CH_2CH_3$

CH_3CHCH_3 (with CH_3 on the central C)

(2) C_5H_{12} $CH_3CH_2CH_2CH_2CH_3$

$CH_3CHCH_2CH_3$ (with CH_3 on C-2)

CH_3CCH_3 (with two CH_3 on the central C)

---

5. We can now see that it is essential to describe molecules accurately. The names given to the linear or unbranched chain alkanes with 1 to 10 carbon atoms are respectively:—*methane, ethane, propane, butane, pentane, hexane, heptane, octane, nonane, decane*. These names must be memorised. They form the basis for naming many other compounds. The simple name is not enough for compounds with more than three carbon atoms. It does not tell us which isomer is involved. In the IUPAC system compounds with branched chains are named as if they were derivatives of shorter chains in which hydrogen atoms have been replaced by *alkyl groups*. The simplest alkyl group is CH_3, known as the *methyl* group. An *ethyl* group would be . . . . (1).

---

(1) CH_3CH_2-

INDEX

6. $CH_3CH_2CH_2CH_2$– is another example of an .... (1) and its specific name is .... (2).

---

(1) alkyl group (2) butyl. The trivial name would be n-butyl. The prefix n- implies that the four carbon atoms are in an unbranched chain.

---

7. The rules for naming branched chain compounds are:—
RULE 1. *Select the longest continuous chain in the molecule and take the name of this alkane.* If the longest continuous chain has six carbon atoms we take the name .... (1); if it has eight carbon atoms we take the name .... (2).

---

(1) hexane (2) octane

---

8. *RULE 2. Name the compound as if it had been derived from the straight chain alkane by replacing hydrogen atoms with alkyl groups.*

$$\begin{array}{l} CH_3CHCH_3 \\ \quad\;\;| \\ \quad\;CH_3 \end{array}$$

The position of the alkyl group is given a number counted from one end of the longest chain. In the compound illustrated the longest chain has .... (1) carbon atoms and the name selected is therefore .... (2). The compound in question has on its second carbon atom a .... (3) group and we call it 2-methylpropane, using the number to show where the methyl group must be placed.

---

(1) three (2) propane (3) methyl

---

9. *RULE 3. Where there is a choice of number to be given to the position of substitution depending on the end from which we count, then the smaller number is given.*

$$\begin{array}{l} CH_3CH_2CHCH_3 \\ \qquad\quad\;| \\ \qquad\;\;CH_3 \end{array}$$

In the compound illustrated, if we numbered from the left the name would be .... (1). If we counted from the right the name would be .... (2). The correct name is .... (3).

---

(1) 3-methylbutane (2) 2-methylbutane (3) 2-methylbutane

---

10. *RULE 4. If more than one group of the same kind is present this is indicated in the name and all the positions are given.*

INDEX

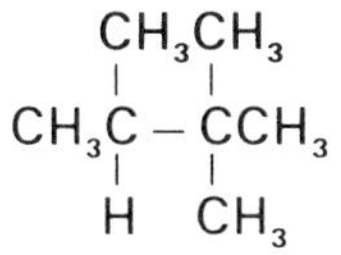

The structure shown is 2,2,3-trimethylbutane. Give the structure of 2,2-dimethylpropane . . . . (1) and 2,3-dimethylbutane . . . . (2).

---

(1) 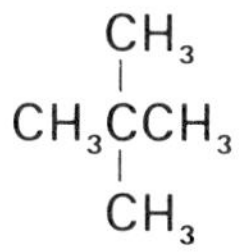

(2) 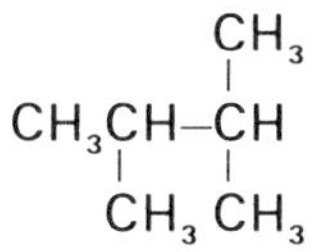 or 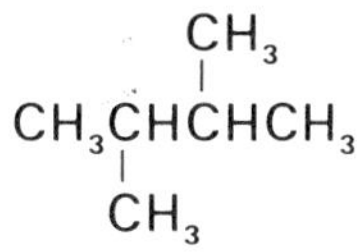 etc.

---

11. *RULE 5. If there are several different alkyl groups present their names are given in alphabetical order.*

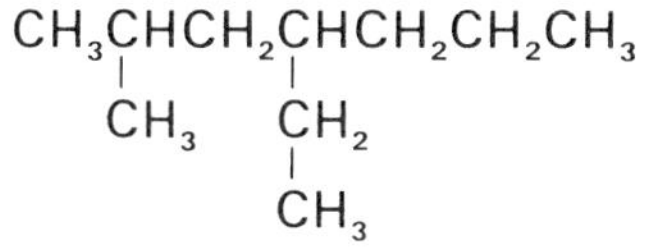

The compound illustrated has . . . . (1) carbon atoms in its longest chain and is named as a derivative of . . . . (2). The substituent . . . . (3) groups are CH_3–, named . . . . (4) and CH_3CH_2–, named . . . . (5). The numbers of the substituent positions will be smaller if we number from the. . . . . (6). The name is . . . . (7).

---

(1) seven (2) heptane (3) alkyl (4) methyl
(5) ethyl (6) left (7) 4-ethyl-2-methylheptane

---

12. Give the structures of:—

2,5-dimethylhexane. . . . . (1), 3-ethyl-2-methylpentane. . . . . (2)
4-ethyl-3,3-dimethylheptane . . . . (3).

Note that in deciding alphabetical order of substituents, the multiplying prefixes di-, tri-, etc., are ignored.

---

(1) 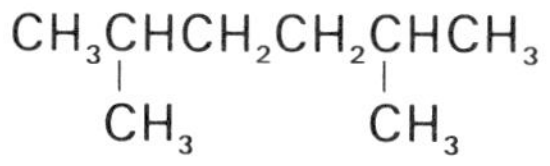

(2) 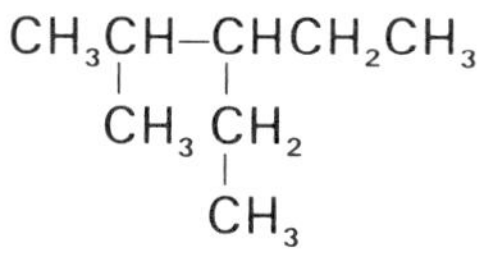

(3) 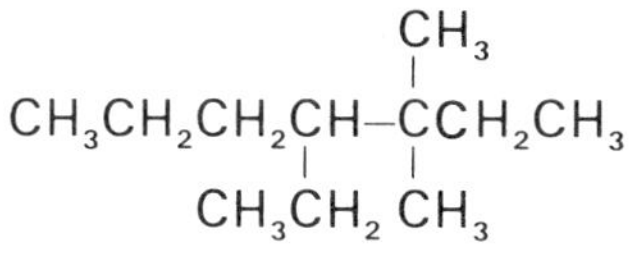

Summarise the synthesis of the racemic form of menthol using the above reaction . . . . (1).

---

(1)

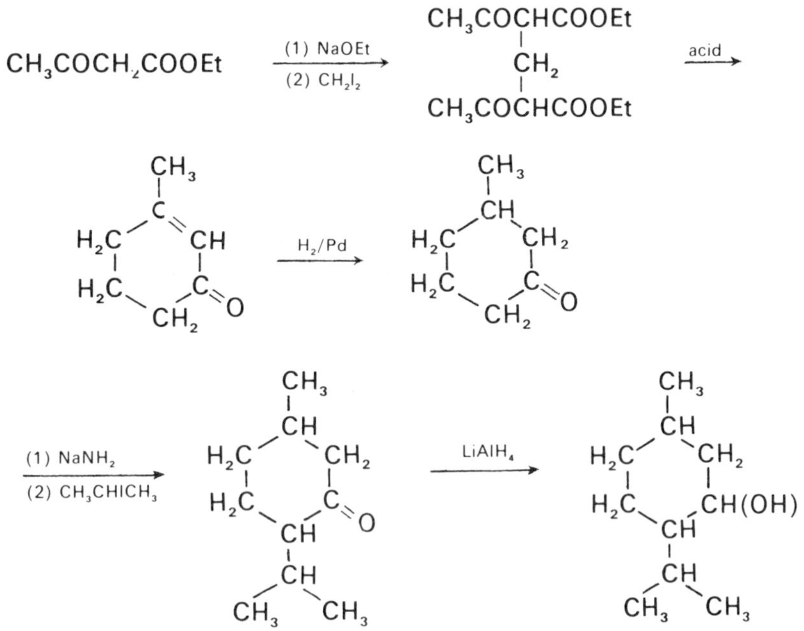

---

61. An alternative route for the synthesis of 3-methylcyclohexanone starts from 3-methylphenol (*m*-cresol). Write down the equations for these reactions . . . . (1).

---

(1)

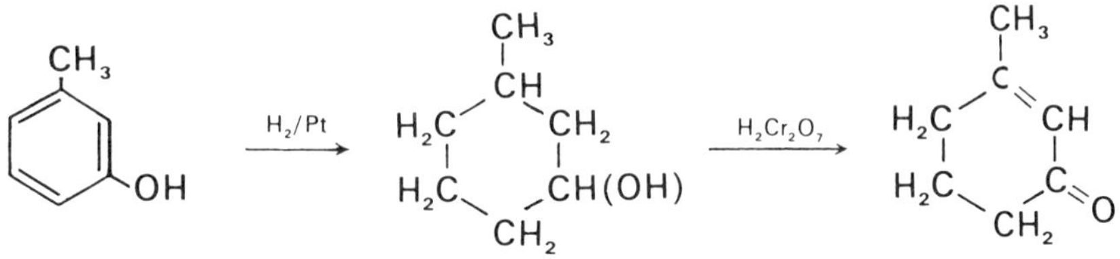

CYCLIC ALKANES

13. The carbon atoms of alkanes can be joined not only in linear and branched open chains but also in ring structures. The names given are e.g. *cyclopropane* for a ring of three carbon atoms, cyclo . . . . (1) for a ring of four carbon atoms and . . . . (2) for a ring of six carbon atoms.

---

(1) cyclobutane 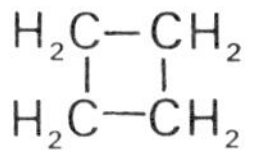 (2) cyclohexane 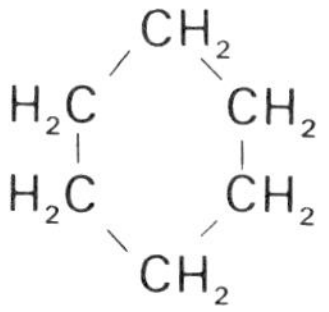

---

14. If we compare a straight chain alkane e.g. hexane, C_6H_{14}, with the corresponding cyclic compound, cyclohexane, it is seen that the cyclic compound has . . . . (1) less hydrogen atoms and its molecular formula is . . . . (2). The general formula for a monocyclic alkane of n carbon atoms is . . . . (3). Write the formulae and names of five cycloalkanes having the formula C_5H_{10} . . . . (4).

---

(1) two (2) C_6H_{12} (3) C_nH_{2n}

(4)

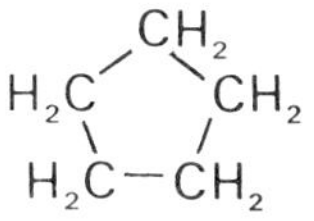

cyclopentane

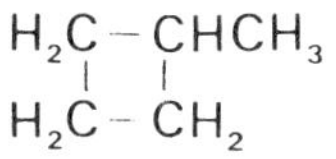

methylcyclobutane

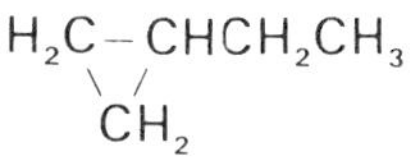

ethylcyclopropane

1,2-dimethylcyclopropane 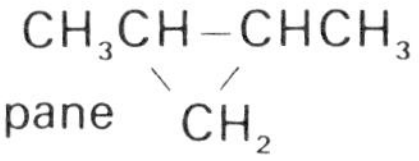

1,1-dimethylcyclopropane 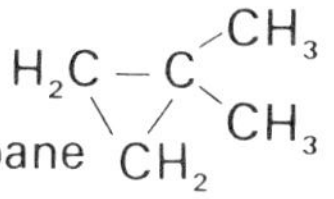

ALKENES

15. Hydrocarbons which contain carbon atoms doubly bonded to each other are called *alkenes* (or olefins). The characteristic functional group of alkenes is >C=C< . For each double bond an alkene has two less hydrogen atoms than the corresponding alkane. The general formula for an alkene with one double bond in an open chain is, therefore . . . . (1).
The names of the alkenes are derived from the name of the corresponding alkane by changing the word ending from *-ane* to *-ene*. The three-carbon hydrocarbon with one double bond has the structure . . . . (2) and is called . . . . (3).

---

57. Consider the reactions of the symmetrical diketone heptane-2,6-dione. Remembering that acetone will condense in the presence of base to give diacetone alcohol, suggest the structure of the compound which is formed when the methyl group at one end of the diketone adds to the carbonyl group at the other. . . . . (1).

(1)

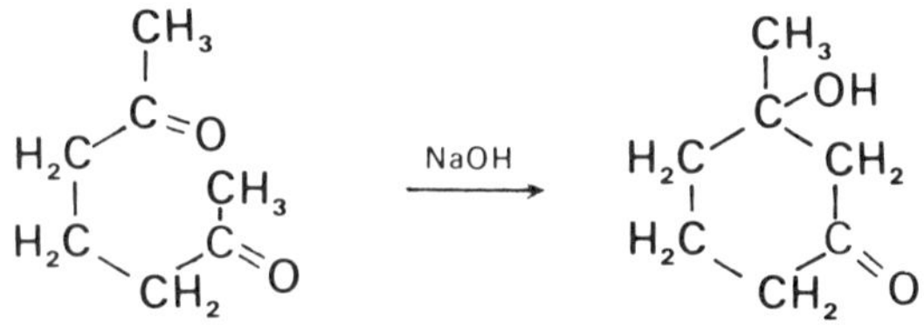

58. Treatment of this compound with acid causes the elimination of water and formation of the product according to the equation. . . . . (1).

(1)

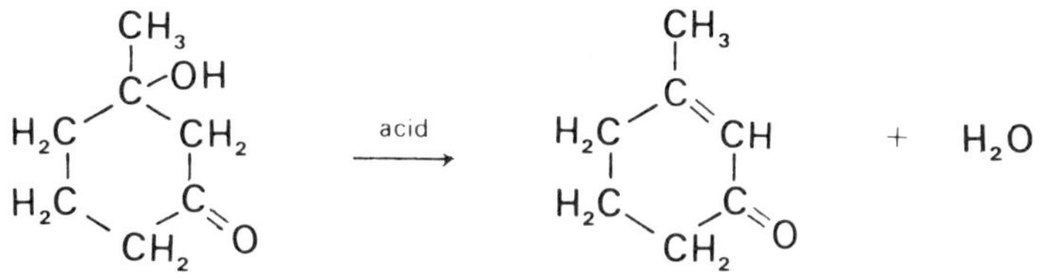

Note that the elimination reaction normally gives a double bond conjugated with the carbonyl group.

59. Careful hydrogenation of this product with hydrogen over palladium results in the reduction of the olefinic bond but not of the ketone according to the equation. . . . . (1).

(1)

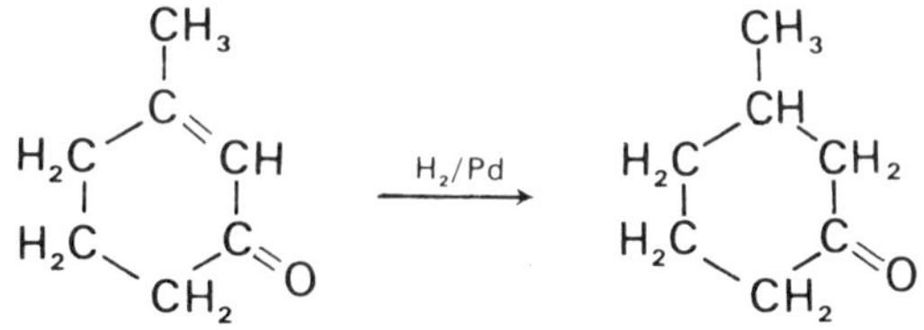

60. The synthesis of 3-methylcyclohexanone and consequently of menthone and menthol is now complete. The number of synthetic steps can be reduced as the intermediate di-ester on treatment with acid undergoes hydrolysis, decarboxylation, acid catalysed aldol addition and elimination of water to give 3-methylcyclohex-2-enone directly.

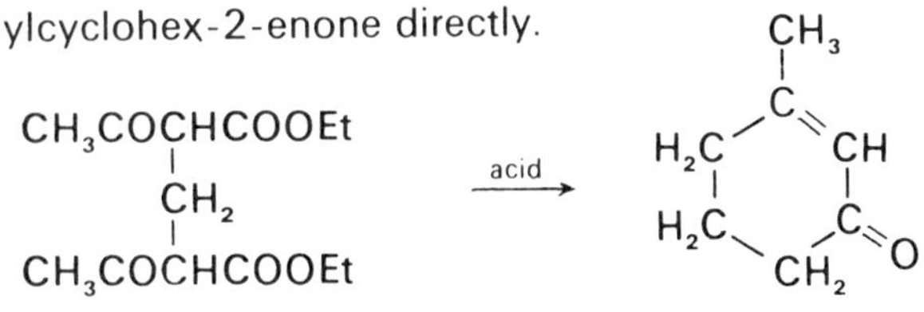

(1) C_nH_{2n} (2) $CH_3CH{=}CH_2$ (3) propene

---

16. Only one structure can be drawn for propene and only one isomer is known. With longer chains more than one position is possible for the double bond and the name must indicate the position. As in naming substituents, we number from the end which gives the smaller number to the position of the double bond and we give the number of the first of the pair of carbon atoms involved in the double bond. $CH_3CH_2CH_2CH{=}CH_2$ is pent-1-ene. When a number of substituents have to be named, the number which gives the position of the double bond is placed next to the *-ene* ending, e.g., 4,4-dimethylpent-1-ene. Draw and name the isomeric alkenes which have four carbon atoms and one double bond . . . . (1).

---

(1) (a) $CH_3CH_2CH{=}CH_2$ but-1-ene (b) $CH_3CH{=}CHCH_3$ but-2-ene (c) 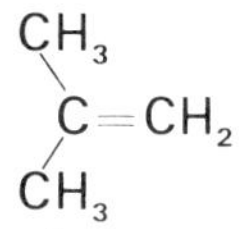 2-methylpropene.

In American usage (a) is 1-butene and (b) is 2-butene.

---

17. The simplest alkene is $H_2C{=}CH_2$ and is called, systematically . . . . (1). The common name of this compound is ethylene.

---

(1) ethene

---

18. A branched chain hydrocarbon which contains a double bond is named as a substituted alkene and *the numbering is determined by the longest chain which contains the double bond.* The chain is numbered from the end which gives the smaller number to the double bond. In the compound illustrated the

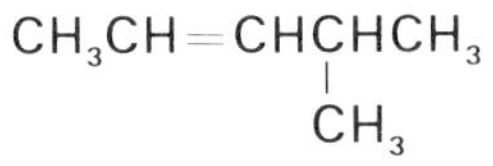

longest straight chain which contains the double bond has . . . . (1) carbon atoms and the name is therefore based on . . . . (2). The smaller number is given to the double bond by numbering from the . . . . (3) and the name of the parent chain is then . . . . (4). We also have to show the substituent alkyl group which is a . . . . (5) group on carbon atom . . . . (6). The name is therefore . . . . (7).

---

(1) five (2) pentene (3) left (4) pent-2-ene
(5) methyl (6) four (7) 4-methylpent-2-ene

form of 3-methylcyclohexanone from ethyl acetoacetate . . . . (1).
(Hint: an intermediate on the route is heptane-2,6-dione).

---

(1)

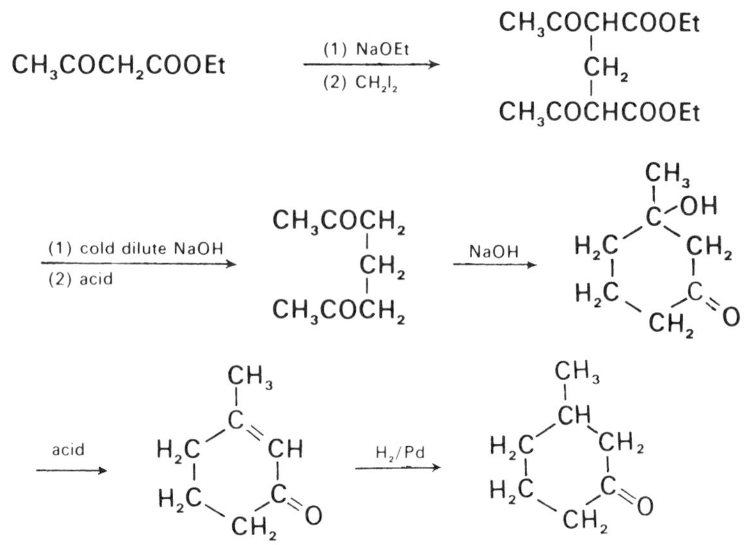

These reactions are discussed in the frames below

---

55. This synthesis of 3-methylcyclohexanone depends on the availability of heptane-2,6-dione. One synthetic application of ethyl acetoacetate results in the conversion RX $\longrightarrow$ RCH_2COCH_3 so that to produce heptane-2,6-dione this conversion has to be done twice on a one-carbon compound.

$$CH_3COCH_2\,\mathbf{CH_2}\,CH_2COCH_3$$

Diiodomethane (methylene iodide) reacts with two moles of sodium ethyl acetoacetate according to the equation . . . . (1).

---

(1)

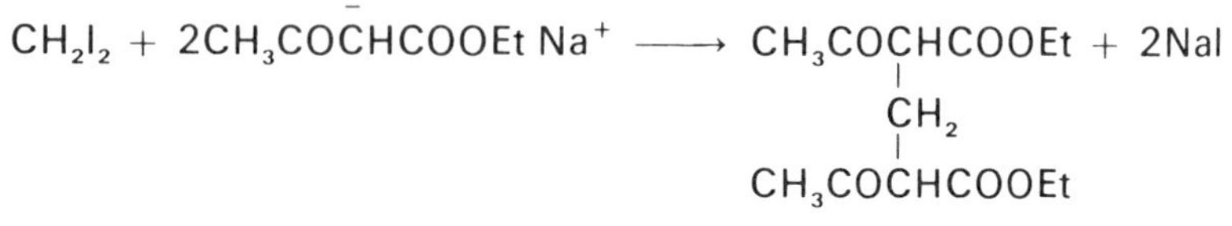

---

56. The resulting diester can be converted into heptane-2,6-dione by treatment with cold dilute sodium hydroxide solution followed by acidification according to the equation . . . . (1).

---

(1)

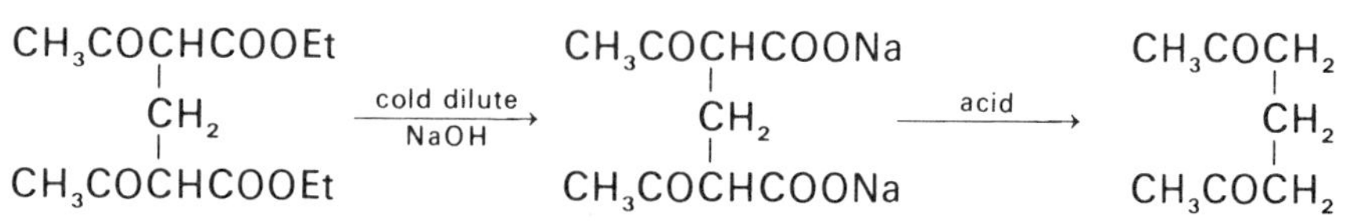

19. Write the name of 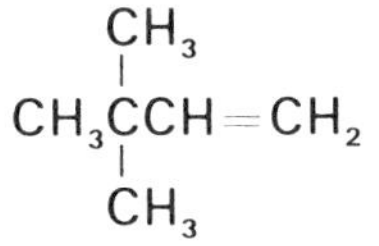 . . . . (1)

---

(1) 3,3-dimethylbut-1-ene

---

20. Give the structure of 2,3-dimethylbut-2-ene . . . . (1) and 3,6-dimethyloct-1-ene . . . . (2).

---

(1) 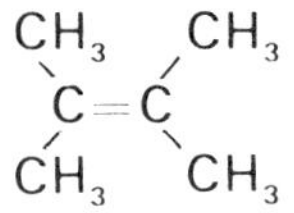

(2) 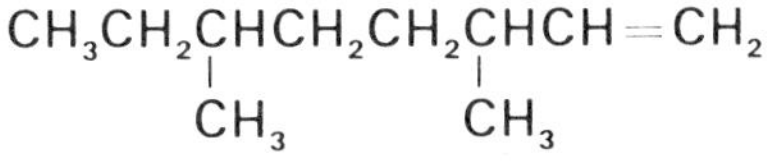

---

21. 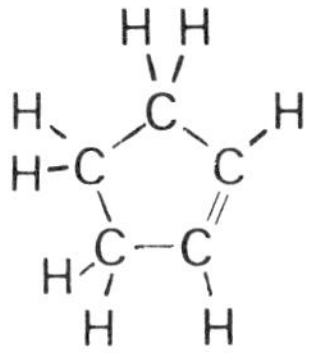 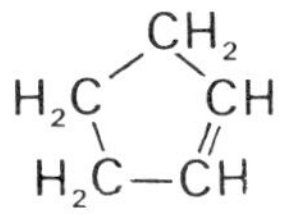

The rules already given can be applied to cyclic alkenes. By analogy with the cyclic alkanes, cyclopentene must be as shown. The compound is named for the C = C functional group and the smallest number which can be given is 1. Carbon atom 2 is the second atom involved in the double bond. Give the structure of 3-methylcyclopent-1-ene . . . . (1).

---

(1) 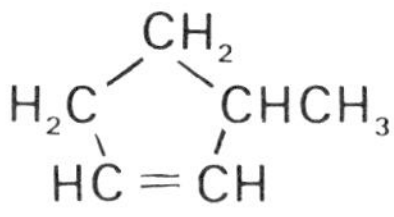

---

22. If two double bonds are present in a compound it is known as a *diene*. A four-carbon chain with 2 double bonds is called *butadiene*. Give the structure of buta-1,3-diene . . . . (1).

---

(1) $CH_2{=}CHCH{=}CH_2$.   If correct go to frame 24.

---

23. The naming of the diene is an extension of the rules already given. The *diene* in the name tells us that there are 2 double bonds and the *buta* that there are 4 carbon atoms in the chain. The two numbers define the positions of the double bonds. Number 1 tells us that there is a double bond between carbon atoms 1 and 2 and number 3 tells us that there is a double bond between carbon atoms 3 and 4. Write the structures of 2-methylpenta-1, 3-diene . . . . (1) and penta-1,4-diene . . . . (2).

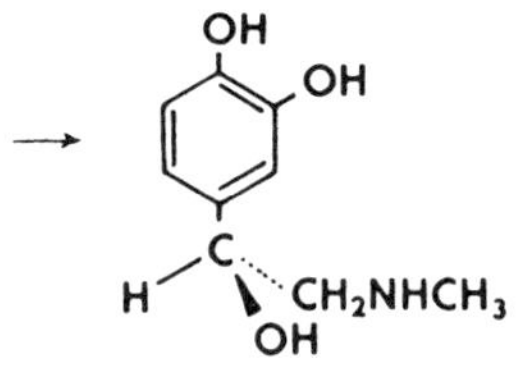

(−)adrenalin

Menthol

53. Menthol is present in peppermint oil and its taste and smell will be familiar to most people. It is related in structure to menthone, also present in peppermint oil, as shown by the equation below. Note that in these formulae the stereochemical complexities are ignored.

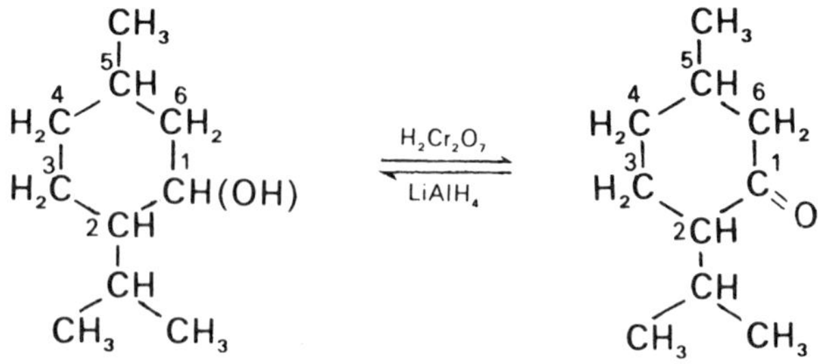

menthol menthone

Menthol (2-isopropyl-5-methylcyclohexanol) contains three asymmetric carbon atoms, namely carbon atoms numbered . . . . (1), while menthone (2-isopropyl-5-methylcyclohexanone) contains . . . . (2) asymmetric carbon atoms.

---

(1) 1, 2, 5 (2) two

---

54. Provided the racemic form of 3-methylcyclohexanone is available, it is possible to synthesise the racemic form of menthone by the following process:

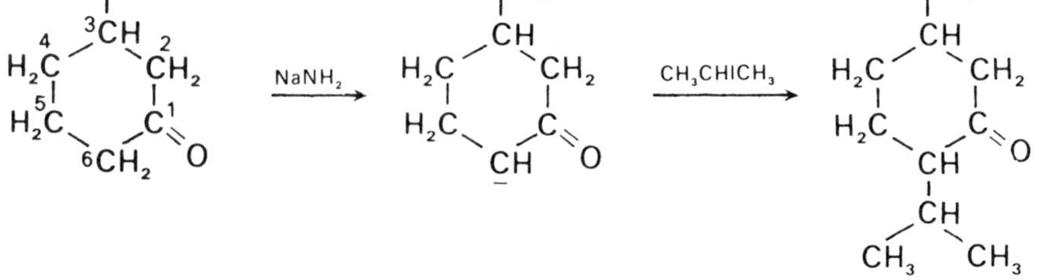

While in theory the sodium amide could also remove a proton from the number 2 carbon atom, it is found in practice that it is possible to direct alkylation largely towards carbon atom 6. Suggest a synthesis of the racemic

(1) $\overset{CH_3}{\overset{|}{CH_2{=}CCH{=}CHCH_3}}$ (2) $CH_2{=}CHCH_2CH{=}CH_2$

---

24. In buta-1,3-diene we have a pattern of alternating single and double bonds (C=C–C=C) which is called a conjugated system. In penta-1,4-diene the double bonds are said to be isolated (i.e., separated by more than one single bond). Penta-1,3-diene shows .... (1) double bonds.

---

(1) conjugated

ALKYNES

25. Alkynes are compounds which contain the functional group –C≡C–. The name ending which indicates the presence of the triple bond is *-yne*. The simplest alkyne is HC≡CH which is called .... (1) in the IUPAC system and has the common name *acetylene*. Draw and name the alkynes with three and four carbon atoms .... (2). The same rules apply for the naming of branched chain alkynes which have been used for alkenes and alkanes.

---

(1) ethyne (2) propyne $CH_3C{\equiv}CH$
but-1-yne $CH_3CH_2C{\equiv}CH$
but-2-yne $CH_3C{\equiv}CCH_3$

TEST FRAMES

A saturated unbranched hydrocarbon chain of ten carbon atoms is called .... (1) and has formula .... (2). A cyclic saturated hydrocarbon of seven carbon atoms is called .... (3) and the molecular formula is .... (4). The name ending for a compound with one double bond is .... (5) and for two double bonds is .... (6). What is the arrangement of conjugated double bonds? .... (7). The characteristic name ending for a compound with a triple bond is .... (8).

---

(1) decane (2) $C_{10}H_{22}$ (3) cycloheptane (4) C_7H_{14}

(5) -ene (6) -diene (7) –C=C–C=C– (8) -yne

Draw and name the two compounds which have the molecular formula C_3H_6 .... (1) and .... (2) and give names to the following

ring. It is consequently necessary to add some functional group to the C_2 unit before acylation. The reagent of choice is chloroacetyl chloride, $ClCH_2COCl$. The structure of the acylation product resulting from step (b) in the synthesis (cf. Frame 55) is. . . . . (1).

---

(1)

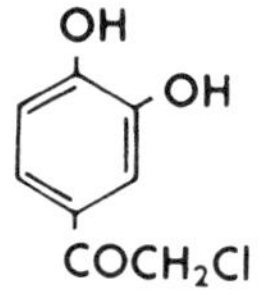

---

50. Reaction of this product with a large excess of methylamine gives. . . . . (1).

---

(1)

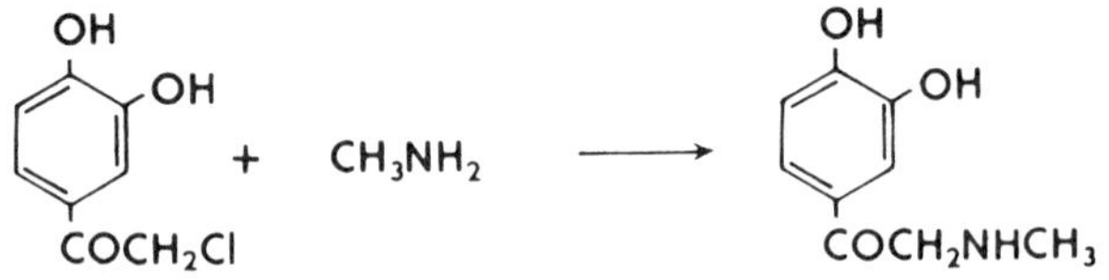

---

51. Finally, reduction of the carbonyl group with hydrogen and platinum proceeds according to the equation. . . . . (1).

---

(1)

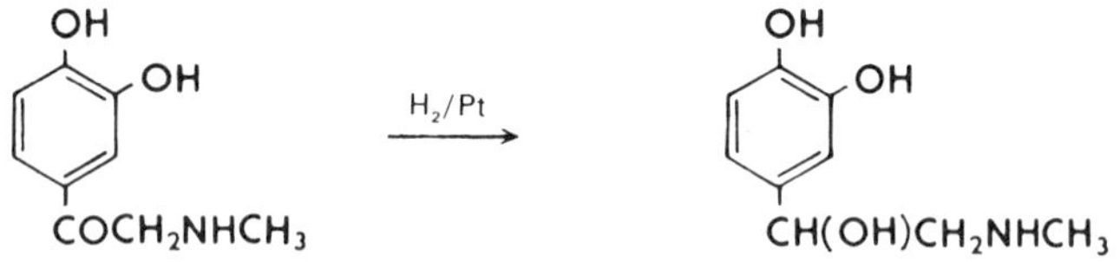

---

52. Summarise the synthesis of (–)adrenalin starting from phenol . . . . (1).

---

(1)

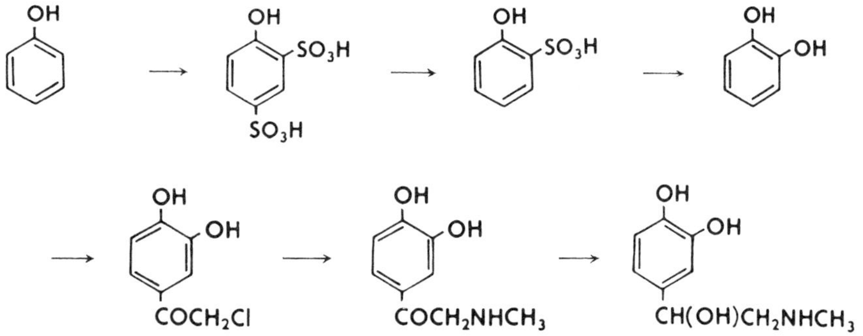

compounds: C_3H_8 . . . . (3), 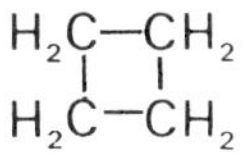 . . . . (4) and

(5) 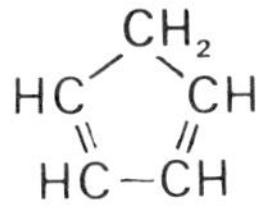

---

(1) $CH_3CH{=}CH_2$ propene (2) 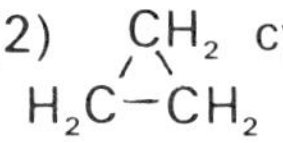 cyclopropane (3) propane

(4) cyclobutane (5) cyclopenta-1,3-diene

Branched chain compounds are named as if the hydrogen atoms of the longest straight chain had been substituted by . . . . (1) groups. CH_3 is a . . . . (2) group. The pentyl group is . . . . (3).

---

(1) alkyl (2) methyl (3) $CH_3CH_2CH_2CH_2CH_2-$

Give the structural formulae of the following compounds:

(1) 5-methylhex-1-yne (2) 2,2,4-trimethylpentane

(3) 4-ethyl-2,4-dimethylheptane (4) 4-methylcyclohexene

---

(1) 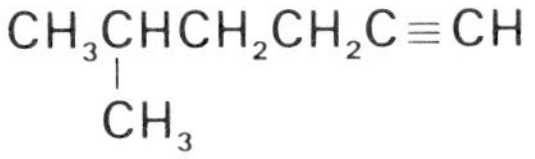

(2) 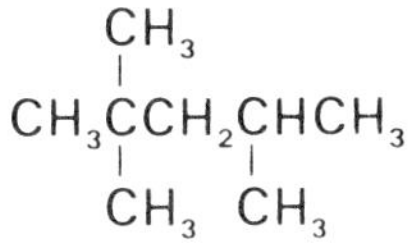

(3) 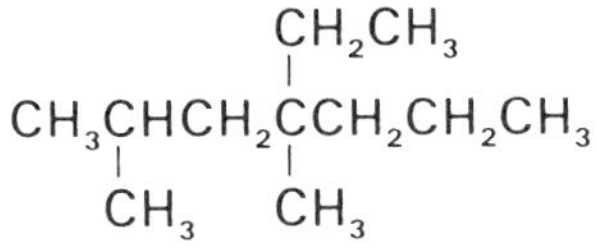

(4) 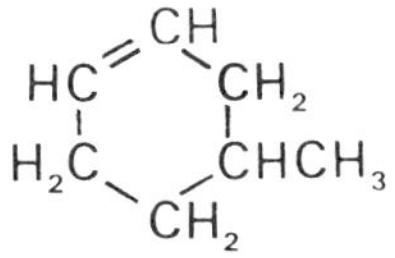

ALKYL GROUPS AND SUBSTITUTED ALKYL GROUPS

26. In this program we have named some of the hydrocarbons as substituted compounds. $(CH_3)_2CHCH_3$, which is isobutane, is named systematically 2-methylpropane. For some alkyl groups which are in frequent use the following abbreviations may be encountered:

CH_3-	methyl	Me
CH_3CH_2-	ethyl	Et
$CH_3CH_2CH_2-$	propyl	Pr
$(CH_3)_2CH-$	isopropyl	i-Pr
$CH_3CH_2CH_2CH_2-$	normal butyl	Bu

47. Separation of this mixture of sulphonic acids is difficult and advantage is taken of the reversibility of sulphonation evident in the following reaction

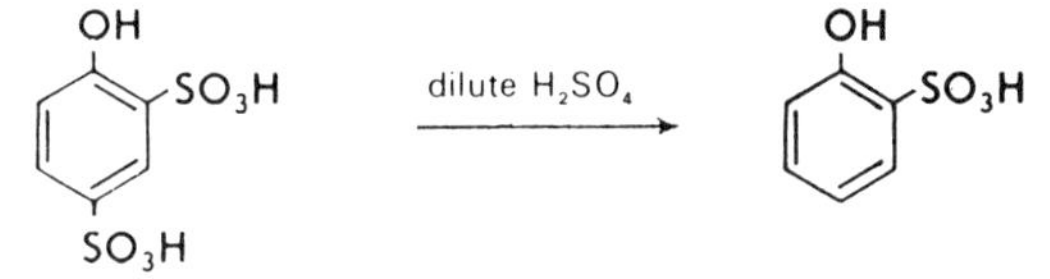

On the basis of this reaction, suggest an efficient synthesis of catechol . . . . (1).

---

(1)

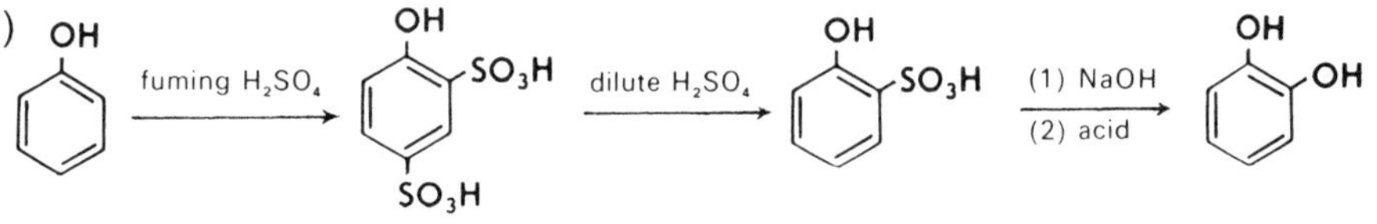

---

48. Methods available for attaching a carbon substituent to a benzene ring include Friedel-Crafts alkylation and acylation. The position of attachment is controlled by the substituents already present. Suggest structures for the products of alkylation and acylation of catechol with the two-carbon reagents $CH_3CH_2Cl/AlCl_3$ and $CH_3COCl/AlCl_3$ respectively. . . . . (1).

---

(1)

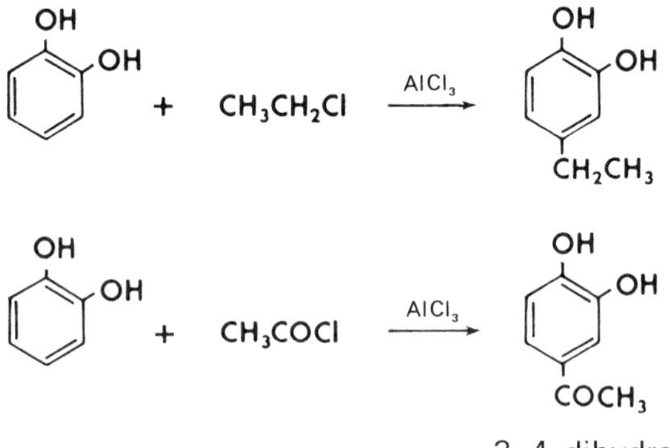

3, 4-dihydroxyacetophenone

The hydroxyl group is *o, p*-directing and the *para*- isomer predominates. Note that attack at the position *para* to either of the two hydroxyl groups leads to the same compound.

---

49. In adrenalin the carbon atom attached to the benzene ring carries a hydroxyl group so that acylation is the obvious choice. Comparison of the structure of 3,4-dihydroxyacetophenone with that of adrenalin shows that a functional group is also needed on the second carbon atom of the side chain. Methyl ketones are susceptible to, say, bromination at the α-carbon atom but any attempt at such a reaction would lead to further substitution in the benzene

$(CH_3)_2CHCH_2-$	isobutyl	i-Bu
$CH_3CH_2CH(CH_3)-$	secondary butyl	s-Bu
$(CH_3)_3C-$	tertiary butyl	t-Bu

Halogen substituents. The hydrogen atoms of hydrocarbons may be substituted by halogen atoms to form compounds of different chemical behaviour. The systematic way of naming these compounds is to use fluoro-, chloro-, bromo- or iodo- as a prefix to the name of the parent hydrocarbon. Thus CH_3Cl (methyl chloride) is named *chloromethane;* CH_2Cl_2 (methylene chloride) is *dichloromethane;* $CHCl_3$ (chloroform) is *trichloromethane* and CCl_4 (carbon tetrachloride) is *tetrachloromethane.* The positions of the halogen atoms must be numbered as soon as any ambiguity can occur. There is only one possible structure for chloroethane but two structures for dichloroethane. Write these two structures and name them . . . . (1).

---

(1) CH_3CHCl_2 1,1-dichloroethane CH_2ClCH_2Cl 1,2-dichloroethane

---

27. The alkyl groups listed in Frame 26 are also used to give semi-systematic names to monohalogen compounds. Tertiary-butyl bromide is . . . . (1) and is named systematically . . . . (2).

---

(1) $(CH_3)_3CBr$ (2) 2-bromo-2-methylpropane

---

28. When a compound has a *nitro* group (NO_2) as a substituent, *nitro-* is used as a prefix to the name of the parent hydrocarbon. CH_3NO_2 is *nitromethane.* Give the structure of 2-methyl-2-nitropropane . . . . (1) and of 1-nitropropene . . . . (2).

---

(1) $CH_3C(CH_3)_2NO_2$ (drawn as CH_3CNO_2 with CH_3 above and below the central C)

(2) $CH_3CH{=}CHNO_2$ or $CH_3CH{=}CH(NO_2)$

Other kinds of functional group which occur as substituents will be introduced later in the program. They include $-OH$, the hydroxy group; $-COOH$, the carboxyl group; $-NH_2$, the amino group; and $-CN$, the cyano group.

properties, can be resolved by treating it with an optically active acid and separating the diastereomeric salts. In this case (+)tartaric acid is a convenient resolving agent and the diastereomers can be separated by virtue of their different solubilities.

Decomposition of the separated diastereomeric salts allows the recovery of (–)adrenalin, (+)adrenalin, and (+)tartaric acid.

---

44. The synthesis of the racemic form of adrenalin can be considered in three steps:
(a) preparation of *o*-dihydroxybenzene (catechol) .... (1);
(b) addition of a two-carbon fragment having substituents suitable for further manipulation; and (c) attachment of the basic nitrogen atom and adjustment of the oxidation level to that required. Suggest a synthesis for adrenalin based on these three steps and check your suggestion against the following frames.

---

(1) 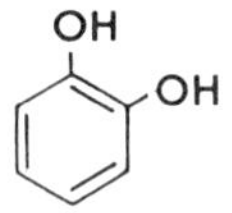

---

45. Catechol can be made either from *o*-chlorophenol by the following reaction

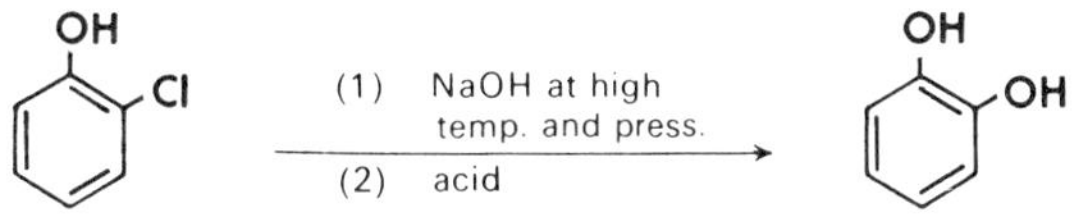

or from *o*-hydroxybenzenesulphonic acid according to the equation .... (1).

---

(1) 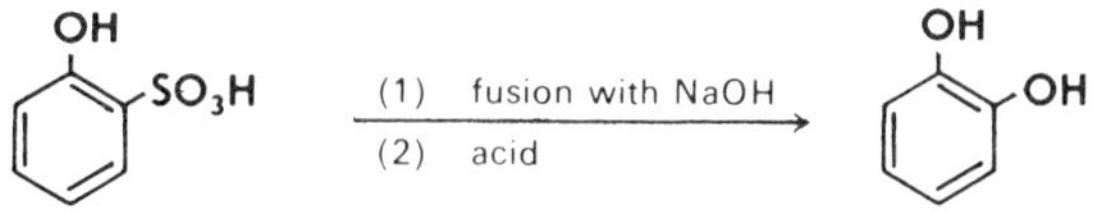

---

46. *o*-Chlorophenol can be prepared by chlorination of phenol although it must be separated from *p*-chlorophenol which is formed in the reaction. Similarly sulphonation of phenol leads to a mixture of isomers according to the equation............ (1).

---

(1) 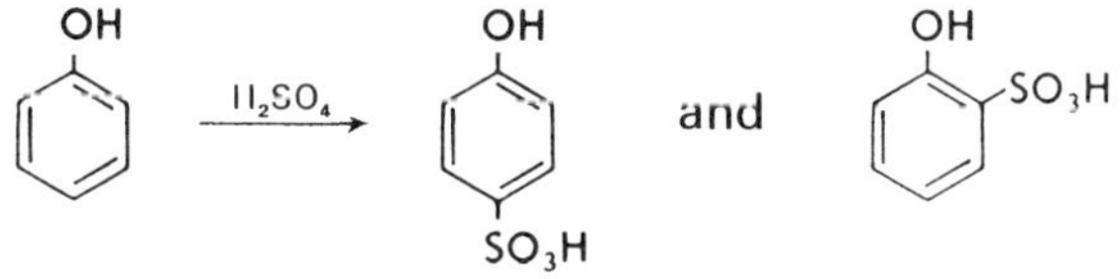

AROMATIC HYDROCARBONS (ARENES)—BENZENE AND BENZENE DERIVATIVES

29. Benzene, C_6H_6, is a stable substance, with chemical properties which distinguish it sharply from unsaturated aliphatic compounds such as alkenes and alkynes. Benzene and other compounds having similar structural features are known as "aromatic" substances. There are two common ways, I and II. of representing benzene.

I

which is shorthand for

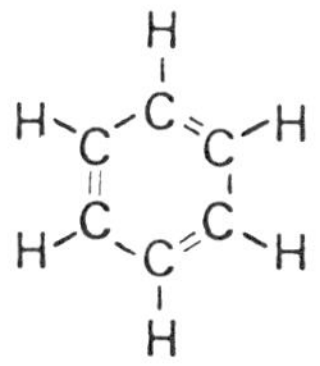

II

which is shorthand for

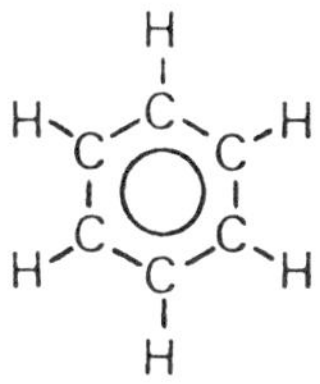

(the circle represents a total of six bonding electrons)

It is beyond the scope of this program to discuss the valency aspects of the structure of benzene. Nevertheless it is essential to an understanding of the nomenclature of benzene compounds to be aware of the symmetry of the benzene molecule. Because the molecule is symmetrical and all the carbon and hydrogen atoms lie in a plane, the replacement of one hydrogen atom by a methyl gives rise to .... (1) isomer(s).

---

(1) One. It is immaterial how we draw this.

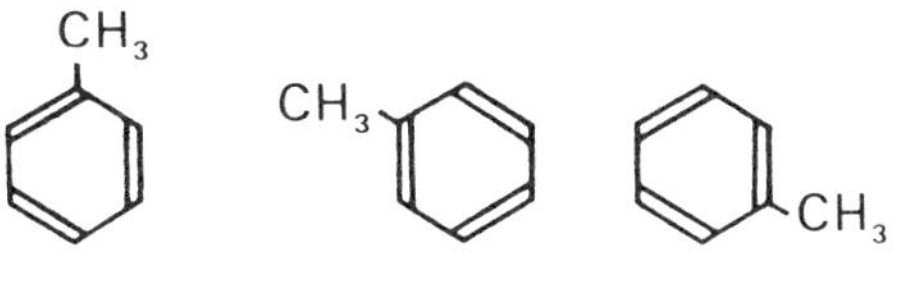

---

30. The systematic name of methyl-substituted benzene is *methylbenzene,* but the name *toluene* is in common use and accepted in the IUPAC system for this arene.
Benzene and all related compounds containing the structural feature of a planar ring of six trigonal carbon atoms, i.e., each carbon atom in the ring attached to three other atoms, are called *arenes*. Arenes are commonly referred to as "aromatic compounds" (cf. Frame 29).
Draw the structure of ethylbenzene .... (1), bromobenzene .... (2) and nitrobenzene .... (3).

---

(8) $C_6H_5CH_2Cl \xrightarrow{(CH_3)_3N} C_6H_5CH_2\overset{+}{N}(CH_3)_3\ Cl^-$

(9) $C_6H_5COCH_3 \xrightarrow{LiAlH_4} C_6H_5CH(OH)CH_3 \xrightarrow{acid} C_6H_5CH{=}CH_2$

$\xrightarrow[\text{peracetic acid}]{CH_3COOOH} C_6H_5CH{-}CH_2$ (epoxide: CH and CH_2 bridged by O)

EXAMPLES OF MORE COMPLEX SYNTHESES

Students completing these programs are not expected to be able to cope with syntheses as complex as the following examples. They are included merely to show how the reactions which have been discussed can be used in sequence to build up complex structures.

Adrenalin

43. The hormone adrenalin is a normal constituent of blood and its formation is controlled by the adrenal gland. An increase in concentration of adrenalin in the blood stream causes dilation of the pupil of the eye, strengthening of the heart beat, and an increase in blood pressure by stimulation of the post-ganglionic sympathetic nerve fibres. Adrenalin is an asymmetric molecule and only one of the enantiomers, (−)adrenalin, possesses the above properties. The structure of adrenalin is

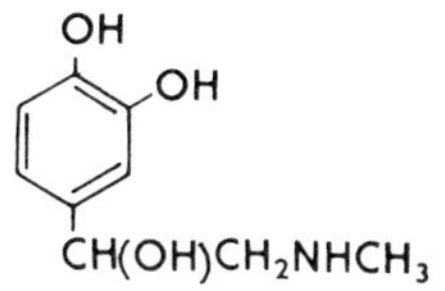

Resolution of the racemic form of the compound can be done as the last stage of the synthesis. Draw the structures of the enantiomers . . . . (1) and suggest a method for their resolution . . . . (2).

---

(1)

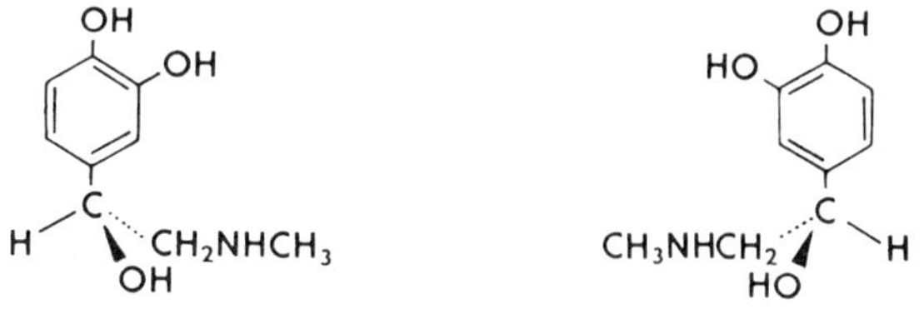

(−)adrenalin (+)adrenalin

Note that rotation of the benzene ring about the C-C bond does not affect the asymmetry.

(2) Racemic adrenalin, which is an amine and therefore possesses basic

(1) 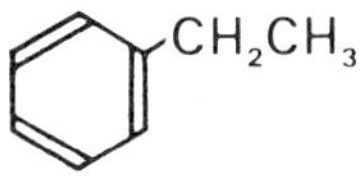

(2) 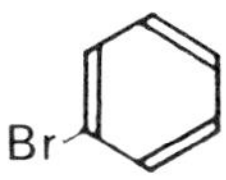

(3) 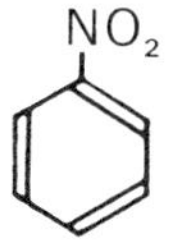

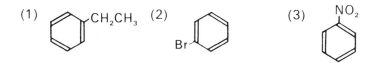

31. When we introduce a second methyl group into toluene, to give rise to what are commonly called xylenes, there are five ring hydrogen atoms which it might substitute but we do not find five isomers. The number 1 is given to the carbon atom which holds the first methyl group. Now, because the

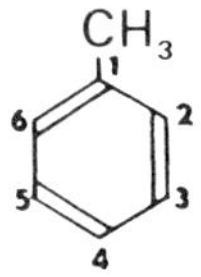

molecule is symmetrical, we cannot distinguish between positions 2 and 6. If the new methyl group substitutes for either of the hydrogen atoms on the carbon atoms 2 and 6 the products are identical and the name given is *ortho*-dimethyl benzene. The term *ortho*- or *o*- thus signifies that the two substituents are on adjacent carbon atoms. *o*-Nitrotoluene has the structure . . . . (1).

---

(1) 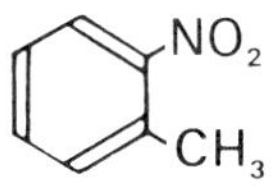

---

32. 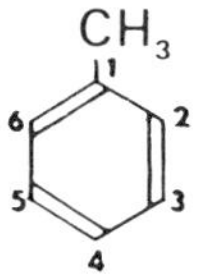

Again in the substitution of toluene we cannot distinguish between the positions numbered 3 and 5 and the 1,3-dimethyl compound is called *meta*-dimethylbenzene or *m*-dimethylbenzene. *m*-Dinitrobenzene has the structure . . . . (1).

---

(1) 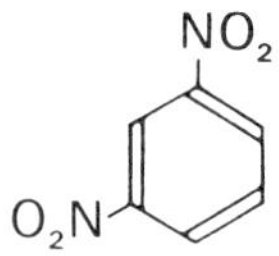

---

33. The third isomer of dimethylbenzene, in which the methyl groups are in the 1 and 4 positions, is called *para*-dimethylbenzene. The structure of *p*-nitrotoluene is . . . . (1).

---

(1) 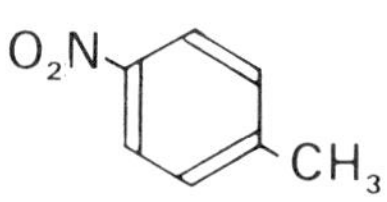

---

34. When we have compounds with three or more substituents on the benzene ring, numbers are used to indicate the positions. Thus 1,2,4-tribromobenzene is . . . . (1).

---

aniline . . . . (4); methyl phenyl ketone (acetophenone) from acetic acid and benzene . . . . (5); benzonitrile from aniline, benzoic acid, or benzaldehyde . . . . (6); 4-phenylbutan-2-one from ethyl acetoacetate . . . . (7); benzyltrimethylammonium chloride from benzyl chloride . . . . (8); phenylethene oxide (styrene oxide) from acetophenone . . . . (9).

---

(1) 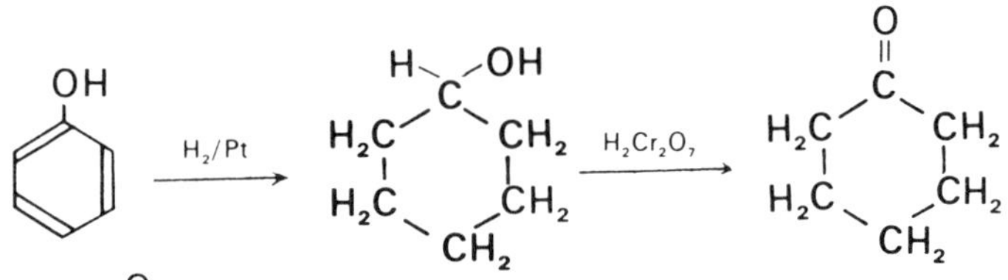

(2) 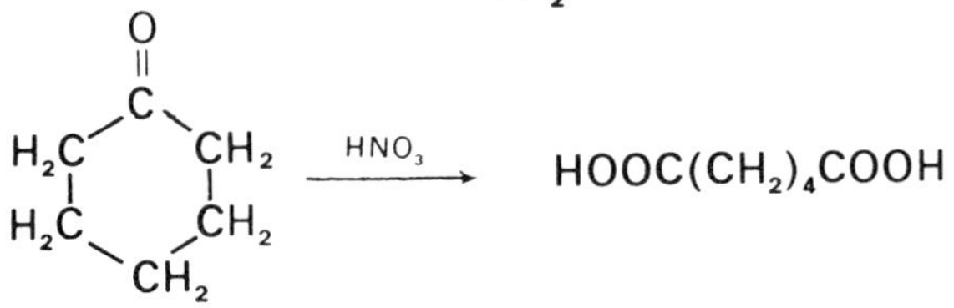

(3) $2CH_3CHO \xrightarrow[\text{dilute}]{\text{NaOH}} CH_3CH(OH)CH_2CHO \xrightarrow{\text{acid}} CH_3CH{=}CHCHO$

(4) 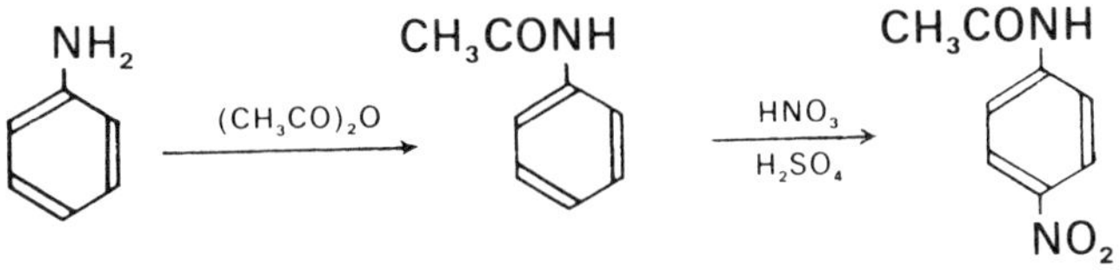

(5) $CH_3COOH \xrightarrow[\text{SOCl}_2]{\text{PCl}_5\text{ or}} CH_3COCl$

$CH_3COCl + C_6H_6 \xrightarrow{\text{AlCl}_3} CH_3COC_6H_5$

(6) $C_6H_5NH_2 \xrightarrow[\text{HCl}]{\text{HONO}} C_6H_5N_2^+Cl^- \xrightarrow{\text{CuCN}} C_6H_5CN$

or

$C_6H_5COOH \xrightarrow{\text{SOCl}_2} C_6H_5COCl \xrightarrow{\text{NH}_3} C_6H_5CONH_2 \xrightarrow{\text{P}_4\text{O}_{10}} C_6H_5CN$

or

$C_6H_5CHO \xrightarrow{\text{H}_2\text{NOH}} C_6H_5CH{=}NOH \xrightarrow{(\text{CH}_3\text{CO})_2\text{O}} C_6H_5CN$

(7) $CH_3COCH_2COOEt \xrightarrow[\text{(2) }C_6H_5CH_2Cl]{\text{(1) NaOEt}} CH_3COCH(CH_2C_6H_5)COOEt$

$CH_3COCH(CH_2C_6H_5)COOEt \xrightarrow[\text{(2) acid}]{\text{(1) cold dilute NaOH}} C_6H_5CH_2CH_2COCH_3$

(1) 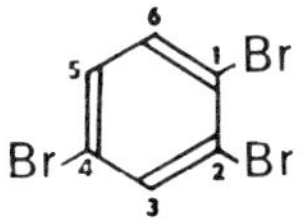

35. A benzene ring substituted with a methyl group and a bromine atom would be named systematically as a bromomethylbenzene. It is more often named as a *bromotoluene*. If the compound is named as a substituted toluene, the number 1 is given to the carbon atom holding the methyl group. Draw the structure of 2,6-dinitrotoluene . . . . (1) and 4-bromo-3-chloronitrobenzene . . . . (2).

(1) 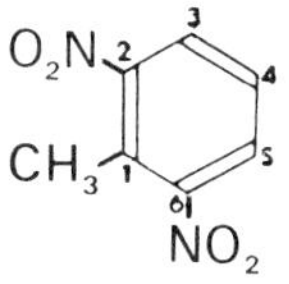

(2) 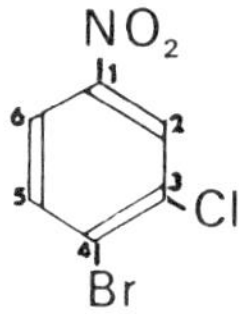

36. 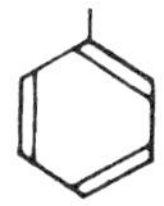

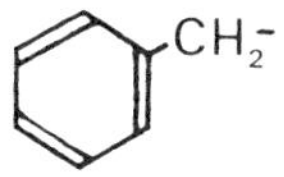

Occasions also arise when the benzene ring is regarded as a substituent of a larger molecule. The substituting group C_6H_5– is called a *phenyl* group. The name *benzyl* is given to $C_6H_5CH_2$–. This nomenclature is a frequent source of confusion but it is the accepted convention and must be learnt. Draw the structures of 2-phenylbutane . . . . (1) and 3-benzylpentane . . . . (2).

(1) 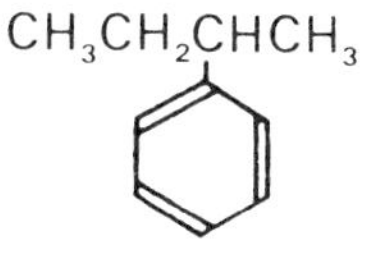

(2) 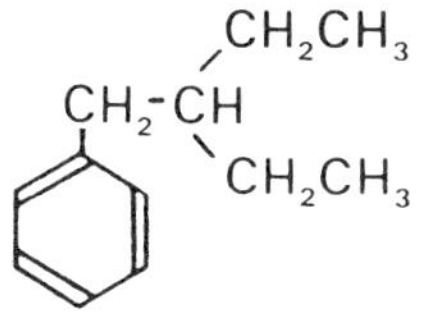

TEST FRAMES

Draw and name the arenes which can be obtained by the disubstitution of benzene with bromine, i.e. two hydrogen atoms are replaced by bromine atoms.

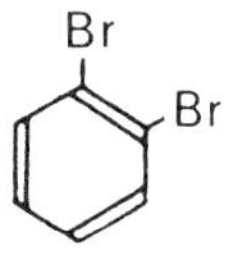

ortho-dibromobenzene

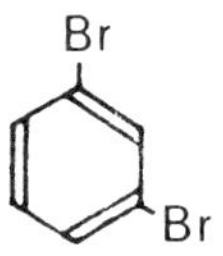

meta-dibromobenzene

para-dibromobenzene

(3) $CH_3CH_2I \xrightarrow{CH_3CO\bar{C}HCOOEt\ Na^+} CH_3COCH(CH_2CH_3)COOEt$

$CH_3COCH(CH_2CH_3)COOEt \xrightarrow[\text{(2) acid}]{\text{(1) hot strong NaOH}} CH_3CH_2CH_2COOH$

(4) $CH_3CH_2I \xrightarrow{Mg} CH_3CH_2MgI \xrightarrow[\text{(2) acid}]{\text{(1) } CO_2} CH_3CH_2COOH$

(5) CH_3CH_2I $\xrightarrow[\text{phthalimide}]{\text{potassium}}$

$$\begin{array}{c} O \\ \| \\ C \\ \diagup \quad \diagdown \\ NCH_2CH_3 \\ \diagdown \quad \diagup \\ C \\ \| \\ O \end{array}$$

$\xrightarrow[\text{hydrolysis}]{\text{alkaline}}$ $CH_3CH_2NH_2$

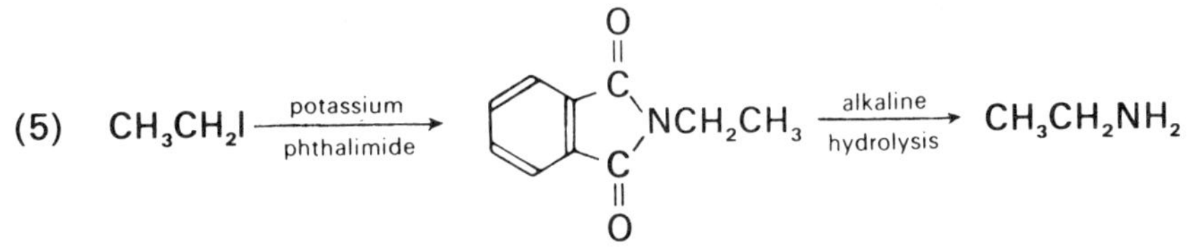

Two or three step syntheses

41. Starting from aniline and any other necessary reagent give methods for the preparation of: iodobenzene . . . . (1); N-ethylaniline . . . . (2); phenylazo-2-naphthol . . . . (3); benzylamine . . . . (4); 2,4,6-tribromobenzene . . . . (5).

(1) $C_6H_5NH_2 \xrightarrow{H_2SO_4/NaNO_2} C_6H_5N_2^+ \xrightarrow{NaI} C_6H_5I$

(2) $C_6H_5NH_2 \xrightarrow{(CH_3CO)_2O} C_6H_5NHCOCH_3 \xrightarrow{LiAlH_4} C_6H_5NHCH_2CH_3$

(3) $C_6H_5NH_2$ $\xrightarrow{HCl/NaNO_2}$ $C_6H_5N_2^+Cl^-$ $\xrightarrow[\text{of 2-naphthol}]{\text{alkaline solution}}$

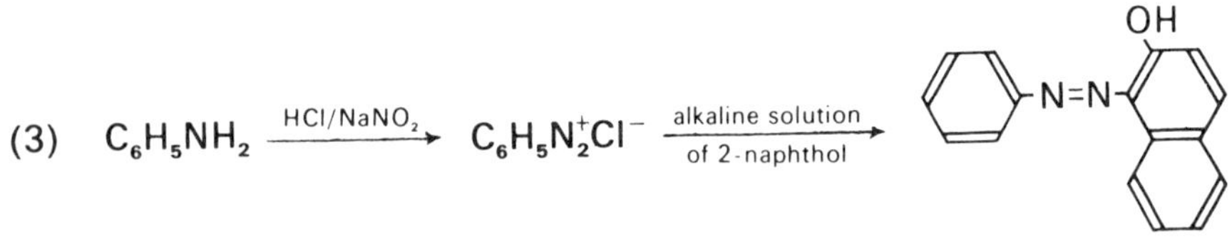

(4) $C_6H_5NH_2 \xrightarrow{H_2SO_4/NaNO_2} C_6H_5N_2^+ \xrightarrow{CuCN} C_6H_5CN \xrightarrow{LiAlH_4} C_6H_5CH_2NH_2$

(5) $C_6H_5NH_2$ $\xrightarrow{Br_2}$

$\xrightarrow[\text{(2) } H_2P(O)OH]{\text{(1) HONO}}$

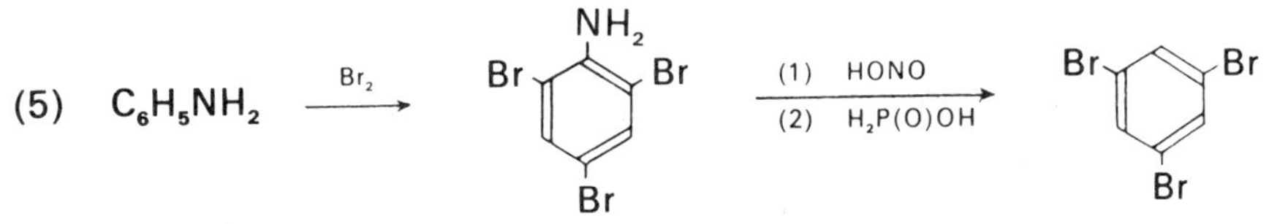

42. Suggest syntheses for the following compounds using the starting materials indicated: cyclohexanone from phenol . . . . (1); hexanedioic acid (adipic acid) from cyclohexanone . . . . (2); but-2-enal (crotonaldehyde) from acetaldehyde . . . . (3); *p*-nitroacetanilide from

If one hydrogen atom of *p*-dibromobenzene is replaced by a nitro group how many isomers are possible? . . . . (1). Give the structure(s) and names(s) . . . . (2). Do the same for *m*-dibromobenzene . . . . (3) and for *o*-dibromobenzene . . . . (4).

---

(1) one (2)

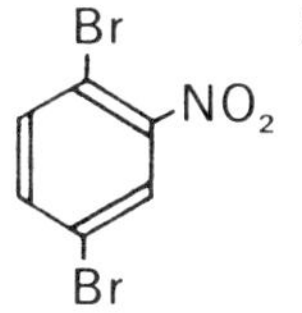

2,5-dibromonitrobenzene

(3) 2,6-dibromonitrobenzene

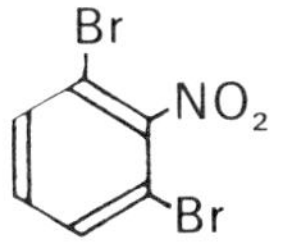

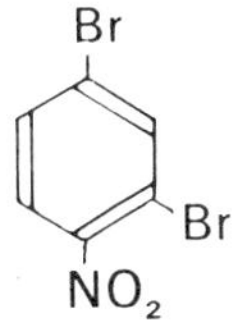

2,4-dibromonitrobenzene

3,5-dibromonitrobenzene

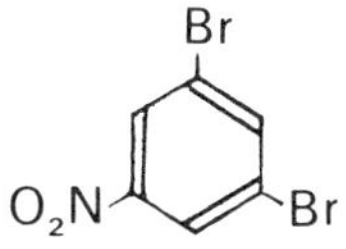

(4) 2,3-dibromonitrobenzene

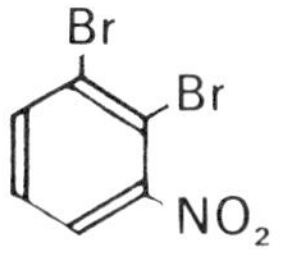

3,4-dibromonitrobenzene

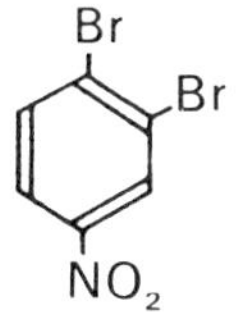

Draw the structures of (1) 2-methyl-3-phenylbut-2-ene
(2) diphenylmethane
(3) *p*-dibenzylbenzene

---

(1)

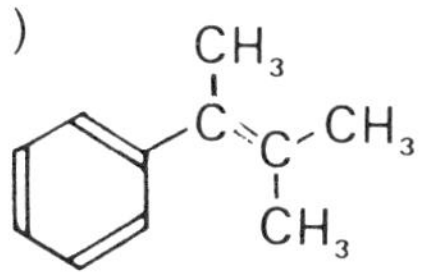

(2)

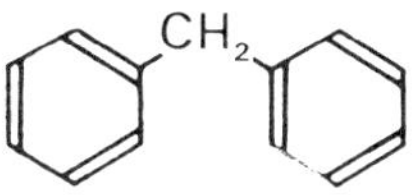

(3)

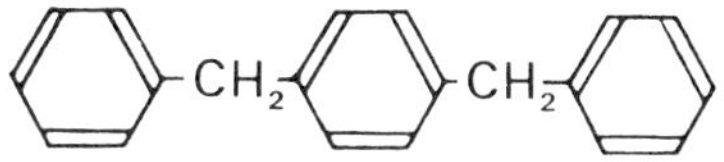

ALCOHOLS

37. The functional group of the alcohols is the hydroxyl group, –OH. These compounds are named in the IUPAC system by replacing the terminal *-e* of the hydrocarbon name by *-ol*. Thus we have *methanol*, CH_3OH, and *ethanol* . . . . (1).

---

(1) CH_3CH_2OH

---

38. Propanol and the longer chain alcohols all have isomeric forms which are chemically distinct. These are named by extension of the rules already given

(1) cyanohydrin (2) three (3) propanal (propionaldehyde)

In the following four frames no explanation is given of the reasoning behind the choice of the reactions.

One step synthesis

39. Procedures involving the formation of intermediate metal complexes which require acidification to yield the product are included under this heading. Starting from acetone and any other necessary reagent give reactions for the preparation of: 2-hydroxy-2-phenylpropane . . . . (1); 4-hydroxy-4-methylpentan-2-one . . . . (2); acetic acid . . . . (3); 4-hydroxy-4-methylpent-2-yne . . . . (4); acetone cyanhydrin . . . . (5).

---

(1) $$CH_3COCH_3 \xrightarrow[\text{(2) acid}]{\text{(1) } C_6H_5MgBr} C_6H_5\underset{\underset{CH_3}{|}}{\overset{\overset{CH_3}{|}}{C}}OH$$

(2) $$2CH_3COCH_3 \xrightarrow{Ba(OH)_2} (CH_3)_2\underset{\underset{OH}{|}}{C}CH_2COCH_3$$

(3) $$CH_3COCH_3 \xrightarrow{I_2 + NaOH} CH_3COOH + CHI_3 \quad \text{or}$$

$$CH_3COCH_3 \xrightarrow{KMnO_4} CH_3COOH$$

(4) $$CH_3COCH_3 \xrightarrow[\text{(2) acid}]{\text{(1) } CH_3C\equiv CNa} (CH_3)_2\underset{\underset{OH}{|}}{C}C\equiv CCH_3$$

(5) $$CH_3COCH_3 \xrightarrow{HCN} (CH_3)_2\underset{\underset{OH}{|}}{C}CN$$

One or two step syntheses

40. Starting from ethyl iodide and any other necessary reagent give reactions for the preparation of: butane . . . . (1); 1-aminopropane . . . . (2); butanoic acid . . . . (3); propanoic acid . . . . (4); ethylamine . . . . (5).

---

(1) $$CH_3CH_2I \xrightarrow{Mg} CH_3CH_2MgI$$

$$CH_3CH_2MgI + CH_3CH_2I \xrightarrow{CoCl_2} CH_3CH_2CH_2CH_3$$

(2) $$CH_3CH_2I \xrightarrow{NaCN} CH_3CH_2CN \xrightarrow[\text{or } LiAlH_4]{H_2/Pt} CH_3CH_2CH_2NH_2$$

for other functional groups. Draw and name the alcohols with the molecular formula C_3H_8O . . . . (1).

---

(1) $CH_3CH_2CH_2OH$ propan-1-ol or propyl alcohol

CH_3CHCH_3 (with OH on the middle carbon) propan-2-ol or isopropyl alcohol

$\quad\;\; OH$

---

39. In the answer to the frame above, both the systematic and semi-systematic names were given. The latter were of the type *alkyl alcohol* using alkyl group names which were given in Frame 26. Draw the structures of the alcohols of formula C_4H_9OH and give the systematic names . . . .(1).

---

(1) $CH_3CH_2CH_2CH_2OH$ butan-1-ol

CH_3CHCH_2OH (with CH_3 on the second carbon) 2-methylpropan-1-ol

$\quad\;\; CH_3$

$CH_3CH_2CHCH_3$ (with OH on the third carbon) butan-2-ol

$\qquad\quad OH$

$\quad\;\; CH_3$

CH_3COH 2-methylpropan-2-ol

$\quad\;\; CH_3$

---

40. Alcohols may be classified as primary, secondary or tertiary according to the number of carbon atoms attached to the carbon which bears the hydroxyl group. Thus we can have

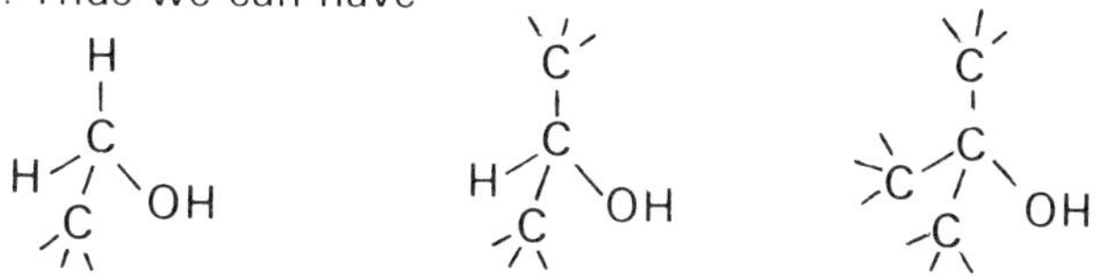

primary secondary tertiary

Ethanol has . . . . (1) carbon atom attached to the carbon atom carrying the –OH group and is therefore a . . . . (2) alcohol.

---

(1) one (2) primary

---

41. Classify in this way butan-1-ol . . . . (1), butan-2-ol . . . . (2), 2-methylpropan-1-ol . . . . (3) and 2-methylpropan-2-ol . . . . (4).

---

36. The synthesis therefore requires that both adipic acid and 1,6-diaminohexane be made separately from butane-1,4-diol and this can be done in the following manner . . . . (1).

---

(1) $HOCH_2CH_2CH_2CH_2OH \xrightarrow{HBr} Br(CH_2)_4Br \xrightarrow{NaCN} NC(CH_2)_4CN$

$$NC(CH_2)_4CN \xrightarrow[\text{hydrolysis}]{\text{acid}} HOOC(CH_2)_4COOH$$

$$NC(CH_2)_4CN \xrightarrow{LiAlH_4} H_2NCH_2(CH_2)_4CH_2NH_2 \text{ or } H_2N(CH_2)_6NH_2$$

2-Hydroxybutanoic acid

37. Suggest a method for synthesising the racemic form of 2-hydroxybutanoic acid and then resolving it into its enantiomers . . . . (1).

---

(1) $CH_3CH_2CHO + HCN \longrightarrow (\pm)CH_3CH_2CH(OH)CN$

$$(\pm)CH_3CH_2CH(OH)CN \xrightarrow[\text{hydrolysis}]{\text{acid}} (\pm)CH_3CH_2CH(OH)COOH$$

This synthesis leads to the racemic form of 2-hydroxybutanoic acid which can be resolved by separating the diastereomeric salts formed by reacting the acid with an optically active amine, shown as e.g., (–)B, (cf. Program 11).

$$(\pm)CH_3CH_2CH(OH)COOH \xrightarrow{(-)B} [(+)CH_3CH_2CH(OH)COO^-(-)BH^+] + [(-)CH_3CH_2CH(OH)COO^-(-)BH^+]$$

The diastereomeric salts can be separated by crystallisation and decomposed with acid.

$$[(+)CH_3CH_2CH(OH)COO^-(-)BH^+] \xrightarrow{\text{acid}} (+)CH_3CH_2CH(OH)COOH + (-)BH^+$$

$$[(-)CH_3CH_2CH(OH)COO^-(-)BH^+] \xrightarrow{\text{acid}} (-)CH_3CH_2CH(OH)COOH + (-)BH^+$$

---

38. Solution of the problem of synthesis of 2-hydroxybutanoic acid required recognition of a method of synthesis of an α-hydroxy acid. Acids can be prepared in a number of ways, one of which is hydrolysis of a nitrile. Another name for an α-hydroxynitrile is a . . . . (1), and such compounds are formed by reacting hydrogen cyanide with an aldehyde or ketone. The sequence of reactions having been decided, the number of carbon atoms in the aldehyde must be . . . . (2) and consequently the starting aldehyde is . . . . (3).

---

(1) primary (2) secondary (3) primary (4) tertiary

---

42. *Cyclic alcohols* are named using the same principles. Cyclohexanol has the structure . . . . (1) and it is a . . . . (2) alcohol. Draw and name the cyclic alcohol $C_5H_{10}O$ having a ring of five carbon atoms . . . . (3).

---

(1)

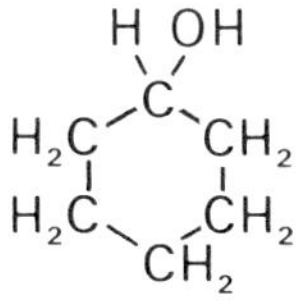

(2) secondary

(3)

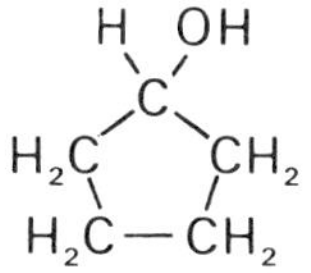

cyclopentanol

---

43. A compound with two –OH groups is called a *diol*, and with three –OH groups a *triol*, and the positions of the –OH groups are given in the name. Give the structure of ethane-1,2-diol, ... (1), propane-1, 2-diol . . . . (2) and propane-1,3-diol . . . . (3).

---

(1) 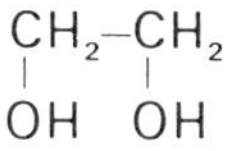 commonly known as ethylene glycol

(2) 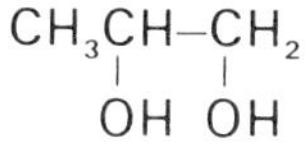 commonly called propylene glycol

(3) 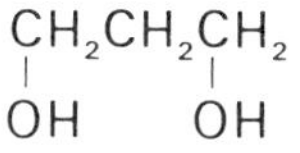 The use of the common name *glycol* is restricted to diols in which the two –OH groups are on adjacent carbon atoms.

---

44. 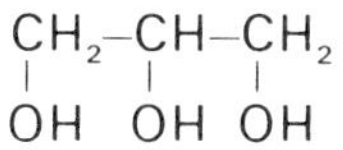 Glycerol, which is a constituent of fats and hence widely distributed in all kinds of biological material, is named in the IUPAC system as . . . . (1).

---

(1) propane-1,2,3-triol

PHENOLS

45. When the hydroxyl group is attached directly to a benzene ring the properties are sufficiently modified from those of an alcohol for a separate family name to be given. *Hydroxybenzene* is known as *phenol* and has the structure . . . . (1). A benzene ring with a methyl group and an hydroxyl group has the common name of *cresol* and is likewise a member of the phenol

$$CH_3CH_2CH_2OH \xrightarrow[250\text{-}300°]{Cu} CH_3CH_2CHO$$

$$CH_3CH_2CH_2MgBr + CH_3CH_2CHO \longrightarrow CH_3CH_2CH_2\underset{\displaystyle OH}{\underset{|}{C}}HCH_2CH_3$$

---

32. The required product has twice as many carbon atoms as the starting material and it contains a linear chain. It can therefore be made by joining two three-carbon units at . . . . (1).

---

(1) the end of each chain

---

33. The method of choice for the preparation of an alcohol is often a Grignard synthesis in which the reactive carbon atoms in the coupling reaction differ in that one is . . . . (1) and the other . . . . (2).

---

(1) electrophilic (2) nucleophilic

Nylon 66

34. Starting from butane-1,4-diol, suggest a synthesis of Nylon 66. This is a linear polymer having the structure $-[\underset{\displaystyle O}{\underset{\|}{C}}(CH_2)_4\underset{\displaystyle O}{\underset{\|}{C}}NH(CH_2)_6NH]_n-$ where n is a large number. Firstly, the bonds holding the nylon molecule together include . . . . (1) bonds which can be cleaved by hydrolysis (cf. Program 8, Frame 49).

---

(1) amide

---

35. The products of acid hydrolysis of nylon are . . . . (1). The industrial synthesis of nylon involves mixing these two substances in equimolecular amounts and heating the resulting salt to form the amide bonds with loss of water. A laboratory synthesis involves the formation of the acid chloride, followed by its reaction with the amine in the presence of base.

---

(1) $HOOC(CH_2)_4COOH$ and $H_2N(CH_2)_6NH_2$
hexanedioic acid or adipic acid; 1,6-diaminohexane or hexamethylenediamine

family. There are . . . . (2) isomers and their formulae are . . . . (3).

---

(1) 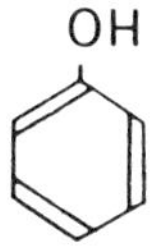   (2) three

(3)

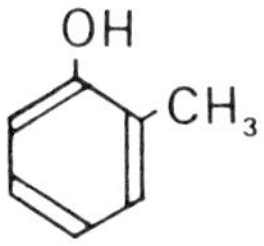

o-

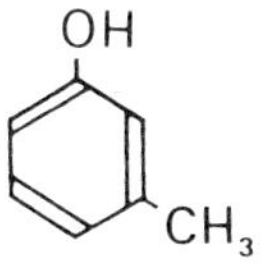

m-

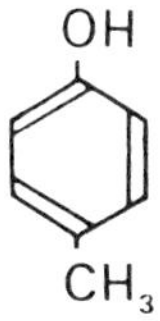

p-

---

46. Draw the structures of *o*-bromophenol . . . . (1), *p*-nitrophenol . . . . (2), and 4-chloro-1,3-dihydroxybenzene . . . . (3).

---

(1) 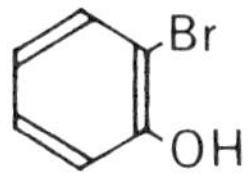

(2) 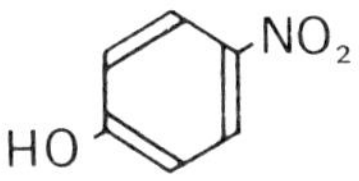

(3) 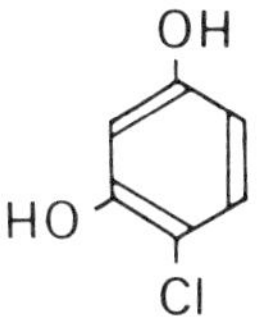

---

47. The name phenol is applied in a general way to all compounds in which an –OH group is attached to a benzene ring.

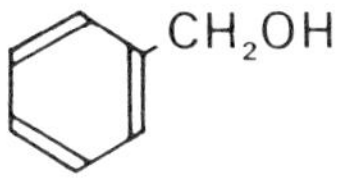

. . . . (is/is not) (1) a phenol because . . . . (2).

---

(1) is not a phenol   (2) because the –OH is not attached to the benzene ring

---

48. 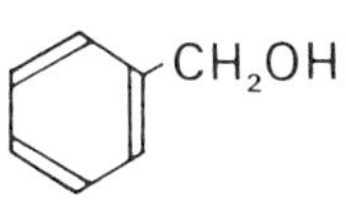

The compound illustrated will be expected to show the properties and reactions of an . . . . (1). It is known as . . . . (2).

---

(1) alcohol   (2) benzyl alcohol

benzanilide gave a lead that benzoic acid and aniline might be useful intermediates. Conversion of benzoic acid into benzoyl chloride according to the equation. . . . . (1) and reaction of this acid chloride with aniline according to equation . . . . (2) clearly provides a synthesis of benzanilide. The problem therefore resolves itself into a synthesis of aniline from benzoic acid.

---

(1) $C_6H_5COOH \xrightarrow[\text{or } PCl_5]{SOCl_2} C_6H_5COCl$

(2) $C_6H_5COCl + C_6H_5NH_2 \xrightarrow{NaOH} C_6H_5CONHC_6H_5$

---

29. Aniline contains one carbon atom less than benzoic acid. Methods available for the removal of a carbon atom include decarboxylation, e.g. $C_6H_5COOH \longrightarrow C_6H_6 + CO_2$. The reaction in this case goes only with difficulty, and solution of the problem would then require a synthesis of aniline from benzene, e.g., by the reaction . . . . (1).

---

(1) $C_6H_6 \xrightarrow{HNO_3/H_2SO_4} C_6H_5NO_2 \xrightarrow{Fe/HCl} C_6H_5NH_2$

---

30. A second possibility is the oxidation of a primary amide with sodium hypobromite to give the amine with one less carbon atom, (cf. Program 5, Frame 38).

$$C_6H_5CONH_2 \xrightarrow{NaOBr} C_6H_5NH_2$$

This synthesis is reasonably efficient since the benzoyl chloride required for the formation of benzanilide can also be used to prepare benzamide. A balanced equation for the formation of benzamide from benzoyl chloride is . . . . (1), and this can be shown in abbreviated form as. . . . . (2).

---

(1) $C_6H_5COCl + 2NH_3 \longrightarrow C_6H_5CONH_2 + NH_4Cl$

(2) $C_6H_5COCl \xrightarrow{NH_3} C_6H_5CONH_2$

Hexan-3-ol

31. Devise a synthesis of hexan-3-ol using propan-1-ol as the only organic starting material . . . . (1).

---

(1) $CH_3CH_2CH_2OH \xrightarrow{HBr} CH_3CH_2CH_2Br \xrightarrow{Mg} CH_3CH_2CH_2MgBr$

The use of the prefix hydroxy- is not restricted to aromatic compounds. An aliphatic molecule which contains both the –COOH and an –OH group, is named as an acid and the presence of the hydroxyl group is indicated by the use of the prefix. Examples are given later.

ALDEHYDES AND KETONES

49. The *carbonyl* group is the functional group of aldehydes and ketones. The carbonyl group is >C=O. In aldehydes one of the remaining bonds must be to hydrogen so an aldehyde always contains the terminal group .... (1). In a ketone the two unassigned bonds are both to carbon atoms, so a ketone always contains the arrangement .... (2).

---

(1) 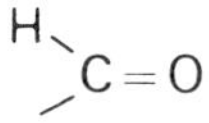

(2) 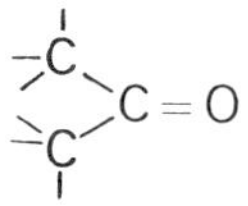

---

50. The aldehydes are named in the IUPAC system by replacing the *e* ending of the alkane name by *-al* but there are common names in use for the simple aldehydes. For example *HCHO* which the IUPAC system would call .... (1) is *formaldehyde* and ethanal, structure .... (2) is known as *acetaldehyde*. Benzaldehyde has the structure .... (3).

---

(1) methanal

(2) 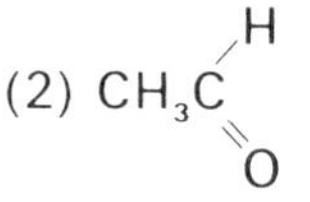

or CH_3CHO

(3) 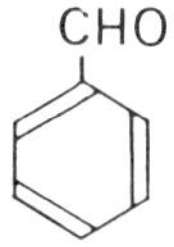

---

51. The IUPAC system is useful for naming more complex aldehydes. A compound which is being named as an aldehyde does not require a number to indicate the position of the carbonyl group because the oxygen atom will always be at the .... (1) of a chain.

---

(1) end (consequently the number 1 is always assumed for the carbonyl carbon)

---

52. Give the structure of propanal .... (1) and of 3-methylpentanal .... (2).

---

(1) CH_3CH_2CHO

(2) $CH_3CH_2CHCH_2CHO$ (with CH_3 on the third carbon, the CH)

Successful completion of this synthesis required the following steps:

(1) The chemical names were translated into. . . . . (1).

(2) By examination of the structure of the desired product, and deciding the. . . . . (2) of compound to which it belongs, one or more possible syntheses of the desired compound were written down.

(3) The nature of each of the possible reactants suggested by step (2) was then considered in relation to the structure of. . . . . (3).

(4) Equations for the preparation of these reactants were then suggested.

(5) The sequence of equations was written down in the order in which they would be carried out in an actual laboratory synthesis.

---

(1) structures (2) class (3) the given starting material

---

26. Problems of this kind should always be solved by working backwards. If the final step of the synthesis is not immediately obvious, it is sometimes useful to consider what compounds might be obtained by, for example, hydrolysis of the desired substance, and then to work out how the hydrolysis products might be recombined. If a problem required the synthesis of N-phenylbenzamide (benzanilide), you could ask whether the substance can be hydrolysed. The answer in this case is yes, and the reaction follows the equation. . . . . (1) (cf. Program 8, Frame 49). You might therefore guess that benzanilide could possibly be prepared from. . . . . (2).

---

(1) $C_6H_5CONHC_6H_5 + H_2O \longrightarrow C_6H_5COOH + C_6H_5NH_2$

(2) benzoic acid, C_6H_5COOH, and aniline, $C_6H_5NH_2$

Benzanilide

27. Devise a synthesis of benzanilide using benzoic acid as the only organic starting material. . . . . (1).

---

(1) $C_6H_5COOH \xrightarrow{SOCl_2} C_6H_5\underset{\underset{O}{\|}}{C}Cl \xrightarrow{NH_3} C_6H_5\underset{\underset{O}{\|}}{C}NH_2 \xrightarrow{NaOBr} C_6H_5NH_2$

benzoyl chloride, benzamide, aniline

$C_6H_5\underset{\underset{O}{\|}}{C}Cl + C_6H_5NH_2 \xrightarrow{NaOH} C_6H_5\underset{\underset{O}{\|}}{C}NHC_6H_5$

benzanilide

---

28. An amide is usually prepared by the reaction of an acid chloride with an amine in the presence of sodium hydroxide or a tertiary base to neutralise the hydrogen chloride produced. Consideration of the hydrolysis of

53. In the IUPAC system a *ketone* is named by selecting the longest chain in which the C=O occurs. The *e* ending of the alkane name is replaced by *-one* and the position of the C=O along the chain is given by number, counting, as usual, from the end which gives the smaller number. Pentan-2-one is . . . . (1) and 3-methylbutan-2-one is . . . . (2). When two or more functional groups are present it may be necessary to define the position of a carbonyl group by a prefix. The oxygen substituent of C=O is then indicated by *oxo-* and the carbon atom is numbered. Thus $ClCH_2COCH_2CHO$ is 4-chloro-3-oxobutanal.

---

(1) $\mathrm{CH_3CH_2CH_2\underset{\underset{\displaystyle O}{\|}}{C}CH_3}$ (2) $\mathrm{CH_3\underset{\underset{\displaystyle CH_3}{|}}{C}H{-}\underset{\underset{\displaystyle O}{\|}}{C}CH_3}$

---

54. Some of the common ketones are named according to the alkyl groups attached to the C=O. The compound illustrated is known as ethyl methyl ketone but would be named systematically as . . . . (1). Diethyl ketone has the structure . . . . (2) and its systematic name is . . . . (3).

$$\mathrm{CH_3\underset{\underset{\displaystyle O}{\|}}{C}CH_2CH_3}$$

---

(1) butanone (2) $\mathrm{CH_3CH_2\underset{\underset{\displaystyle O}{\|}}{C}CH_2CH_3}$ (3) pentan-3-one

---

55. Two ketones which are usually known by their trivial names are:

acetophenone and benzophenone

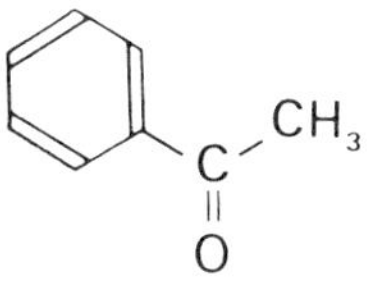

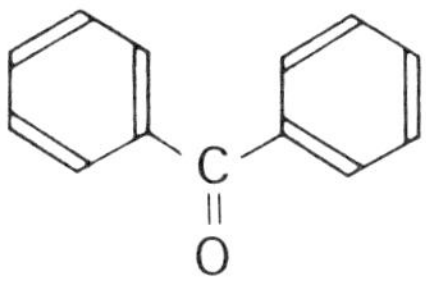

o-Hydroxyacetophenone has the structure . . . . (1).

While not part of the IUPAC system, it is sometimes useful to designate the carbon atoms of an alkyl chain attached to a carbonyl group by the Greek letters α, β, γ etc.

$$\mathrm{-\overset{\gamma}{C}-\overset{\beta}{C}-\overset{\alpha}{C}-\underset{\underset{\displaystyle O}{\|}}{C}-}$$

Hydrogen atoms attached to these carbons are then known respectively as α-hydrogens, β-hydrogens, etc. (cf. Program 7, Frame 16).

---

21. Benzyl benzoate could be prepared by several methods available for ester formation. Thus it could be made by reaction of benzyl alcohol either with .... (1) in the presence of an .... (2) catalyst, or by reaction with .... (3) in the presence of a base.

---

(1) benzoic acid (2) acid (3) benzoyl chloride

---

22. The equations of these two reactions leading to benzyl benzoate are .... (1) and .... (2).

---

(1) $C_6H_5COOH + C_6H_5CH_2OH \xrightarrow[\text{catalyst}]{\text{acid}} C_6H_5COOCH_2C_6H_5 + H_2O$

(2) $C_6H_5COCl + C_6H_5CH_2OH \xrightarrow{\text{base}} C_6H_5COOCH_2C_6H_5$

---

23. Both methods require benzyl alcohol. The first thing to determine is whether or not this alcohol has the same number and arrangement of carbon atoms as the starting material, which in the present case is benzoic acid. On inspection it is seen that there is no variation of the carbon skeleton but only a change in oxidation level. Benzyl alcohol can be prepared from benzoic acid by .... (reduction/oxidation) (1) with a reagent such as .... (2).

---

(1) reduction (2) $LiAlH_4$

---

24. The synthesis of benzyl benzoate from benzoyl chloride requires preparation of the latter compound from benzoic acid. This can be done by heating the acid with .... (1).

---

(1) thionyl chloride, $SOCl_2$, or phosphorus pentachloride, PCl_5

---

25. The reactions can now be summarised:

$$C_6H_5COOH \xrightarrow[\text{(2) acid}]{\text{(1) LiAlH}_4} C_6H_5CH_2OH$$

$$C_6H_5COOH \xrightarrow[\text{or PCl}_5]{\text{SOCl}_2} C_6H_5COCl$$

$$C_6H_5COOH + C_6H_5CH_2OH \overset{\text{acid}}{\rightleftharpoons} C_6H_5COOCH_2C_6H_5 + H_2O$$

$$C_6H_5COCl + C_6H_5CH_2OH \xrightarrow{\text{base}} C_6H_5COOCH_2C_6H_5$$

(1) 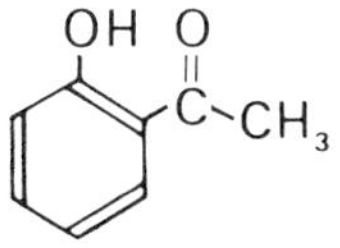

TEST FRAMES

Give the structures of the following alcohols and state whether they are primary, secondary or tertiary.

(1) 3-phenylpropan-1-ol (2) hexan-3-ol
(3) 3-phenylpentan-3-ol (4) benzyl alcohol
(5) 2-methylpropan-1-ol

---

(1) 

a primary alcohol

(2) $CH_3CH_2CH(OH)CH_2CH_2CH_3$

a secondary alcohol

(3) 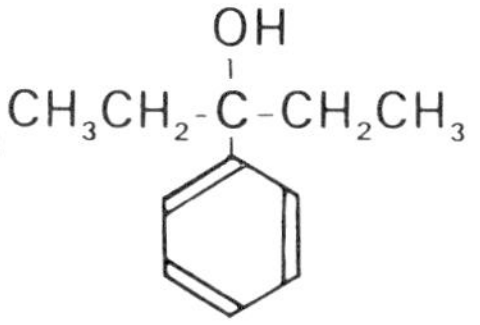

a tertiary alcohol

(4) 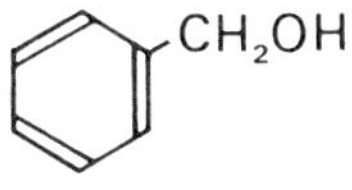

a primary alcohol

(5) $CH_3CH(CH_3)CH_2OH$ a primary alcohol.

Give the structural formulae of the following compounds:

(1) 2,4-dinitrophenol (2) picric acid, 2,4,6-trinitrophenol
(3) *m*-chlorophenol (4) cyclohexanone
(5) phenylacetaldehyde (6) 3-hydroxypentanal

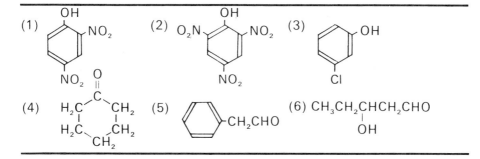

---

(1) 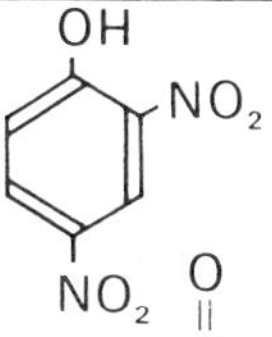

(2) 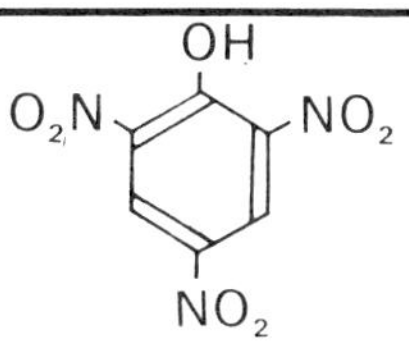

(3) 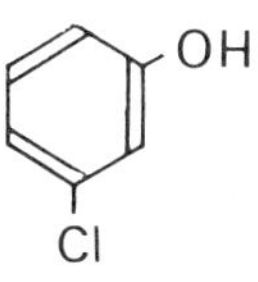

(4) 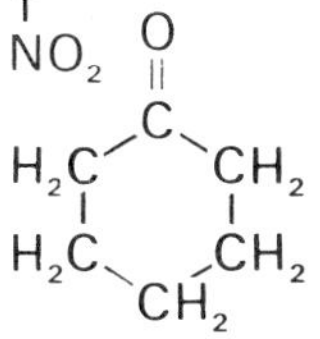

(5) 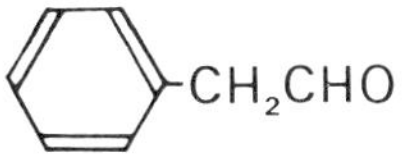

(6) $CH_3CH_2CH(OH)CH_2CHO$

CARBOXYLIC ACIDS

56. The functional group of the carboxylic acids is –COOH or –C(=O)OH. Trivial names have been retained for the first four homologues and thereafter the systematic names are more frequently used. In the IUPAC system, the name of the acid is derived from the name of the hydrocarbon with the same number of carbon atoms. Thus *formic acid*, HCOOH, is

(7, 8) $RCHO + HCN \longrightarrow RCH(OH)CN$

(7, 8) $RCOR' + HCN \longrightarrow RR'C(OH)CN$

Amines:

(4, 5) $RX + NaCN \longrightarrow RCN \longrightarrow RCH_2NH_2$

(5, 8) $RCONH_2 \xrightarrow{NaOBr} RNH_2 + CO_2$

Mechanisms

The reactions listed above in which a carbon-carbon bond is formed can be considered in terms of the reaction of nucleophilic carbon atoms with electrophilic carbon atoms. In most instances the nucleophilic carbon atom is a carbanion $\rangle C^-$ which provides the pair of electrons required to form the new bond with an electrophilic carbon atom such as is present in $\rangle C—X$ or $\rangle C=O$. In general terms the fundamental reactions are:

1. Nucleophilic substitution at a saturated carbon atom

2. Nucleophilic addition at an unsaturated carbon atom.

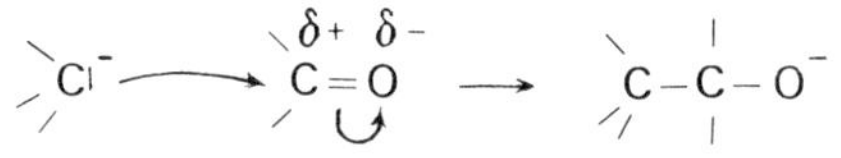

In many cases the way in which the reaction proceeds can be thought of in terms of these simple electron shifts. It should be remembered, however, that such representations are oversimplifications of complex reaction mechanisms, many of which are poorly understood at the present time. The mechanisms of the reactions of acyl chlorides, RCOCl, are not considered.

SYNTHESIS OF ORGANIC COMPOUNDS

You should try to complete each synthesis before looking at the answer and the explanatory frames. In this way you will gain confidence in your ability to apply the knowledge of organic reactions you have acquired from the earlier programs. It may be helpful when considering each problem to refer to the list of carbon-carbon bond-forming and bond-breaking reactions.

Benzyl benzoate

20. Let us suppose that you are required to devise a synthesis of benzyl benzoate using benzoic acid as the only starting material. The required product has the structure. . . . . (1) and it is a member of the general class of compounds called . . . . (2).

---

(1) $C_6H_5COOCH_2C_6H_5$ (2) esters

methanoic acid; acetic acid, CH_3COOH, is *ethanoic acid; propionic acid,* CH_3CH_2COOH, is .... (1) butyric acid, $CH_3CH_2CH_2COOH$ is .... (2).

---

(1) propanoic acid (2) butanoic acid

---

57. I 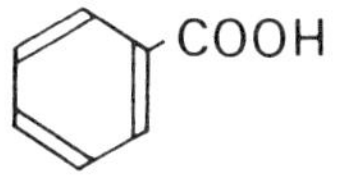

II 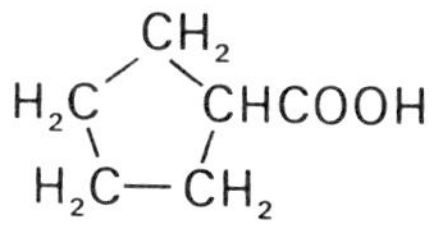

A carboxyl group may be attached to a ring. The name benzoic acid is given to structure (I). In other compounds the attachment of –COOH to a ring is shown by giving the name of the hydrocarbon followed by *carboxylic acid.* Compound II is *cyclopentanecarboxylic acid.* In a similar fashion attachment of –CHO to a ring is shown by adding *-carboxaldehyde* to the name of the hydrocarbon.
Give the structures of *m*-bromobenzoic acid .... (1), 2,4-dinitrobenzoic acid .... (2) and cyclohexanecarboxylic acid............... (3).

---

(1) 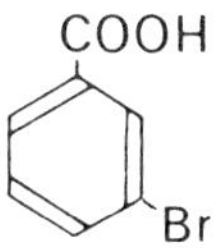

(2) 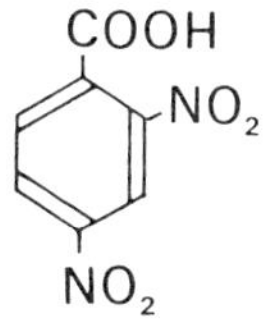

(3) 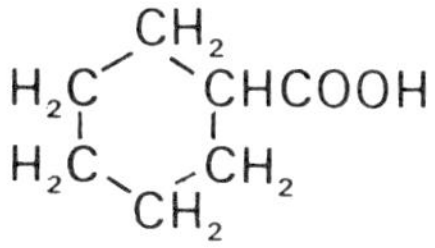

---

58. 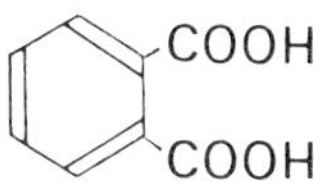
I

A compound with two –COOH groups is a *dicarboxylic* acid. There are .... (1) isomeric compounds having a benzene ring substituted with two carboxyl groups. Compound I, 1,2-benzenedicarboxylic acid, has the trivial name phthalic acid. 1,3-Benzenedicarboxylic acid has structure .... (2) and is known as isophthalic acid. 1,4-Benzenedicarboxylic acid (terephthalic acid), has structure .... (3).

---

(1) three

(2) 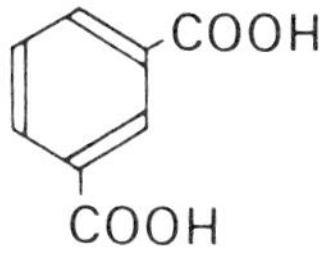

(3) 

---

59. In the open chain aliphatic dicarboxylic acids the carbon atoms of both carboxyl groups are counted in selecting the name of the parent hydrocarbon. A monocarboxylic acid is named by adding *-oic acid* to the name of the hydrocarbon and a dicarboxylic acid is named by the addition of *-dioic acid.*

(7) $2CH_3CHO \xrightarrow{base} CH_3CH(OH)CH_2CHO$

(7) $2CH_3COCH_3 \xrightarrow{base} (CH_3)_2C(OH)CH_2COCH_3$

Aldehydes:

(7) $2CH_3CHO \xrightarrow{base} CH_3CH(OH)CH_2CHO \xrightarrow{acid} CH_3CH{=}CHCHO$
cf. *alkenes* above

(6) $RCH(OH)CH(OH)R' \xrightarrow{HIO_4} RCHO + R'CHO$

Ketones:

(4, 12) $CH_3COCH_2COOEt \xrightarrow[(2)\ RX]{(1)\ NaOEt} CH_3COCHRCOOEt \longrightarrow CH_3COCH_2R$

(4, 12) $CH_3COCHRCOOEt \xrightarrow[(2)\ R'X]{(1)\ NaOEt} CH_3COCRR'COOEt \longrightarrow CH_3COCHRR'$

(7) $2CH_3COCH_3 \xrightarrow{base} (CH_3)_2C(OH)CH_2COCH_3 \xrightarrow{acid} (CH_3)_2C{=}CHCOCH_3$
cf. *alkenes* above

(7) $RCOCl + Cd(CH_3)_2 \longrightarrow RCOCH_3$

(9) $RCOCl + C_6H_6 \xrightarrow{AlCl_3} RCOC_6H_5$

Acids:

(4, 8) $RX + NaCN \longrightarrow RCN \longrightarrow RCOOH$

(10) $ArN_2^+\ Cl^- + NaCN \xrightarrow{CuCN} ArCN \longrightarrow ArCOOH$

(4, 12) $CH_3COCH_2COOEt \xrightarrow[(2)\ RX]{(1)\ NaOEt} CH_3COCHRCOOEt \longrightarrow RCH_2COOH$

(4, 12) $CH_3COCHRCOOEt \xrightarrow[(2)\ R'X]{(1)\ NaOEt} CH_3COCRR'COOEt \longrightarrow RR'CHCOOH$

(8) $RMgX + CO_2 \longrightarrow RCOOH$

(7, 8) $RCHO + HCN \longrightarrow RCH(OH)CN \longrightarrow RCH(OH)COOH$

Nitriles:

(4, 8) $RX + NaCN \longrightarrow RCN$

(10) $ArN_2^+\ Cl^- + NaCN \xrightarrow{CuCN} ArCN$

It is not necessary to specify the carbon atoms of the carboxyl groups by number since the parent hydrocarbon is taken as the chain which contains both the carboxyl groups and hence they occur at the two ends of this chain. HOOC-COOH would be named systematically as ethanedioic acid but is often known as oxalic acid. Propanedioic acid (malonic acid) has the structure . . . . (1). Hexanedioic acid (adipic acid) is . . . . (2) and 2-ethyl-3-methylbutanedioic acid has the structure . . . . (3).

---

(1) $HOOCCH_2COOH$

(2) $HOOCCH_2CH_2CH_2CH_2COOH$ or $HOOC(CH_2)_4COOH$

(3) 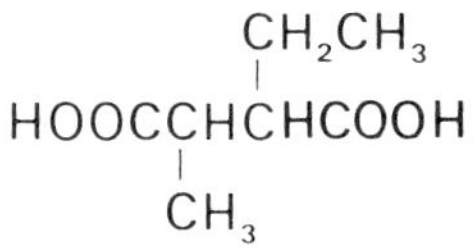

---

60. Draw the structures of *p*-hydroxybenzoic acid . . . . (1), 3,5-dinitrobenzoic acid . . . . (2) and phenylacetic acid . . . . (3).

---

(1) 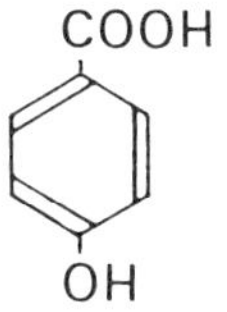

(2) 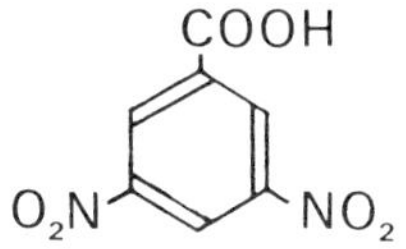

(3) 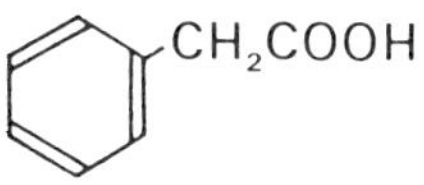

DERIVATIVES OF CARBOXYLIC ACIDS

61. Carboxylic acids can be converted into a number of derivatives:— acid chlorides, acid anhydrides, amides and esters. The names of all these compounds are derived from the name of the corresponding acid.

Acid chlorides. The general structure illustrated can also be written as RCOCl, where R is used to represent the rest of the molecule which may be an alkyl group, a benzene ring etc. The systematic name of the acid chloride corresponding to ethanoic acid is *ethanoyl chloride*. The name acetyl chloride is commonly used and is derived from the trivial name of the acid, acetic acid. The structure of acetyl chloride is . . . . (1).

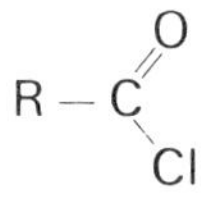

---

(1) 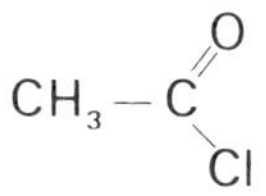

---

62. The acid chloride which can be obtained from benzoic acid is called . . . . (1) and has the structure . . . . (2). It should not be

FORMATION OF CARBON-CARBON BONDS

Before considering the synthesis of organic compounds, it might be helpful to summarise the reactions available for the formation of carbon-carbon bonds and also a few of those available for their cleavage. Both procedures involve a change in the size of the carbon skeleton and this must be borne in mind. Where a sequence of reactions is indicated in the following list the first generally involves the formation of a carbon-carbon bond and the second shows reactions such as reduction, elimination, hydrolysis, or hydrolysis and decarboxylation which are necessary to obtain the desired compound. The numbers preceding the equations show which program contains a more detailed discussion of the reaction concerned.

Alkanes:

(12) $RMgX + R'X \xrightarrow{CoCl_2} RR'$

(3, 4) $RCOOH \xrightarrow{heat} RH + CO_2$

(3, 4) $RX + NaC{\equiv}CR' \longrightarrow RC{\equiv}CR' \xrightarrow{H_2/Pt} RCH_2CH_2R'$

Alkenes:

(3, 4) $RX + NaC{\equiv}CR' \longrightarrow RC{\equiv}CR' \xrightarrow{H_2/Pd} RCH{=}CHR'$
cf. *aldehydes* and *ketones* below

Alkynes:

(3, 4) $RX + NaC{\equiv}CH \longrightarrow RC{\equiv}CH$

(3, 4) $RX + NaC{\equiv}CR' \longrightarrow RC{\equiv}CR'$
cf. *alcohols* below

Arenes:

(9) $RX + C_6H_6 \xrightarrow{AlCl_3} RC_6H_5$

Alcohols:

(6, 7) $RMgX + HCHO \longrightarrow RCH_2OH$

(6, 7) $RMgX + R'CHO \longrightarrow RR'CHOH$

(6, 7) $RMgX + R'COR'' \longrightarrow RR'R''COH$

(6) $2RMgX + R'COOC_2H_5 \longrightarrow R_2R'COH$

(6, 7) $RC{\equiv}CNa + R'CHO \longrightarrow RC{\equiv}CCH(OH)R'$
cf. *alkynes* above

(6, 7) $RC{\equiv}CNa + R'COR'' \longrightarrow RC{\equiv}CC(OH)R'R''$

(7, 8) $RCHO + HCN \longrightarrow RCH(OH)CN \longrightarrow RCH(OH)COOH$

(7, 8) $RCOR' + HCN \longrightarrow RR'C(OH)CN \longrightarrow RR'C(OH)COOH$

confused with benzyl chloride which has the structure . . . . (3).

---

(1) benzoyl chloride (2) 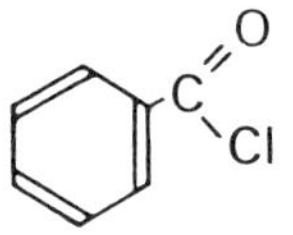 (3) 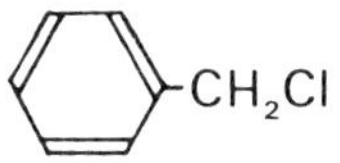

---

63. If the acid chloride, ester etc. of an alkanoic acid carries a substituent in the alkyl chain, its position is indicated by a number. Since the carboxyl function must be at the end of the chain, numbering commences there. Thus $C_6H_5CH_2CH_2COCl$ is 3-phenylpropanoyl chloride. 4-Methylhexanoic acid has the structure . . . . (1).

---

(1) $CH_3CH_2CHCH_2CH_2COOH$ (with CH_3 on the third carbon)

---

64. The general structure of a primary *amide* can be written $RCONH_2$, 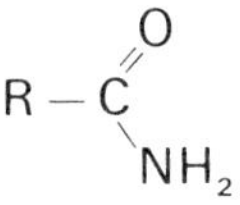 Such compounds can be made by the action of ammonia on acid chlorides. The name is based on the name of the acid which is obtained on hydrolysis of the amide. CH_3CONH_2 has the trivial name acetamide derived from the name acetic acid. The systematic name, being based on ethanoic acid, is . . . . (1). Benzamide is . . . . (2).

---

(1) ethanamide (2) 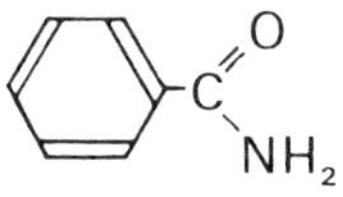

---

65. If, instead of reacting an acid chloride with ammonia, it is allowed to react with an amine of general formula $R'NH_2$, the product is a secondary amide, RCONHR′, while a secondary amine R′R″NH yields a tertiary amide RCONR′R″. CH_3COCl, which is called . . . . (1) reacts with CH_3NH_2 to give $CH_3CONHCH_3$ to which the name N-methylethanamide (N-methylacetamide) is given. The prefix N- shows that one of the hydrogen atoms initially attached to the nitrogen has been replaced by the methyl group. Write the structures of N-ethylbenzamide . . . . (2) and N,N-dimethylethanamide (N,N-dimethylacetamide) . . . . (3).

---

(1) ethanoyl chloride or acetyl chloride (2) 

(3) $CH_3CON(CH_3)_2$

17. Hydrolysis of a monoalkylated or dialkylated diethyl malonate with aqueous sodium hydroxide followed by acidification leads to the substituted malonic acid, e.g.,

$$RCH(COOEt)_2 + 2NaOH \longrightarrow RCH(COONa)_2 + 2EtOH$$

$$RCH(COONa)_2 + 2HCl \longrightarrow \ldots\ldots\ldots\ldots \ (1).$$

---

(1) $RCH(COONa)_2 + 2HCl \longrightarrow RCH(COOH)_2 + 2NaCl$

---

18. When malonic acid or a substituted malonic acid is heated alone or in acid solution it decarboxylates (loses carbon dioxide) and a monocarboxylic acid is formed.

$$RCH(COOH)_2 \xrightarrow{heat} \ldots\ldots\ldots\ldots \ (1) + CO_2$$

---

(1) $RCH(COOH)_2 \xrightarrow{heat} RCH_2COOH + CO_2$

---

19. The hydrolysis and decarboxylation steps can be represented by the equation

$$RR'C(COOEt)_2 \xrightarrow[(2)\ acid]{(1)\ NaOH} \ldots\ldots\ldots\ldots \xrightarrow{heat} \ldots\ldots\ldots\ldots \ (1)$$

---

(1) $RR'C(COOEt)_2 \xrightarrow[(2)\ acid]{(1)\ NaOH} RR'C(COOH)_2 \xrightarrow{heat} RR'CHCOOH$

TEST FRAME

Using abbreviated forms of the equations outline a synthesis of 2-methylpentanoic acid from diethyl malonate .... (1).

---

(1) $CH_2(COOEt)_2 \xrightarrow[(2)\ CH_3I]{(1)\ NaOEt} CH_3CH(COOEt)_2$

$$CH_3CH(COOEt)_2 \xrightarrow[(2)\ CH_3CH_2CH_2Br]{(1)\ NaOEt} CH_3CH_2CH_2\underset{\displaystyle CH_3}{\underset{|}{C}}(COOEt)_2$$

(Note that the alkylations could be done in the reverse order).

$$CH_3CH_2CH_2\underset{\displaystyle CH_3}{\underset{|}{C}}(COOEt)_2 \xrightarrow[(2)\ acid]{(1)\ NaOH} CH_3CH_2CH_2\underset{\displaystyle CH_3}{\underset{|}{C}}(COOH)_2$$

$$CH_3CH_2CH_2\underset{\displaystyle CH_3}{\underset{|}{C}}(COOH)_2 \xrightarrow{heat} CH_3CH_2CH_2\underset{\displaystyle CH_3}{\underset{|}{C}}HCOOH$$

66. *Esters* are obtained by the reaction of a carboxylic acid with an alcohol in the presence of a strong acid according to the general equation:—

$$R-C(=O)OH \quad + \quad R'OH \quad \underset{\text{catalyst}}{\overset{\text{acid}}{\rightleftharpoons}} \quad R-C(=O)OR' \quad + \quad H_2O$$

The name of the ester indicates both the acid and the alcohol from which it was formed. The ester derived from methanol and ethanoic acid (acetic acid) is called methyl ethanoate (methyl acetate). In this name the first word indicates the name of the . . . . (1) and the second the name of the . . . . (2). Methyl benzoate is formed by the reaction of . . . . (3) and . . . . (4) and it has the structure . . . . (5).

---

(1) alcohol (2) acid (3) methanol (4) benzoic acid

(5) 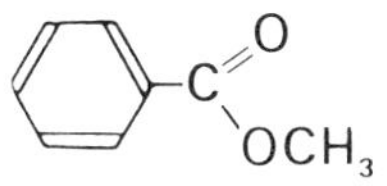

---

67. Benzyl acetate is derived from . . . . (1) and . . . . (2) and has the structure . . . . (3) The insect repellent, dimethylphthalate is derived from . . . . (4) and . . . . (5), and has the structure . . . . (6).

---

(1) benzyl alcohol (2) acetic acid

(3) 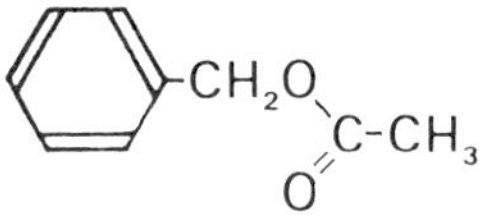

(4) methanol

(5) phthalic acid

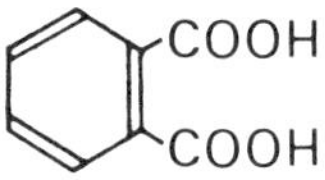

(cf. Frame 58)

(6) 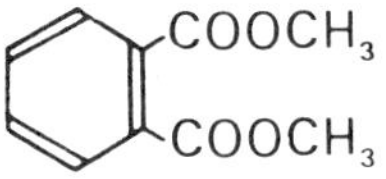

---

68. *Acid anhydrides* are derived from carboxylic acids by the elimination of water between two carboxyl groups. The commonest anhydride is acetic anhydride or ethanoic anhydride, formed from two moles of acetic acid. In phthalic acid two carboxyl groups are present and dehydration leads to intramolecular loss of water with formation of phthalic anhydride.

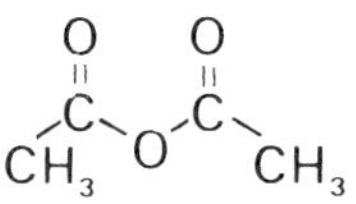

acetic anhydride

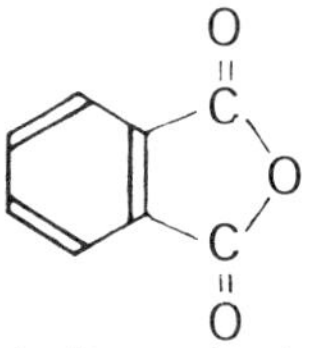

phthalic anhydride

Draw the structure of the acid anhydride which is obtained on dehydration of the general acid RCOOH . . . . (1).

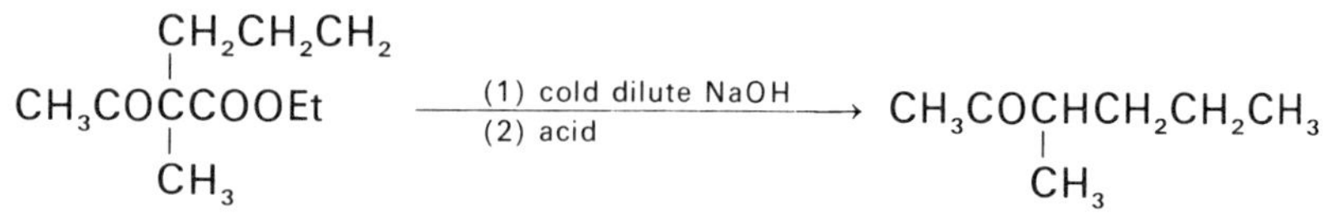

3-methylhexan-2-one

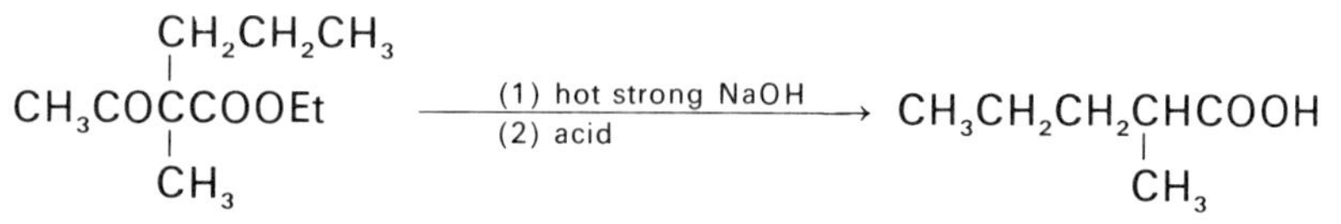

2-methylpentanoic acid

DIETHYL MALONATE

14. Diethyl malonate, $CH_3CH_2OC(=O)CH_2C(=O)OCH_2CH_3$, is often loosely called ethyl malonate or malonic ester and the structure can be written in a number of ways, e.g. $EtOOCCH_2COOEt$ or $CH_2(COOEt)_2$. It is an ester of malonic acid, (propanedioic acid) which has the structure . . . . (1).

---

(1) $HOOCCH_2COOH$

---

15. Diethyl malonate is prepared by reacting sodium chloroacetate with sodium cyanide to form sodium cyanoacetate which on reaction with excess sulphuric acid and ethanol is converted into diethyl malonate.

$$ClCH_2COONa + NaCN \longrightarrow NCCH_2COONa + NaCl$$

$$NCCH_2COONa + 2HOEt + H_2SO_4 \longrightarrow CH_2(COOEt)_2 + NH_4^+ + Na^+ + SO_4^{2-}$$

Diethyl malonate belongs to the class of . . . . (1).

---

(1) β-diesters

---

16. Diethyl malonate can be both monoalkylated and dialkylated as was described for ethyl acetoacetate. The monoalkylation of diethyl malonate can be shown by the equation

$$CH_2(COOEt)_2 \xrightarrow[(2)\ RX]{(1)\ NaOEt} RCH(COOEt)_2$$

Write the equation for the alkylation of diethyl malonate with 1-bromobutane . . . . (1).

---

(1) $CH_2(COOEt)_2 \xrightarrow[(2)\ CH_3CH_2CH_2CH_2Br]{(1)\ NaOEt} CH_3CH_2CH_2CH_2CH(COOEt)_2$

(1) 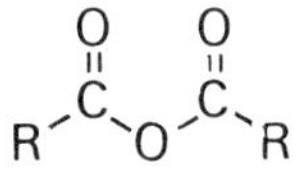

NITRILES

69. The functional group of a nitrile is $-C{\equiv}N$. In older usage these compounds were named for the acid they give on hydrolysis e.g. CH_3CN hydrolyses to .... (1) and is known as *acetonitrile*. The systematic name adds the ending *nitrile* to the name of the parent hydrocarbon, the CN carbon atom being regarded as part of the hydrocarbon chain. Acetonitrile becomes .... (2). The carbon atom of the CN group is numbered 1 as in the carboxylic acids, so that 3-chlorobutanenitrile is .... (3).

---

(1) CH_3COOH acetic acid (2) ethanenitrile (3) CH_3CHCH_2CN (with Cl on the CH)

---

70. If a compound with more than one functional group contains a –CN group but is not being named as a nitrile, then the presence of the –CN group is indicated by the prefix cyano- and a number is used to indicate its position on the open chain or ring. Cyanoethanoic acid (cyanoacetic acid) is .... (1) and 2-chloro-4-cyanotoluene is .... (2).

---

(1) $NCCH_2COOH$ (cyanoacetic acid) (2) 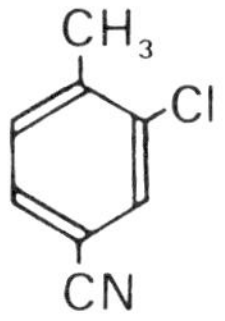

AMINES

71. Amines can be regarded as derivatives of ammonia in which one, two or three hydrogen atoms have been replaced by organic groups. A *primary amine* has one hydrogen replaced and the general formula is RNH_2; a *secondary amine* has .... (1) hydrogens replaced and the general formula is .... (2). The general formula of a *tertiary amine* is .... (3). Note that this use of primary, secondary and tertiary differs from that used with alcohols (Frame 40).

---

(1) two (2) RR′NH (3) RR′R″N

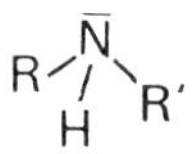

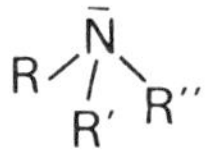

Synthesis of substituted acetic acids—hydrolysis with hot strong alkali.

12. When monosubstituted or disubstituted ethyl acetoacetates are treated with hot strong alkali they are cleaved between the α and β carbon atoms and in addition the ester group is hydrolysed, e.g.,

$$\underset{\underset{R}{|}}{\overset{\overset{R'}{|}}{CH_3COCCOOEt}} + 2NaOH \longrightarrow CH_3COONa + RR'CHCOONa + EtOH$$

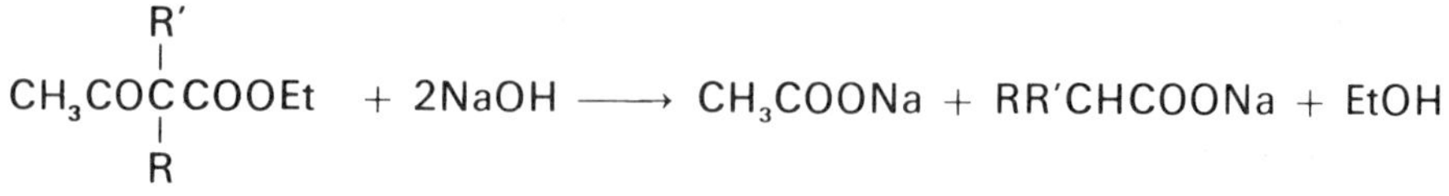

On acidification the free acids are liberated and the required substituted acetic acid . . . . (1) can be isolated and separated from the acetic acid. Note that in the last step of this synthesis a carbon-carbon bond is *broken*.

---

(1) RR'CHCOOH

---

13. The reaction can be written in abbreviated form:

$$\underset{\underset{R}{|}}{\overset{\overset{R'}{|}}{CH_3COCCOOEt}} \xrightarrow[\text{(2) acid}]{\text{(1) hot strong NaOH}} \text{. . . . . (1)}$$

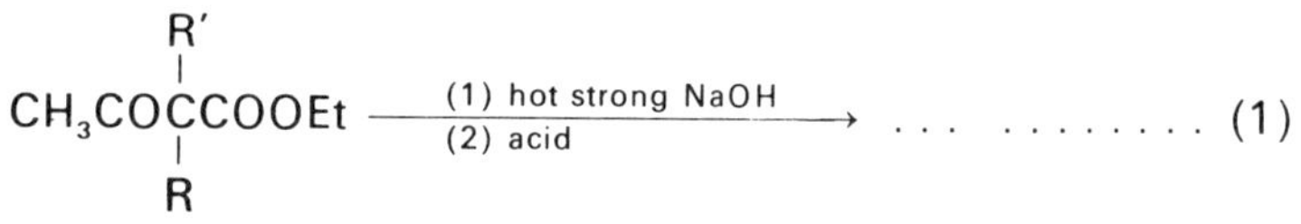

Indicate with an arrow the carbon-carbon bond which is broken in step (2).

---

(1) $\underset{\underset{R}{|}}{\overset{\overset{O}{\|}\ \overset{R'}{|}}{CH_3C-CCOOEt}} \xrightarrow[\text{(2) acid}]{\text{(1) hot strong NaOH}} RR'CHCOOH$

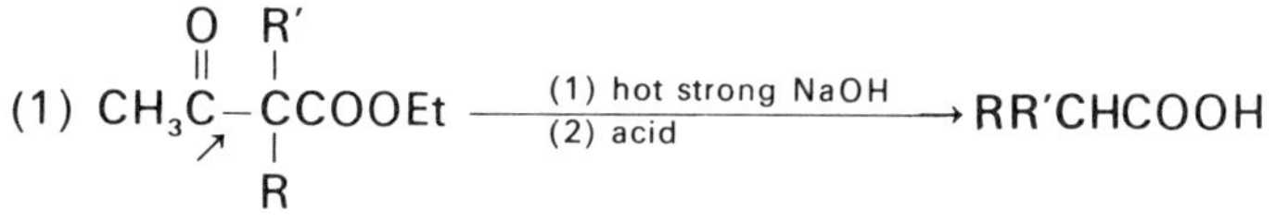

TEST FRAME

Using abbreviated forms of the equations, outline syntheses of 3-methyl-hexan-2-one and 2-methylpentanoic acid from ethyl acetoacetate . . . . (1).

---

(1) $CH_3COCH_2COOEt \xrightarrow[\text{(2) CH}_3\text{I}]{\text{(1) NaOEt}} \underset{\underset{CH_3}{|}}{CH_3COCHCOOEt}$

$\underset{\underset{CH_3}{|}}{CH_3COCHCOOEt} \xrightarrow[\text{(2) CH}_3\text{CH}_2\text{CH}_2\text{Br}]{\text{(1) NaOEt}} \underset{\underset{CH_3}{|}}{\overset{\overset{CH_2CH_2CH_3}{|}}{CH_3COCHCOOEt}}$

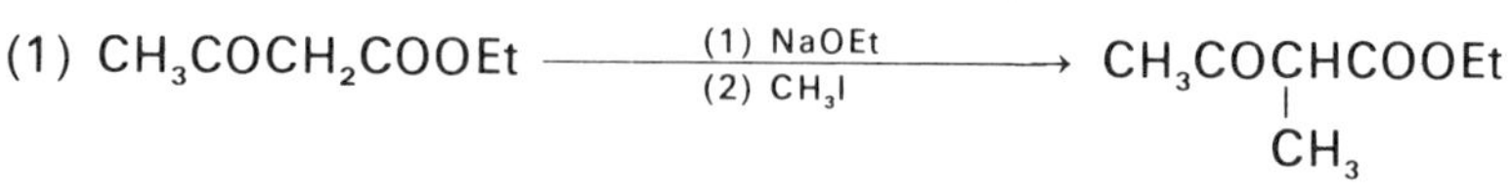

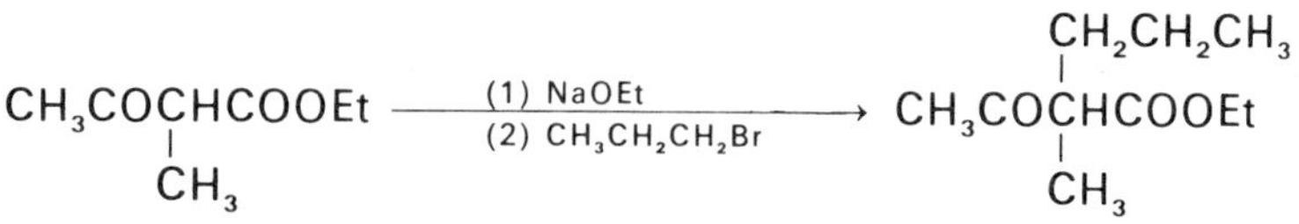

(Note that the alkylations could be done in the reverse order).

72. The amines are bases and, like ammonia, they can form salts with acids e.g. $RNH_2 + HCl \rightleftharpoons RNH_3^+ \ Cl^-$

$CH_3NHCH_2CH_3 + CH_3COOH \rightleftharpoons$ . . . . (1)

$(CH_3)_3N + C_6H_5COOH \rightleftharpoons$ . . . . (2)

---

(1) $CH_3\overset{+}{N}H_2CH_2CH_3 \ CH_3COO^-$ (2) $(CH_3)_3\overset{+}{N}H \ C_6H_5COO^-$

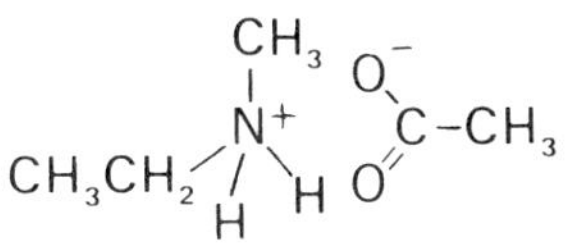

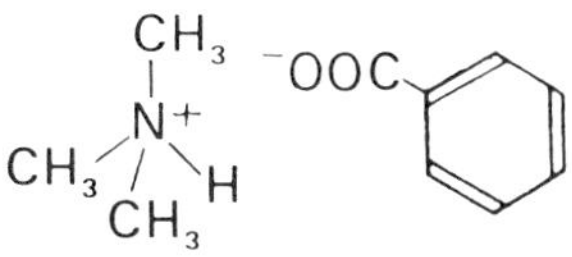

---

73. These salts are named as if they were derivatives of ammonia, substituted with alkyl or aryl groups. The simplest amine, CH_3NH_2, is called *methylamine* and the salt it forms with hydrogen chloride is called *methylammonium chloride*. Ethylamine is . . . . (1) and the salt it forms with acetic acid is . . . . (2), called . . . . (3).

---

(1) $CH_3CH_2NH_2$ (2) $CH_3CH_2\overset{+}{N}H_3 \ CH_3COO^-$ (3) ethylammonium acetate

---

74. 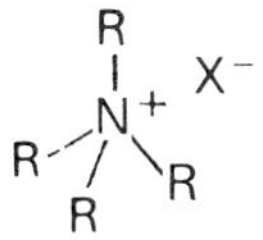 It is also possible to obtain compounds of the type illustrated where the groups R may be the same or different and X^- represents either an inorganic or an organic anion. These are called quaternary ammonium salts. Tetramethylammonium iodide has the structure . . . . (1).

---

(1) $(CH_3)_4N^+I^-$

---

75. Ethylmethylamine and diphenylamine are both examples of . . . . (1) amines and their structures are . . . . (2) and . . . . (3) respectively.

---

(1) secondary (2) $CH_3CH_2NHCH_3$

(3)

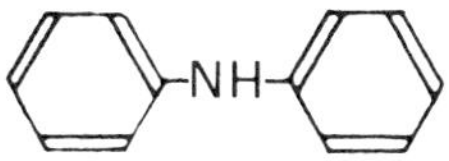

$$(1)\ CH_3COCH_2COOEt \xrightarrow[(2)\ C_2H_5X]{(1)\ NaOEt} CH_3CO\underset{\substack{| \\ C_2H_5}}{C}HCOOEt$$

$$CH_3CO\underset{\substack{| \\ C_2H_5}}{C}HCOOEt \xrightarrow[(2)\ CH_3X]{(1)\ NaOEt} CH_3CO\overset{\substack{CH_3 \\ |}}{\underset{\substack{| \\ C_2H_5}}{C}}COOEt$$

The order of addition of the alkyl groups can be reversed.

Synthesis of methyl ketones—hydrolysis with cold dilute alkali.

9. The ester group of a monoalkylated or dialkylated ethyl acetoacetate can be saponified (hydrolysed) on treatment with cold dilute aqueous alkali, e.g.,

$$CH_3CO\underset{\substack{| \\ R}}{C}HCOOEt + NaOH \longrightarrow \ldots\ldots\ldots\ldots\ (1)$$

$$(1)\ CH_3CO\underset{\substack{| \\ R}}{C}HCOOEt + NaOH \longrightarrow CH_3CO\underset{\substack{| \\ R}}{C}HCOO^-\ Na^+ + EtOH$$

---

10. On acidification, the free β-keto acid is formed which readily loses carbon dioxide (decarboxylates) to give a methyl ketone.

$$CH_3CO\underset{\substack{| \\ R}}{C}HCOO^-\ Na^+ + HCl \longrightarrow CH_3CO\underset{\substack{| \\ R}}{C}HCOOH$$

$$CH_3CO\underset{\substack{| \\ R}}{C}HCOOH \longrightarrow \ldots\ldots\ldots\ldots\ (1) + CO_2$$

$$(1)\ CH_3CO\underset{\substack{| \\ R}}{C}HCOOH \longrightarrow CH_3COCH_2R + CO_2$$

---

11. These reactions can be written in the general form

$$CH_3CO\overset{\substack{R' \\ |}}{\underset{\substack{| \\ R}}{C}}COOEt \xrightarrow[(2)\ acid]{(1)\ cold\ dilute\ NaOH} \ldots\ldots\ldots\ldots\ (1)$$

$$(1)\ CH_3CO\overset{\substack{R' \\ |}}{\underset{\substack{| \\ R}}{C}}COOEt \xrightarrow[(2)\ acid]{(1)\ cold\ dilute\ NaOH} CH_3COCHRR'$$

76. 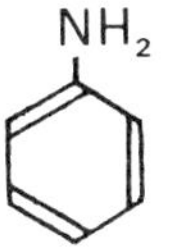 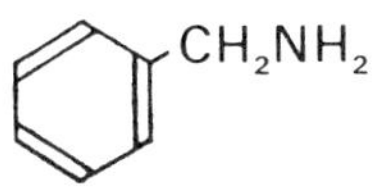

I II

Structure I which might be given the names phenylamine or aminobenzene is commonly known as *aniline*. Structure II is named normally and is . . . . (1). Both are . . . . (2) amines.

---

(1) benzylamine (2) primary

---

77. The benzene derivatives which have one methyl group and one *amino group* ($-NH_2$) both separately attached to the ring are commonly called toluidines. Draw and name the isomeric toluidines . . . . (1). Toluidines can be named systematically as aminotoluenes.

---

(1)

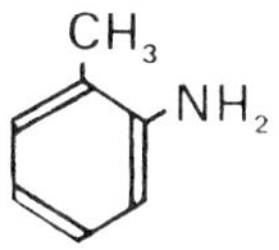

o-toluidine

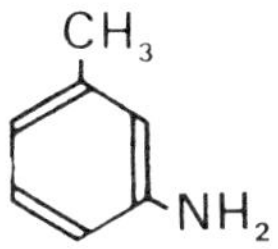

m-toluidine

p-toluidine

---

78. Sometimes, in naming a secondary or tertiary amine, it is necessary to indicate clearly that the substituent group is attached to the nitrogen and not to some part of the carbon chain. As already shown with the amides, this is done by writing N- before the name of the R group. N,N-dimethylaniline has the structure . . . . (1) and it is a . . . . (2) amine.

---

(1) 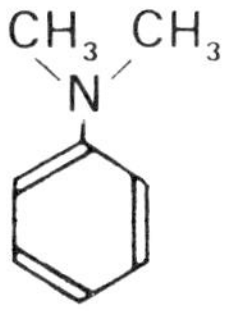

(2) tertiary

---

79. An amino group may be present in a molecule which contains in addition another functional group. Many amino acids are known and these contain both $-COOH$ and $-NH_2$ groups. In naming such compounds the $-COOH$ group has priority over the $-NH_2$ group and the compound is named as an acid with an amino group as substituent. The amino acid known as alanine has structure I and is named systematically as 2-aminopropanoic acid.

$CH_3CHCOOH$ (with NH_2 on the CH) — I

$H_2NCH_2CH_2OH$ — II

Structure II is named as an alcohol and has the systematic name . . . . (1) and the common name ethanolamine.

Alkylation of ethyl acetoacetate

6. Many useful compounds can be made by processes in which the first step is the alkylation of ethyl acetoacetate. The procedure is as follows:
Ethyl acetoacetate is added to a solution of sodium ethoxide in ethanol and then an alkyl halide is added. There is a slow precipitation of sodium halide and the mixture is heated until reaction is complete. Ethanol is removed by distillation, water is added, and the mixture is neutralised. The product is usually isolated by extraction with ether. The β-keto ester is first converted into the sodium salt, that is it forms the conjugate base.

$$CH_3COCH_2COOEt + Na^+ \rightleftharpoons {}^-OEt\ CH_3CO\bar{C}HCOOEt\ Na^+ + EtOH$$

The carbanion then reacts with the alkyl halide causing a nucleophilic substitution and a new carbon-carbon bond is formed (cf. Program 4).

$$CH_3CO\bar{C}HCOOEt\ Na^+ + RX \longrightarrow \ldots\ldots\ldots\ldots \quad (1).$$

---

(1) $CH_3CO\bar{C}HCOOEt\ Na^+ + RX \longrightarrow CH_3CO\underset{R}{\underset{|}{C}}HCOOEt + Na^+\ X^-$

---

7. The overall reaction consists in the substitution of an α-hydrogen atom by an alkyl group and it is called. . . . . (1). It can be written in the abbreviated form:

$$CH_3COCH_2COOEt \xrightarrow[(2)\ RX]{(1)\ NaOEt} CH_3CO\underset{R}{\underset{|}{C}}HCOOEt$$

Write a similar equation for the alkylation of ethyl acetoacetate with methyl iodide. . . . . (2).

---

(1) alkylation

(2) $CH_3COCH_2COOEt \xrightarrow[(2)\ CH_3I]{(1)\ NaOEt} CH_3CO\underset{CH_3}{\underset{|}{C}}HCOOEt$

---

8. The product of this reaction in which one α-hydrogen has been substituted by an alkyl group is called a monoalkylated product. This substance can be further alkylated by a similar process, shown in abbreviated form

$$CH_3CO\underset{R}{\underset{|}{C}}HCOOEt \xrightarrow[(2)\ R'X]{(1)\ NaOEt} CH_3CO\overset{R'}{\overset{|}{\underset{R}{\underset{|}{C}}}}COOEt$$

Write equations for the synthesis of ethyl 2-ethyl-2-methyl-3-oxobutanoate from ethyl acetoacetate. . . . . (1).

---

(1) 2-aminoethanol

POLYFUNCTIONAL COMPOUNDS

80. When more than one functional group is present the IUPAC system gives priority to the groups which inevitably occur at the end of the chain, such as –COOH, –CHO and –CN. The carboxyl group also has priority over the other two and so we named $NCCH_2COOH$ as cyanoethanoic acid. For the common substituents the decreasing order of priority is as follows: ammonium group, carboxylic acid, sulphonic acid, acid chloride, amide, aldehyde, ketone, nitrile, alcohol, phenol, thiol, amine, alkyl ether, alkyl sulphide, and the halogen and nitro groups. Lactic acid $CH_3CHCOOH$ (with OH on the CH) which occurs in sour milk, is called .... (1). Pyruvic acid CH_3CCOOH (with =O on the C) is called .... (2) and ethyl 3-oxobutanoate (ethyl acetoacetate) has the structure .... (3).

---

(1) 2-hydroxypropanoic acid (2) 2-oxopropanoic acid (cf. Frame 53).

(3) $CH_3CCH_2COOCH_2CH_3$ (with =O on the second C)

TEST FRAMES

The compounds which follow are all polyfunctional compounds, but the application of the nomenclature rules already given should allow the structures to be deduced from the names, or names to be given to structures. Draw structures for 2-aminobutan-1-ol .... (1), 3-cyanopropanal .... (2), N-methyl-N-phenylbenzamide .... (3) and cyclopropane-1,2-dicarboxylic acid .... (4).

---

(1) $CH_3CH_2CHCH_2OH$ (with NH_2 on the third C)

(2) $NCCH_2CH_2CHO$

(3)

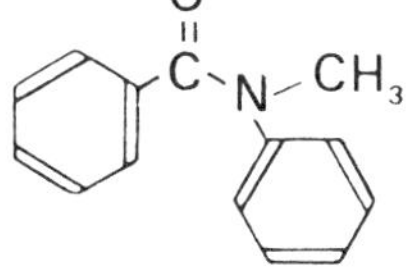

(4)

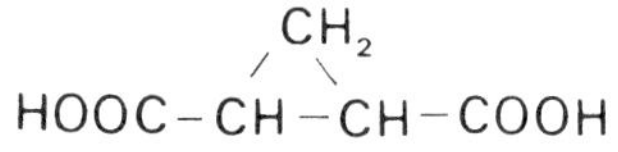

Give names to

(5) $CH_3CH{-}CHCN$ (with H_2N on the second C and Cl on the third C)

(6)

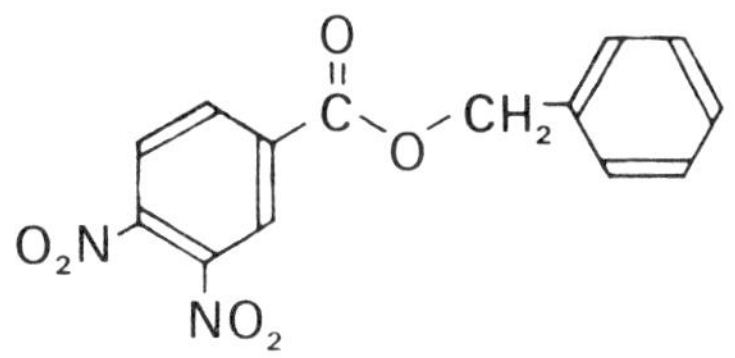

contains two functional groups, an . . . . (1) and a . . . . (2). It is a member of the general class of compounds called . . . . (3).

---

(1) ester (2) carbonyl (as a ketone) (3) β-keto esters

---

3. Ethyl acetoacetate is prepared by reacting ethyl acetate containing a trace of ethanol with metallic sodium. The resulting mixture is neutralised with acid and the product is isolated. The overall reaction is

$$2CH_3COOEt \longrightarrow CH_3COCH_2COOEt + EtOH$$

Sodium ethoxide, formed by reaction of sodium with ethanol, causes condensation of two molecules of ethyl acetate. The sodium salt of ethyl acetoacetate is formed together with ethanol which further reacts with the metallic sodium to form sodium ethoxide and so continues the reaction.

$$2CH_3COOEt + \overset{+}{Na}\ \bar{O}Et \longrightarrow CH_3CO\bar{C}HCOOEt\ Na^+ + 2EtOH$$

Acidification of the end product proceeds according to the equation

$$CH_3CO\bar{C}HCOOEt\ \ Na^+ + HCl \longrightarrow \text{. . . . . (1).}$$

---

(1) $CH_3CO\bar{C}HCOOEt\ \ Na^+ + HCl \longrightarrow CH_3COCH_2COOEt + NaCl$

---

4. The usefulness of ethyl acetoacetate in synthesis depends on the fact that the α-hydrogens are weakly acidic. The anion (conjugate base) is formed on addition of the ester to sodium ethoxide in ethanol.

$$CH_3COCH_2COOEt + Na^+\ {}^-OEt \longrightarrow \text{. . . . . (1)}$$

---

(1) $CH_3COCH_2COOEt + Na^+\ {}^-OEt \longrightarrow CH_3CO\bar{C}HCOOEt\ Na^+ + EtOH$

---

5. The negative charge on the ethyl acetoacetate anion can be delocalised over a number of atoms. Draw the structures of the three principal mesomeric anions . . . . (1).

---

(1)

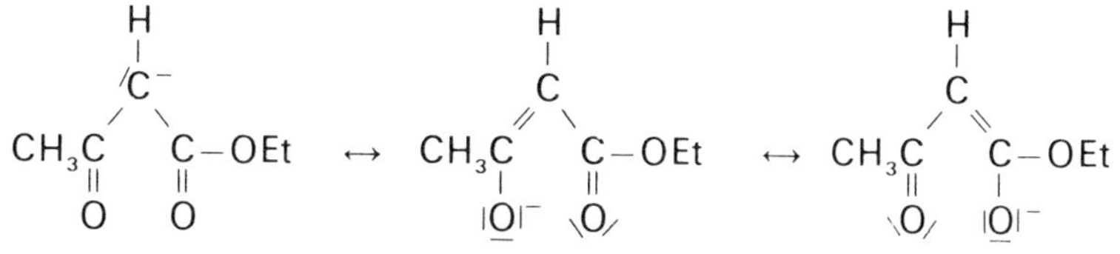

Note that the remaining α-hydrogen atom could be replaced by an alkyl group without affecting the mesomeric structures.

(7) $H_2C{=}CHCH_2COOH$ (8) $CH_3\underset{\displaystyle O}{\underset{\|}{C}}CH_2CH_2COOC_2H_5$

---

(5) 3-amino-2-chlorobutanenitrile (6) benzyl 3,4-dinitrobenzoate
(7) but-3-enoic acid (8) ethyl 4-oxopentanoate

Draw structures for: N,N-dimethylaniline .... (9), N,N-dimethylanilinium acetate .... (10), dec-5-ynedioic acid .... (11) and 2-oxocyclohexanecarboxaldehyde .... (12).

---

(9) 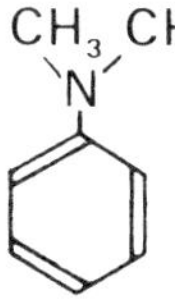

(10) 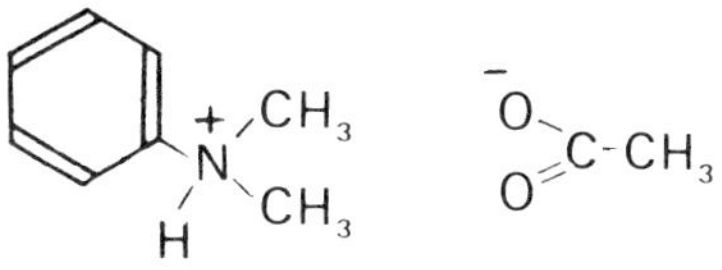

(11) $HOOCCH_2CH_2CH_2C{\equiv}CCH_2CH_2CH_2COOH$

(12) 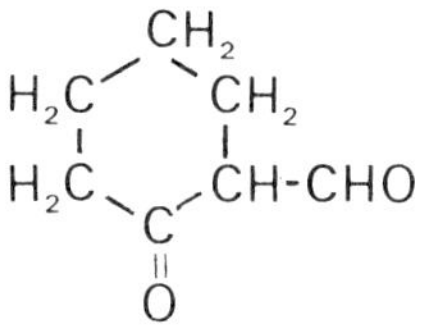

Give names to:

(13) 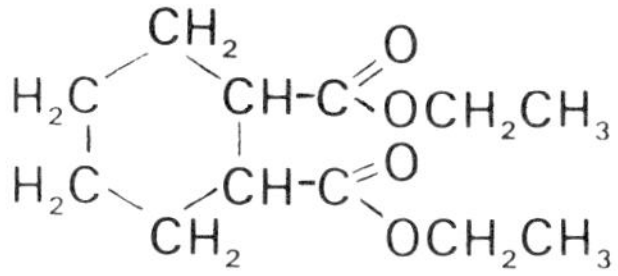

(14) $CH_3CH_2CH_2CH_2CH_2\underset{\displaystyle O}{\underset{\|}{C}}CH_2NH_2$

(15) 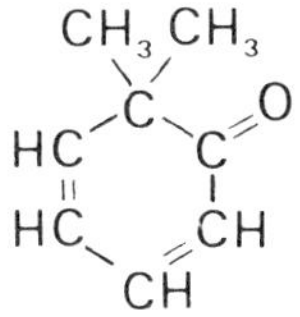

---

(13) diethyl cyclohexane-1,2-dicarboxylate (14) 1-aminoheptan-2-one
(15) 6,6-dimethylcyclohexa-2,4-dienone

SYNTHESIS OF ORGANIC COMPOUNDS

Program 12

The first section of this program is similar to the previous programs in that new information is being imparted. Several synthetic procedures involving the formation of carbon-carbon bonds are discussed. The second section is concerned with application of the chemical reactions discussed in Programs 3-11 to problems of chemical synthesis. It is not sufficient in this latter section to know of a reaction; you must be able to see its applicability and to make use of it in a chemical situation.

GRIGNARD REAGENTS (cf. Program 4, Frames 66-70; Program 6, Frames 33-41; Program 7, Frame 49; and Program 9, Frame 20)

1. Grignard reagents, RMgX, are prepared by the reaction of alkyl or aryl halides with magnesium metal in anhydrous ether solution.

$$RX + Mg \xrightarrow{\text{diethyl ether}} RMgX$$

The structure of the Grignard reagent can be thought of in terms of mesomeric contributions of the type

$$R-Mg-X \leftrightarrow R-Mg^{+}X^{-} \leftrightarrow R^{-}Mg^{+}-X \leftrightarrow R^{-}Mg^{2+}X^{-}$$

The polarisation of a Grignard reagent can be indicated by the structure $\overset{\delta-}{R}-\overset{\delta+}{Mg}-\overset{\delta-}{X}$, and the group R has nucleophilic character. Grignard reagents react with the electrophilic carbon atoms of reactive alkyl halides. The reaction usually requires the presence of cobalt(II) chloride as a catalyst and the mechanism is complex.

$$RMgX + R'X \xrightarrow{CoCl_2} R-R' + MgX_2$$

The equation for the reaction of ethyl magnesium bromide with 1-bromobutane is. . . . . (1), and the hydrocarbon formed is called. . . . . (2).

---

(1) $CH_3CH_2MgBr + CH_3CH_2CH_2CH_2Br \xrightarrow{CoCl_2} CH_3(CH_2)_4CH_3 + MgBr_2$

(2) hexane

ETHYL ACETOACETATE

2. See Program 7 for a discussion of the formation of enolate ions and the phenomenon of tautomerism.

The ester ethyl acetoacetate CH_3COCH_2COOEt or $CH_3\overset{O}{\overset{\|}{C}}CH_2\overset{O}{\overset{\|}{C}}OCH_2CH_3$ is also called acetoacetic ester or ethyl 3-oxobutanoate. The compound

SOME COMMENTS ON SYSTEMS OF NOMENCLATURE

Nomenclature is a tool for the use of chemists. "Correct" nomenclature like "correct" spelling, is a consensus of opinion. The IUPAC (1957) rules of nomenclature are accepted by most chemists as the basis of sound practice, but even with these rules there are allowable variations. As a simple example, $CH_3CH=CHCH_2CH_2OH$ would be named pent-3-en-1-ol in current British journals but 3-penten-1-ol in American. Likewise, the order of naming substituent groups can be done alphabetically or according to increasing complexity. The former method has been used throughout the programs. It must also be realised that the IUPAC system differs markedly from the systems used in two major reference works, Beilstein's *Handbuch der organischen Chemie* and *Chemical Abstracts*.

While we have used IUPAC nomenclature as far as possible in these programs, we have also shown in many cases one or more alternative names that are frequently found in current literature. We have done this because first-year text books in organic chemistry are usually indexed on the basis of semi-systematic and trivial names. For example if only the name but-*trans*-2-enedioic acid were to be used in these programs, there would be no way of knowing that this compound is indexed (and indeed discussed) under the name fumaric acid in text books. Eradication of all trivial names could thus make communication between chemists more difficult, and students who had learnt only systematic nomenclature would not be properly equipped to read the existing chemical literature. In studying the programs, the basic chemistry can be learnt without the need to memorise trivial names. The learning process in organic chemistry follows from an understanding of the chemical reactions and not from a mastery of names. Advanced students of organic chemistry find that a knowledge of systematic, semi-systematic and trivial names is essential, but this additional learning is made easier if the IUPAC system is mastered first.

Program 12: SYNTHESIS OF ORGANIC COMPOUNDS

CONTENTS

Program 2: FORMULAE, STRUCTURES, EQUATIONS AND REACTION MECHANISMS

CONTENTS

THE SEQUENCE RULE AND THE NAMING OF STEREOISOMERS

The sequence rule is a method of specifying substituents in an order of priority to allow unambiguous naming of stereoisomers. In assigning precedence to substituents on a carbon atom, the atoms directly bonded are considered first. The basic rule is that higher atomic number precedes lower, e.g. $I > Br > Cl > F > OH > NH_2 > CH_3 > H$. If substituents contain a common atom the further atoms must be taken into account, e.g. $CH_2Cl > CH_2OH$. The decisions are taken by working outward concurrently along the groups in question, atom by atom, until the first point of difference is found, thus $CH_2CH_2F > CH_2CH_2NH_2$. Atoms doubly or triply bonded are considered to be bonded to the multiple of the atom concerned, e.g. $COOH[C(O, O, OH)] > COCH_3[C(O, O, CH_3)] > CHO[C(O, O, H)] > CH_2OH[C(OH, H, H)] > C \equiv N[C(N, N, N)] > C_6H_5[C(CH, C, CH)]$.

cis-trans-*Isomerism around double bonds*. For each carbon atom in a double bond the attached atom or group having the higher priority is selected. Then, the two possible arrangements are either higher priority groups *cis*—designated by *Z* (zusammen, together), or higher priority groups *trans*—designated by *E* (entgegen, opposite).

e.g. 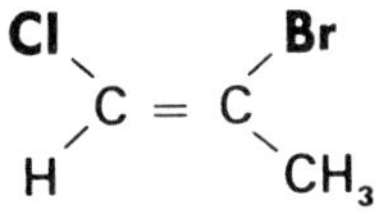

For each pair of substituents, the group of higher priority is shown bold

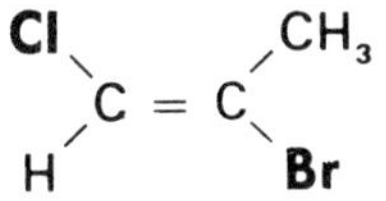

(*Z*)-2-bromo-1-chloropropene (*E*)-2-bromo-1-chloropropene

Optical isomerism. For tetrahedral carbon atoms the sequence rule can be used to specify the handedness or chirality of a molecule. In order to name the enantiomers of a compound such as glyceraldehyde (2, 3-dihydroxypropanal, $HOCH_2CH(OH)CHO$) the order of precedence of the four groups, $OH > CHO > CH_2OH > H$, is determined and the molecule is viewed from the side *remote* from the atom of lowest precedence. In this example, H (lowest priority) is concealed behind the carbon atom.

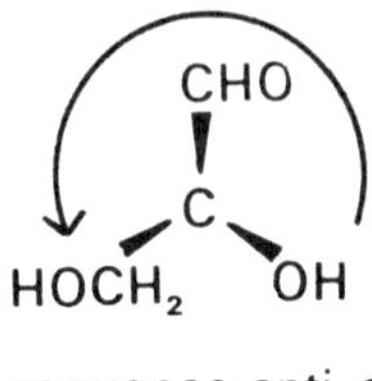

group sequence anti-clockwise
S = sinister, left-handed

(2*S*)-2, 3-dihydroxypropanal

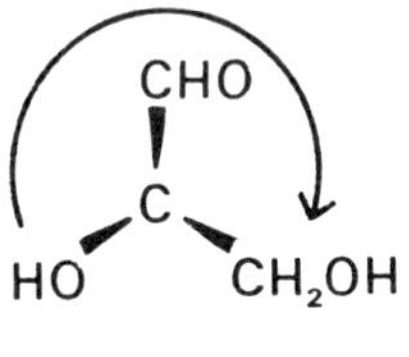

group sequence clockwise
R = rectus, right-handed

(2*R*)-2, 3-dihydroxypropanal

FORMULAE, STRUCTURES, EQUATIONS AND REACTION MECHANISMS

Program 2

MOLECULAR FORMULAE

1. Molecular formulae define the number and types of atoms in a molecule. Thus methane is CH_4, acetone C_3H_6O, glucose $C_6H_{12}O_6$. Valency rules were developed from molecular formulae when it was observed that the number of atoms of a given element present in any stable molecule .... (is/is not) (1) dependent on the number and type of the other elements present.

---

(1) is

---

2. Molecular formulae are obtained from a combination of elemental analysis and measurement of molecular weight. Suppose that the molecular weight of a given compound is determined as 113 ± 2 and that within the limits of experimental error in the analysis, the elemental composition can be calculated as $C_6H_9O_2$ (mol. wt. 113) or $C_6H_{10}O_2$ (mol. wt. 114) Since the valencies of carbon and of oxygen are both even, the number of (univalent) hydrogen atoms must also be even, so that the correct molecular formula is .... (1).

---

(1) $C_6H_{10}O_2$

---

3. In general for any molecular formula the sum of the odd-valence atoms must be even. Indicate by writing "yes" or "no" whether or not the following molecular formulae are possible:
C_5H_7OCl .... (1); $C_{11}H_{22}O_3N$ .... (2); $C_8H_6O_2NSCl$ .... (3).

---

(1) yes—the sum of H and Cl atoms is even
(2) no—the sum of H and N atoms is odd
(3) yes—the sum of H, N and Cl atoms is even

STRUCTURAL FORMULAE

4. The simple hydrides of the elements carbon, nitrogen, oxygen, and chlorine can be represented by the following *structural formulae*

46. Certain *cis-trans* or geometrical isomers can also be viewed as optical isomers. *cis-trans*-Isomers have the same substituents differently arranged in space on a rigid framework (compare Frame 12). This framework can alternatively be a ring of carbon atoms. Thus *cis*-1, 2-dibromocyclopentane has the two bromine atoms on the same side of the 5-membered ring, while the *trans* isomer has them on opposite sides of the ring. The *cis* form has two asymmetric carbon atoms, while the molecule as a whole has a plane of symmetry, so that the compound is a . . . . (1) form, and is optically

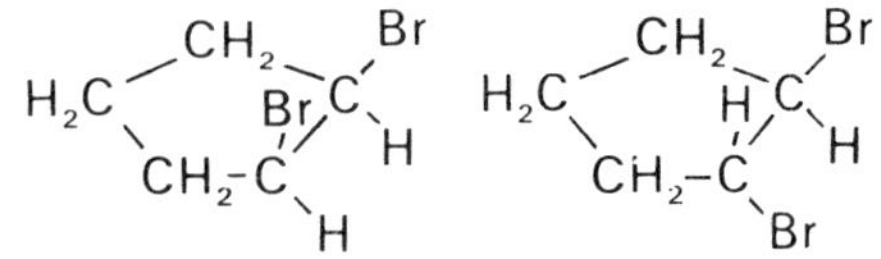

inactive. The *trans* form again possesses two asymmetric carbon atoms but these have. . . . . (the same/opposite) (2) configuration. The question of internal compensation does not arise, and the molecule. . . . . (can/cannot) (3) be resolved into two enantiomeric forms.

(1) meso (2) the same (3) can

47. Less common forms of molecular asymmetry are those analogous to the spiral staircase and the two-bladed propeller. The molecule "hexahelicene" has a spiral shape and has been resolved into two enantiomeric forms, one consisting of left-handed and the other of right-handed spiral molecules.

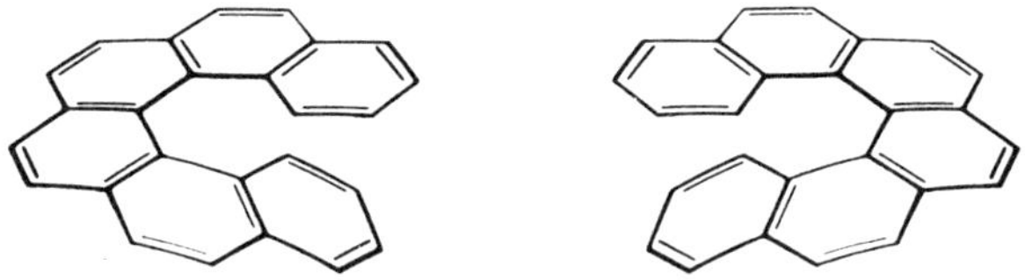

In the substance biphenyl (I), free rotation is possible around the carbon-carbon bond joining the two rings. In biphenyl molecules having bulky substituents in the *ortho*- positions e.g. II, free rotation around the central bond is hindered because of the steric interference between the opposed substituent groups. The molecule then adopts a twisted configuration.

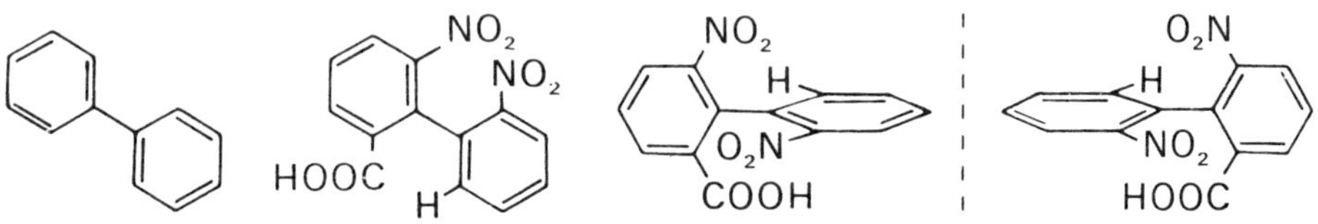

I II IIa IIb

The mirror images (IIa and IIb) . . . . (are/are not) (1) superimposable. If the carboxyl group of II is replaced by a nitro group, the mirror images. . . . . (are/are not) (2) superimposable.

(1) are not (2) are

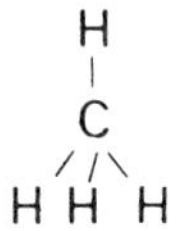 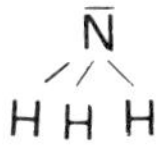 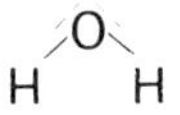 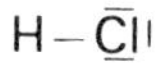

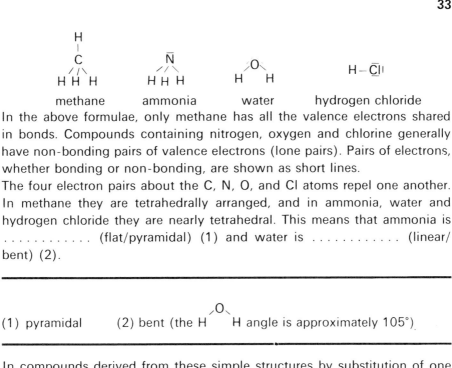

methane ammonia water hydrogen chloride

In the above formulae, only methane has all the valence electrons shared in bonds. Compounds containing nitrogen, oxygen and chlorine generally have non-bonding pairs of valence electrons (lone pairs). Pairs of electrons, whether bonding or non-bonding, are shown as short lines.

The four electron pairs about the C, N, O, and Cl atoms repel one another. In methane they are tetrahedrally arranged, and in ammonia, water and hydrogen chloride they are nearly tetrahedral. This means that ammonia is .... (flat/pyramidal) (1) and water is .... (linear/ bent) (2).

---

(1) pyramidal (2) bent (the H–O–H angle is approximately 105°)

---

5. In compounds derived from these simple structures by substitution of one or more hydrogen atoms by alkyl groups, the central atom tends to retain its tetrahedral electron distribution. Thus the C-O-H bonds in alcohols and the C-O-C bonds in ethers are .... (1) and the number of non-bonding electron pairs on oxygen is .... (2).

---

(1) bent (2) two

---

6. The following structural formulae show double and triple bonding (multiple bonding) between like and unlike atoms

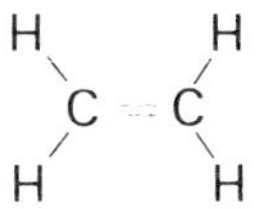 $H-C\equiv C-H$ 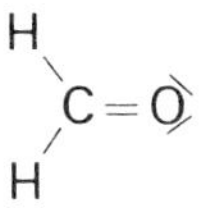 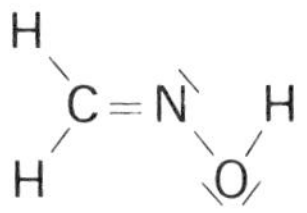 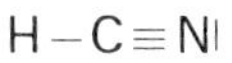

ethene (ethylene) ethyne (acetylene) methanal (formaldehyde) methanal oxime (formaldoxime) hydrogen cyanide

Molecules in which the atoms are joined by single bonds only are said to be saturated; when multiple bonds are present the molecule is said to be unsaturated. Lone pairs can be found on oxygen and nitrogen atoms in both saturated and unsaturated molecules. In formaldoxime there are .... (1) lone pairs of electrons, ... (2) on the nitrogen atom and .... (3) on the oxygen atom.

---

(1) three (2) one pair (3) two pairs

However, the resulting carbanion is not usually able to retain its "quasi-tetrahedral" shape, but like the ammonia molecule, undergoes rapid inversion.

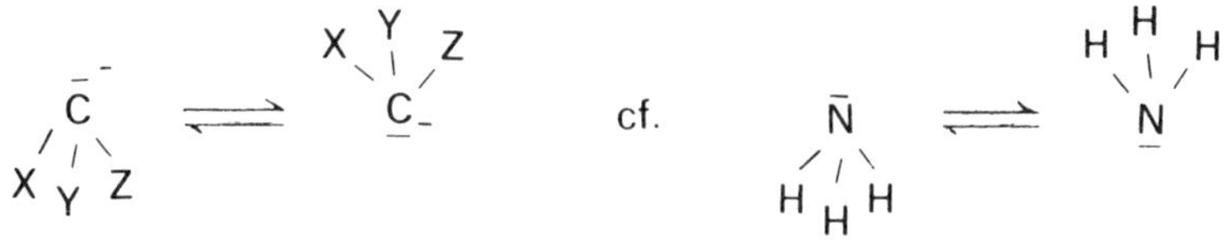

Protonation of the carbanion (the right to left step of the acid-base equilbrium) then gives rise not only to the original asymmetric substance, but to a statistically equal amount of the. . . . . (1).

---

(1) enantiomer (cf. Frames 25-27).

---

45. *Glucose*. In Frame 31 a linear form of the glucose molecule was shown. It is now known that glucose can exist in the solid state as one or other of two cyclic structures, α-glucose and β-glucose, having different melting points, solubilities and optical rotations. In aqueous solution both forms come into equilibrium with each other and with the linear structure (II). These relationships can be shown in 3-dimensional projection as

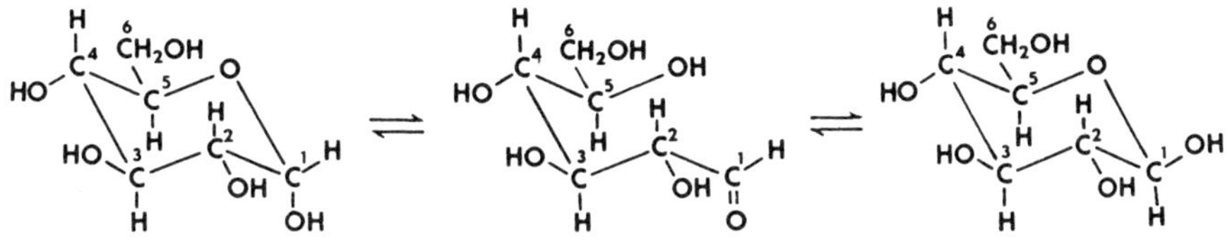

I α-Glucose   II   III β-Glucose

In forming the cyclic compounds, the hydroxyl group attached to C-5 in II adds across the carbonyl C=O double bond, to give a hemiacetal, with production of a new centre of asymmetry. Thus

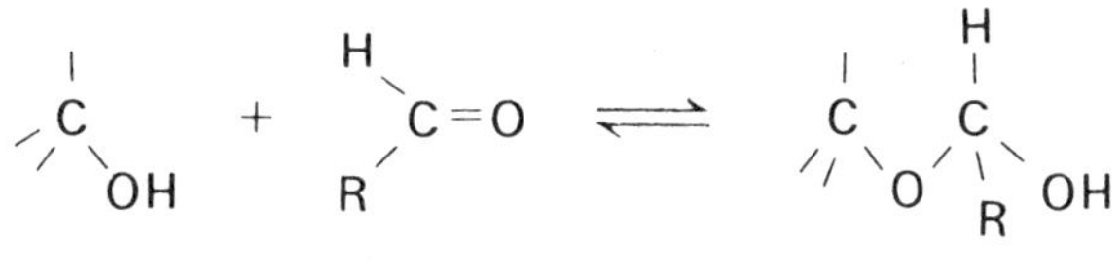

(cf. Program 7, Frame 59)

The two forms I and III differ only in the configuration of the four (different) groups around carbon atom number 1, while the configurations around carbon atoms 2, 3, 4 and 5 are the same in each case. Hence compounds I and III are. . . . . (enantiomers/diastereomers) (1). By considering each atom in turn write structure II as a Fischer projection. . . . . (2).

---

(1) diastereomers—i.e., the compounds are stereoisomers but are not mirror image forms.

(2) Refer to Frame 31 for D-glucose.

7. The disposition of the four electron pairs around a carbon, nitrogen, or oxygen atom is considerably modified if the element is participating in a multiple bond. Given that ethene (ethylene) is a planar molecule, write down the atoms in methanal which are in the same plane . . . . (1).

---

(1) 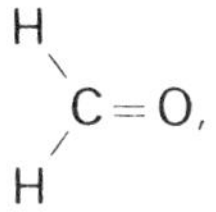 i.e. this molecule is also planar

---

8. Now write down the atoms in methanal oxime which are in the same plane . . . . (1).

---

(1) 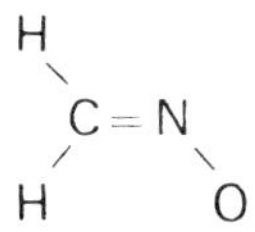

Due to free rotation about the N-O bond as a result of thermal energy, the hydrogen atom attached to oxygen describes a circular path. It passes through the plane of the remaining atoms twice each revolution.

---

9. In certain cases, e.g. benzene, C_6H_6, it is not possible to write a convincing structure for the molecule using single bonds, multiple bonds, and lone pairs. Physical evidence shows that the benzene molecule is hexagonal, with all the bond angles identical, and all the carbon-carbon bonds of equal length. Since single and double bonds are not equal in length, the representation of benzene as I cannot be correct. Despite this, a convenient shorthand structural formula for benzene is II, or the equivalent structure III, and these abbreviations are used extensively in these programs.

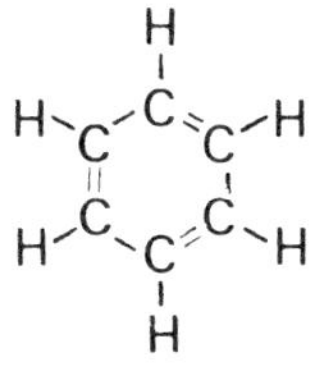

I

II

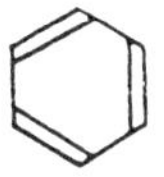

III

Write molecular formulae corresponding to the following abbreviated structures:

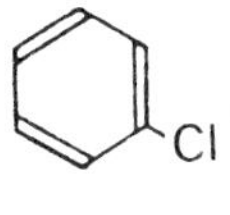 . . . . (1) 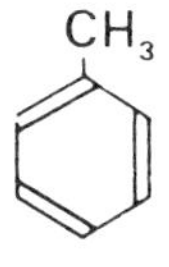 . . . . (2) 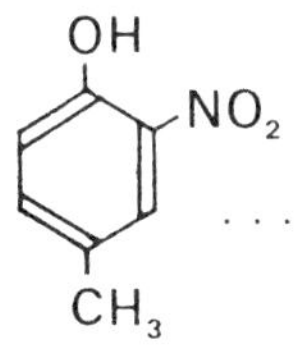 . . . . (3)

---

(1) C_6H_5Cl (2) C_7H_8 (3) $C_7H_7NO_3$

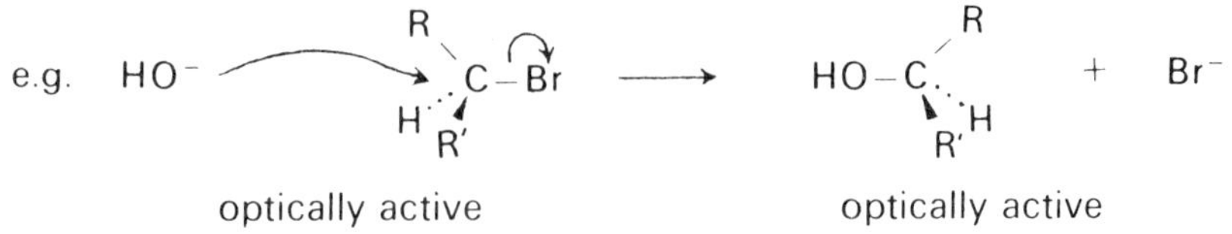

e.g.

$$HO^- \longrightarrow \quad \underset{H}{\overset{R}{\underset{R'}{C}}}-Br \longrightarrow \quad HO-\underset{R'}{\overset{R}{\underset{H}{C}}} \quad + \quad Br^-$$

optically active   optically active

The fact that inversion has occurred in this process of forming an optically-active secondary alcohol from an optically-active alkyl bromide is not easy to prove.

---

(1) inversion   (2) enantiomer, or mirror image form

Racemisation

43. It is sometimes observed that when an optically-active compound is subjected to certain chemical treatments, it loses its optical activity and gives rise to the racemic form of the same compound. The original substance is then said to have been racemised, or to have undergone racemisation.

Inversion occurs when optically-active 2-iodo-octane (e.g. the (+)-enantiomer) is heated with sodium iodide in aqueous acetone. The substitution of one iodine atom by another can be shown:

$$I^- \quad \underset{H \quad C_6H_{13}}{\overset{CH_3}{C}}-I \quad \rightleftharpoons \quad \left[I--\underset{H \quad C_6H_{13}}{\overset{CH_3}{C}}--I \right]^- \quad \rightleftharpoons \quad I-\underset{C_6H_{13}}{\overset{CH_3}{C}}{}_H \quad I^-$$

$$(-) \qquad\qquad\qquad\qquad\qquad\qquad\qquad (+)$$

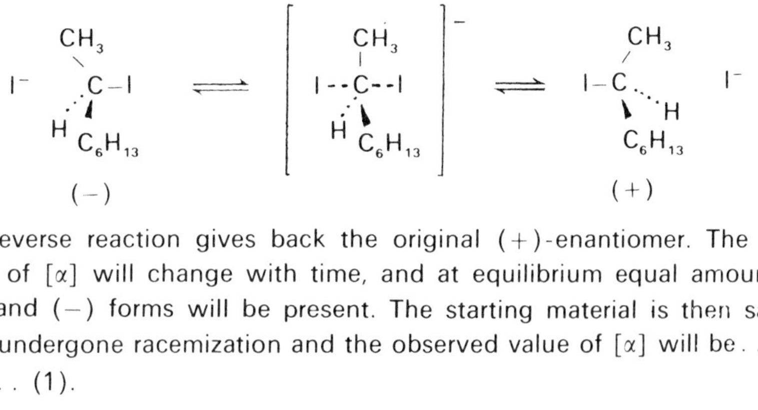

The reverse reaction gives back the original (+)-enantiomer. The initial value of [α] will change with time, and at equilibrium equal amounts of (+) and (−) forms will be present. The starting material is then said to have undergone racemization and the observed value of [α] will be . . . . (1).

---

(1) zero

---

44. Suppose we have an optically-active compound HCXYZ, $[\alpha]_D^{20} = +16°$, in which the hydrogen atom attached to the asymmetric carbon is weakly acidic (cf. Program 7, Frames 16-20). When this acidic compound is dissolved in alkali an acid-base equilibrium will be established:

$$\underset{X \quad Y \quad Z}{\overset{H}{\overset{|}{C}}} \quad + \quad OH^- \quad \rightleftharpoons \quad \underset{X \quad Y \quad Z}{\overset{-}{C}} \quad + \quad H_2O$$

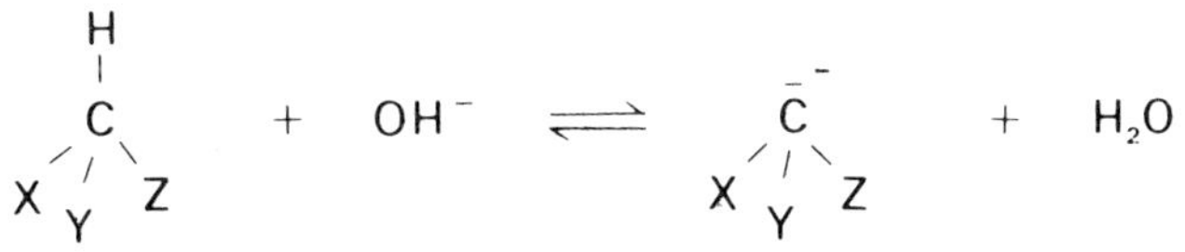

10. *Mesomerism.* A structure for benzene could be written in which the atoms are held together by single (2-electron) bonds, but this would leave a further six electrons unaccounted for. It could then be suggested that these electrons are accommodated in bonds which are effectively smeared out over all six carbon atoms. A formula which endeavours to convey this idea is shown in IV, and an abbreviated form is shown in V.

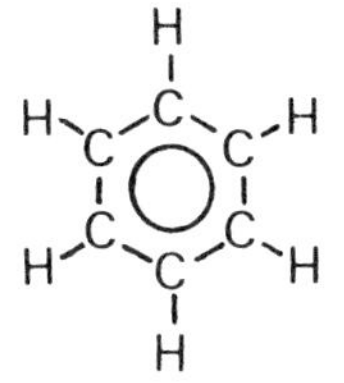

IV

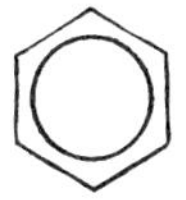

V

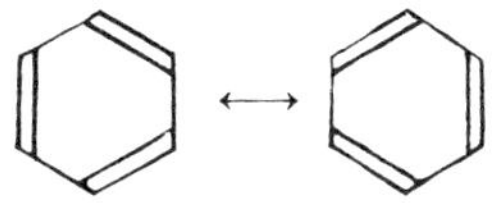

VI

In the representation VI the double-headed arrow is meant to imply that the structure lies somewhere between the two extreme forms shown. The arrow does NOT mean that each molecule is interchanging rapidly from one structure to another. The representation VI is equivalent to V, and makes use of the principle of mesomerism: Whenever a molecule or ion can be represented by structures having the same arrangement of atomic nuclei but different distributions of valence electrons, the actual electronic structure is a weighted average of these different distributions.

It can be shown by physical methods that the carbon-oxygen bonds in the acetate ion, CH_3COO^-, are of equal length. Suggest a mesomeric structure for the carboxylate group, $-COO^-$, in the acetate anion to account for the equality in bond length . . . . (1).

---

(1) 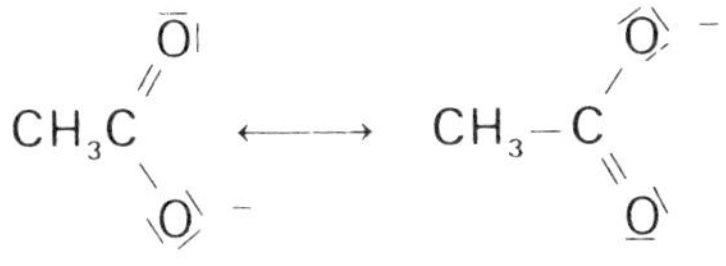

These are the principal contributors to the mesomeric structure but others, such as (i), can be envisaged. The carboxylate ion is sometimes shown as in formula (ii).

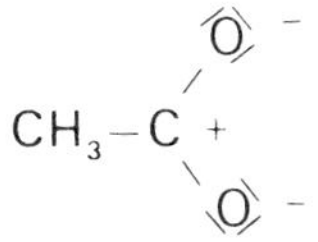

(i)

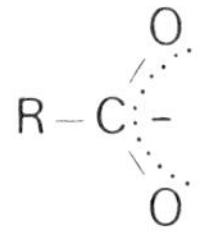

(ii)

Conventions in writing structural formulae

11. The full structural formula of an organic substance is a specification of which atoms are joined to which in the molecule, the way in which the atoms are arranged relative to each other in space, and the disposition of electron lone pairs, if any. It is not easy to show all these features by symbols drawn on

(1) inactive (because two of the groups attached to the initially asymmetric carbon atom are now identical)
(2) D-valine (because the L-valine initially present in equal amount to the D-valine will have undergone decarboxylation)
(3) will. An enzymic reaction of this kind thus constitutes another method of resolution (with destruction of one of the enantiomers).

---

41. *Inversion.* A right-hand glove can be converted into a left-hand glove by turning it inside out. This process of inversion has its counterpart in optical isomerism. Like right- and left-hand gloves, enantiomers are often quite stable with respect to each other, but inversion can be achieved by chemical means. The inversion of L-alanine would give. . . . . (1), inversion of D-glyceraldehyde would give. . . . . (2).

---

(1) D-alanine (2) L-glyceraldehyde

---

42. The two models shown below can be easily made with small pieces of plasticine and some matches.

These models represent mirror image forms of an asymmetric substance, i.e., they are enantiomers. If we consider one of these models, there are several ways of converting it into the other. Thus (i) any two matches can be interchanged; (ii) the markings on two of the matches can be interchanged (iii) one match can be removed and the remaining three matches "turned inside out" by squeezing the plasticine:

and the first match then replaced on the side of the plasticine *opposite* its original position. In chemical terms it is this last method which is the most important. When a molecule is "turned inside out" in this way it is said to have undergone. . . . . (1), and the new molecule so produced is the. . . . . (2) of the original. A process of this kind involves the breaking and making of chemical bonds.
Such inversions are often observed in nucleophilic substitution reactions (cf. Program 4, Frame 40).

a two-dimensional surface. Various ways of representing the structure of ethane, C_2H_6, are shown below.

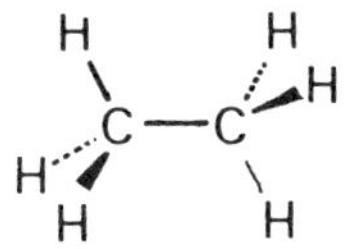

I

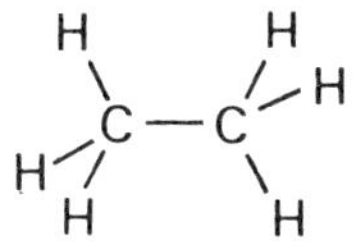

II

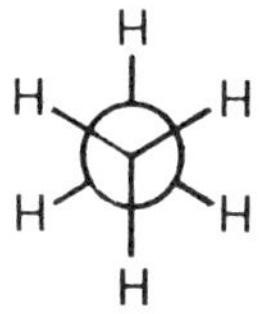

III

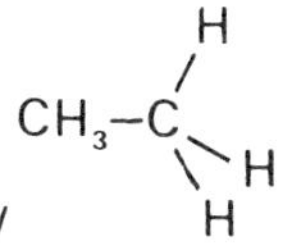

IV

CH_3CH_3

V

The disposition of the four bonds around each carbon atom in ethane is .... (1), and formula I endeavours to give a three dimensional view of the molecule. The thickened lines represent bonds which project forwards out of the plane of the paper, and the dotted lines bonds which project backwards behind the plane of the paper. Formula II is a simplified version of I. Formula III shows a view of ethane along the carbon-carbon bond, so that the bonds radiating from the centre of the circle are projecting .... (towards/away from) (2) the viewer. Formulae IV and V are common abbreviations of I.

---

(1) tetrahedral (2) towards

---

12. In abbreviated formulae such as CH_3CH_3 for ethane, or $CH_3CH_2CH_3$ for propane, the atoms attached to a particular atom are generally written to the right of that atom, and so far as possible, carbon and hydrogen atoms are written alternately. Write the structure of normal hexane in this manner .... (1), and suggest a further possible contraction based on grouping repetitive parts of the chain .... (2).

---

(1) $CH_3CH_2CH_2CH_2CH_2CH_3$ (2) $CH_3(CH_2)_4CH_3$

---

13. Ethylene, shown in Frame 6 as 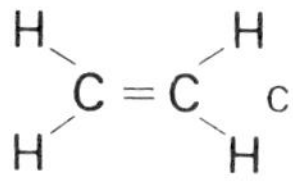 can be abbreviated to $CH_2{=}CH_2$, or sometimes $H_2C{=}CH_2$. Acetylene, $H{-}C{\equiv}C{-}H$, is abbreviated to $HC{\equiv}CH$. In abbreviated forms of the molecular structures of alkenes and alkynes, the double and triple bonds are always shown. Give abbreviated structures for 1-butene and 1-butyne, first with the multiple bonds on the right .... (1), and then with the multiple bonds on the left .... (2).

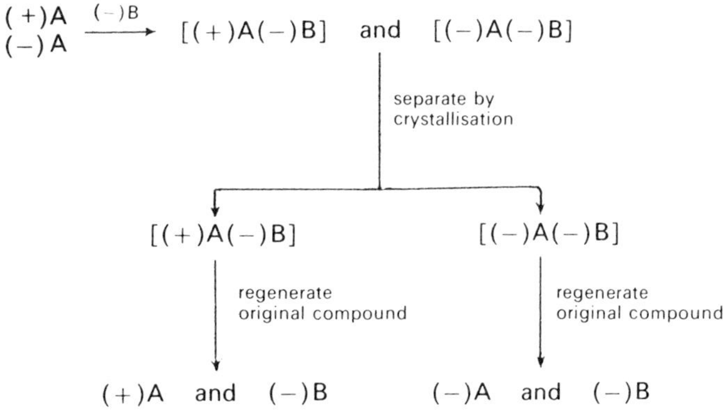

In this scheme the racemic form is shown . . . . (1), the resolving agent is . . . . (2), the mixture of diastereomers is . . . . (3) and the two enantiomers are . . . . (4).

---

(1) (+)A (−)A (2) (−)B

(3) [(+)A(−)B] and [(−)A(−)B] (4) (+)A, (−)A

---

40. *Naturally-occurring asymmetric molecules.* Living matter is a rich source of optically-active substances. If an asymmetric substance occurs naturally, it will nearly always be found in one or other of the two possible enantiomeric forms. Most α-amino acids of general formula $RCH(NH_2)COOH$ isolated from plants or animals are found to be optically active and to belong to the L-family. All sugars and other carbohydrates, and proteins (including enzymes), are optically active. If a microorganism can digest, say, the D-form of a given sugar, it is usually found that it is not able to digest the enantiomer, the L-form.

It is possible to isolate from liver an enzyme which catalyses the decarboxylation of L-α-amino acids (but not D-α-amino acids). In the case of L-valine we can write

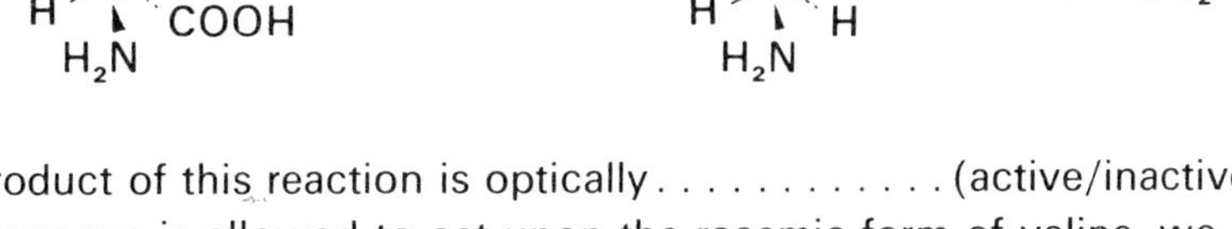

The product of this reaction is optically . . . . (active/inactive) (1). If the enzyme is allowed to act upon the racemic form of valine, we expect that at the end of the reaction, the unchanged amino acid present will be largely . . . . (2), and this . . . . (will/will not) (3) be optically active.

---

(1) $CH_3CH_2CH{=}CH_2$, $CH_3CH_2C{\equiv}CH$
(2) $CH_2{=}CHCH_2CH_3$, $HC{\equiv}CCH_2CH_3$

---

14. Formulae I and II are shorthand abbreviations of the same structure

I 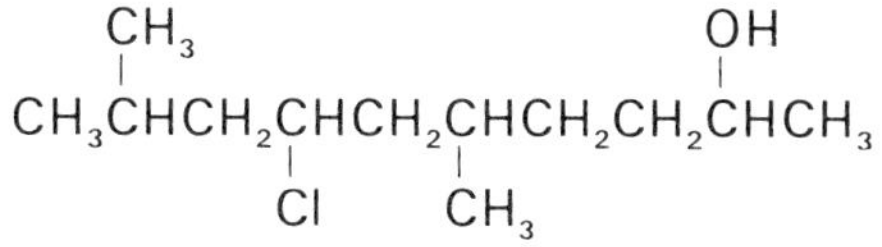

II $(CH_3)_2CHCH_2CHClCH_2CH(CH_3)CH_2CH_2CH(OH)CH_3$

Using the above as a guide, write a structure of type II for 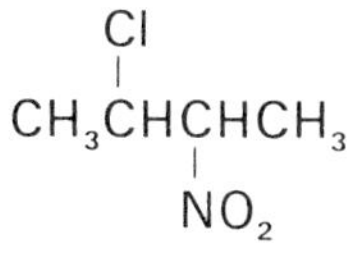

---

(1) $CH_3CHClCH(NO_2)CH_3$ or $CH_3CHClCH(CH_3)NO_2$
Note that a substituent having more than one atom such as NO_2, CH_3, OH is put into brackets if it is enumerated other than at the end of the linear formula. Lone pairs are normally not shown.

---

15. Now write abbreviated structures for

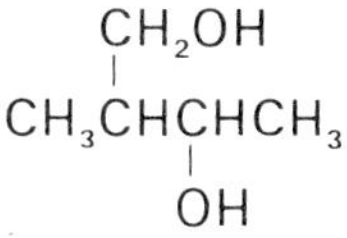 . . . . (1) 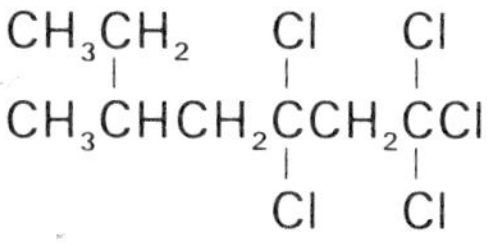 . . . . (2)

---

(1) $CH_3CH(CH_2OH)CH(OH)CH_3$ or $HOCH_2CH(CH_3)CH(OH)CH_3$

(2) $CH_3CH_2CH(CH_3)CH_2CCl_2CH_2CCl_3$

---

16. The structures in the above frame could equally well be drawn from right to left. Then (1) becomes . . . . (1), and (2) becomes . . . . (2).

---

(1) $CH_3CH(OH)CH(CH_2OH)CH_3$ or $CH_3CH(OH)CH(CH_3)CH_2OH$

(2) $CCl_3CH_2CCl_2CH_2CH(CH_3)CH_2CH_3$

---

17. Whereas carbon-carbon double bonds are always shown, carbon-oxygen double bonds are not. Ketones, aldehydes, esters etc. may be written in linear form having the substituent above or below the principal chain, e.g., ethyl 3-oxobutanoate can be written

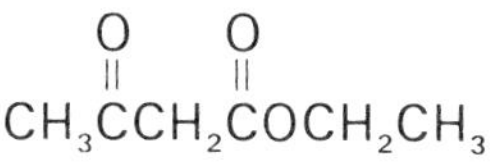

(1) will not

---

38. The separation of a racemic form into two enantiomers is known as resolution. Methods for resolving a racemic form are largely due to Pasteur, and the most important method is described below. Suppose we have a 50/50 mixture of a large number of right-handed and left-handed spirals. This is analogous to a racemic form—for each right-handed spiral there will be a left-handed spiral which is its non-superimposable mirror image.

Now suppose we attach to each spiral a second, different spiral which is, say, left-handed. The above mirror image pair then becomes

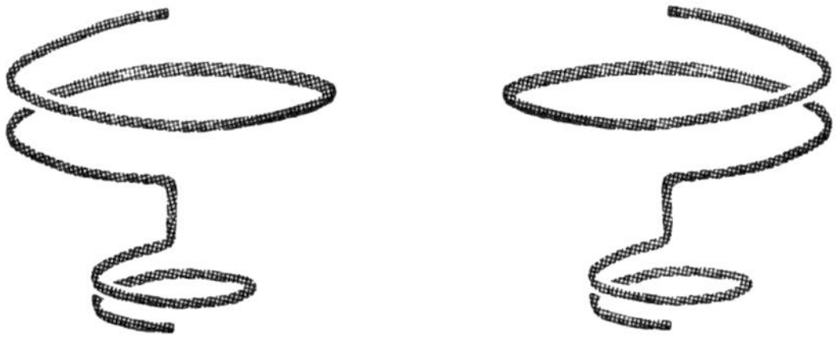

These two combined spirals are *no longer mirror images of each other*. Hence they are no longer enantiomers, but are .... (1).

---

(1) diastereomers (cf. Frame 29).

---

39. While two enantiomers have identical solubilities in all common solvents, this is no longer true of diastereomers. Such isomers are *not* related as object to mirror image, and have different melting points, solubilities, etc. It is often quite easy to separate diastereomers by crystallisation. There are many chemical analogies for the linking together of spirals as discussed above. For example, the racemic form of lactic acid can be treated with a naturally-occurring *optically active* amine such as the alkaloid (−)strychnine. The two salts which form, the (−)strychnine salt of (+)lactic acid, and the (−)strychnine salt of (−)lactic acid, will be diastereomers and can be separated by crystallisation. The two salts can then be decomposed by hydrochloric acid giving the separated (+) and (−) forms of lactic acid. In general, if (+)A and (−)A represent two enantiomers present in a racemic form and (−)B (for example) represents a single optically active form of another compound that will combine in some way with A, then the process of resolution may be shown.

Alternatively the substituent is placed on the line after the carbon atom concerned and the double bond is omitted. The above structure then becomes $CH_3COCH_2COOCH_2CH_3$. Note that with a terminal group such as an aldehyde the hydrogen comes before the oxygen, e.g., CH_3CHO. Rewrite the following formulae into the corresponding linear form:

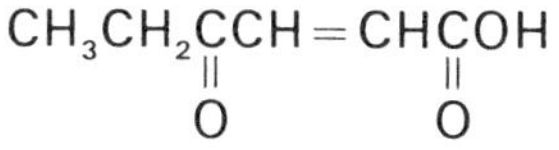 . . . . (1)

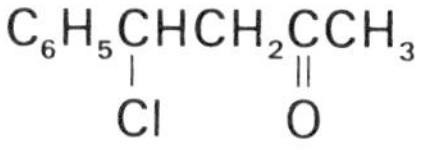 . . . . (2)

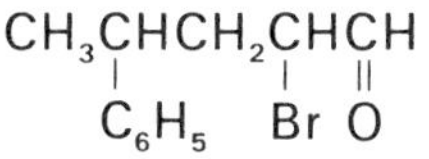 . . . . (3)

---

(1) $CH_3CH_2COCH{=}CHCOOH$ (2) $C_6H_5CHClCH_2COCH_3$

(3) $CH_3CH(C_6H_5)CH_2CHBrCHO$

Use of the symbols R and Ar

18. To concentrate attention on a particular functional group it is often useful to represent the remainder of the molecule by the symbol R. Thus alcohols can be written as ROH, primary amines as RNH_2 and carboxylic acids as . . . . (1); etc.

---

(1) RCOOH

---

19. The symbol R may represent an aliphatic or aromatic group as in say a primary amine RNH_2. There are, however, many differences in the chemistry of aliphatic and aromatic compounds and it is sometimes useful to use the abbreviation Ar to indicate that the functional group is attached directly to an aromatic ring (aryl group). The symbol ArOH represents a . . . . (1).

---

(1) phenol

---

20. Different alkyl groups can be indicated by the symbols R, R′, R″ etc. Thus a general structure representing any dialkyl ether would be ROR′, while a diaryl ether could be shown . . . . (1).

---

(1) ArOAr′

(1) two

---

35. Compounds I and II in the above frame are enantiomers, each molecule being the non-superimposable mirror image of the other. The compounds are known respectively as D- and L- tartaric acid. On the other hand the molecules III and IV, which are different from I and II, are also mirror images of each other, but close examination shows that these mirror images . . . . (are/are not) (1) superimposable.

---

(1) are. A Fischer representation may be rotated through 180° in the plane of the paper without change of configuration. Thus if we rotate the structure III in this way we arrive at the "upside-down" figure, IIIA.

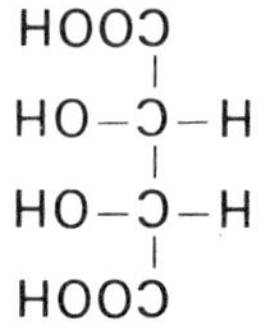

IIIA

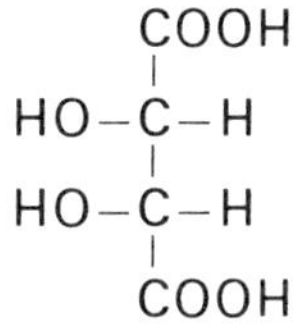

IIIB

This figure can be immediately redrawn as IIIB, which is identical with IV. The use of 3-dimensional models made with plasticine and matches (cf. Frame 18) will help to clarify these ideas.

---

36. Structure III above gives rise to a superimposable mirror image because it is a symmetrical substance, having a plane of symmetry between carbon atoms 2 and 3, at right angles to the C—C bond. Hence there are only three stereoisomers of tartaric acid, the D- form, the L- form and the . . . . (1) form. The latter form is optically inactive because of the phenomenon of internal . . . . (2).

---

(1) meso (2) compensation

---

37. *Resolution*

Given that in a racemic form we have a mixture of two enantiomers, how do we go about separating the optical isomers one from the other? Enantiomers differ in the way they interact with polarised light, but in most properties such as melting point, boiling point, and solubility in common solvents, they are indistinguishable. If two isomeric substances have the same solubility in, say, chloroform, it . . . . (will/will not) (1) be possible to separate these isomers by crystallisation from that solvent.

CHEMICAL EQUATIONS

21. A stoichiometric equation involves quantitative relationships between all the elements involved in the reaction. Complete the stoichiometric equation

$5CH_3CH_2OH + 4MnO_4^- + 12H^+ \longrightarrow$ . . . . (1)

given that the products are acetic acid (ethanoic acid) and manganese(II) ion.

---

(1) $5CH_3CH_2OH + 4MnO_4^- + 12H^+ \longrightarrow 5CH_3COOH + 4Mn^{2+} + 11H_2O$

---

22. The arrow shows the direction in which the reaction proceeds and a single arrow means that the reaction is irreversible or virtually so. A reaction which at equilibrium has significant amounts of both reactants and products present is shown as

$$A + B \rightleftharpoons C + D$$

A reversible reaction may also be shown as e.g., $A + B \rightleftarrows C + D$ which means that the position of equilibrium lies far to the . . . . (right/left) (1) and the concentrations of A and B are . . . . (2).

---

(1) right (2) small

---

23. Acetyl chloride, CH_3COCl, reacts with aqueous ammonia to form acetamide, CH_3CONH_2. The yield of acetamide is not quantitative as some of the acetyl chloride reacts with water to form acetic acid. We may, however, write a stoichiometric equation for the particular reaction of interest, while ignoring the side reaction.

$CH_3COCl + 2NH_3 \longrightarrow CH_3CONH_2 +$ . . . . (1)

---

(1) $CH_3COCl + 2NH_3 \longrightarrow CH_3CONH_2 + NH_4Cl$

This reaction is chosen for illustrative purposes only; the preparation of amides is discussed in detail in Program 8.

---

24. A stoichiometric equation such as $CH_3COCl + 2NH_3 \longrightarrow CH_3CONH_2 + NH_4Cl$ may be simplified to $CH_3COCl \xrightarrow{NH_3} CH_3CONH_2$. Write a simplified equation . . . . (1) for the reaction

$$CH_3COOH + PCl_5 \longrightarrow CH_3COCl + POCl_3 + HCl$$

---

(1) $CH_3COOH \xrightarrow{PCl_5} CH_3COCl$

---

25. In catalytic reactions, the catalyst, temperature or pressure may need to be specified

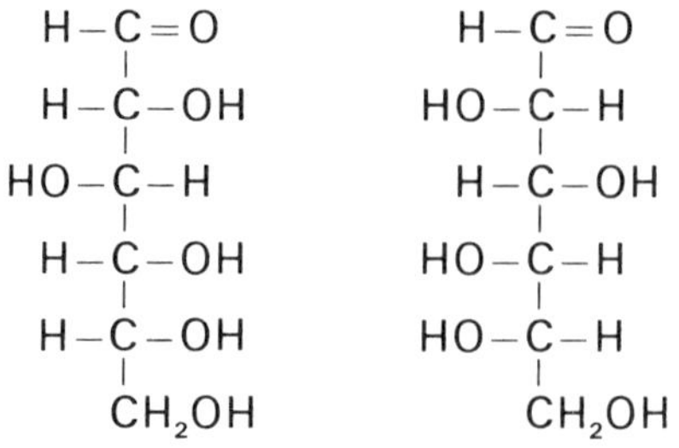

D-glucose L-glucose

32. It is possible for a stereoisomer to have two (or even more) asymmetric carbon atoms and yet not show optical activity. This can happen when the molecule has a plane of symmetry or mirror plane dividing two separate *asymmetric* units. If one half of the molecule is to be the mirror image of the other half, then the two asymmetric units must have .... (the same/opposite) (1) configurations.

(1) opposite

33. If a molecule has a plane of symmetry dividing two asymmetric units having opposite configurations, then the optical rotatory effects of each of the two asymmetric units will cancel out (provided always that free rotation around the bonds joining the units is possible). This phenomenon is called internal compensation, and a substance which is optically inactive because of internal compensation is called a *meso* compound. The measured optical rotation of the meso compound will be .... (1) at all wavelengths.

(1) zero.

34. The above phenomenon is shown by tartaric acid. This compound, HOOCCH(OH)CH(OH)COOH, has .... (1) asymmetric carbon atoms and by analogy with isoleucine (Frame 28) it might be thought that four optical isomers would be possible, as shown below

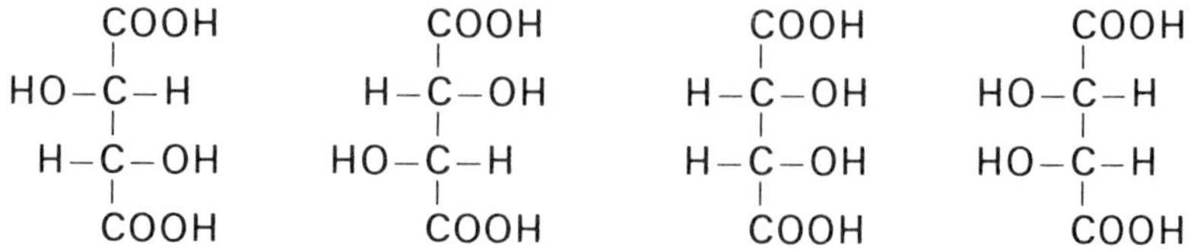

I II III IV

e.g.

$$CH_3CHO \xrightarrow[\text{room temp.}]{H_2/Pt} CH_3CH_2OH$$

$$C_6H_5Cl \xrightarrow[300° \ 150 \text{ atm}]{\text{aq. NaOH}} C_6H_5OH$$

$$CH_3CH_2OH \xrightarrow[250\text{-}300°]{Cu} CH_3CHO$$

$$CH_3CH_2OH \xrightarrow[350\text{-}400°]{Al_2O_3} CH_2{=}CH_2$$

Write balanced equations for the latter two reactions, ... (1) . . . . (2).

---

(1) $CH_3CH_2OH \xrightarrow[250\text{-}300°]{Cu} CH_3CHO + H_2$

(2) $CH_3CH_2OH \xrightarrow[350\text{-}400°]{Al_2O_3} CH_2{=}CH_2 + H_2O$

---

26. Many preparations require two or more separate reactions. Thus acetamide can be prepared from acetic acid by first converting the acid into acetyl chloride and then reacting this substance with ammonia (cf. Frame 24). This can be shown

$$CH_3COOH \xrightarrow{PCl_5} CH_3COCl \xrightarrow{NH_3} CH_3CONH_2$$

or

$$CH_3COOH \xrightarrow[(2)\ NH_3]{(1)\ PCl_5} CH_3CONH_2$$

The following equations represent a synthesis of ethyl magnesium bromide from ethanol:

$$3CH_3CH_2OH + PBr_3 \longrightarrow 3CH_3CH_2Br + H_3PO_3$$

$$CH_3CH_2Br + Mg \xrightarrow{\text{dry ether}} CH_3CH_2MgBr$$

Write abbreviated equations for this synthesis . . . . (1).

---

(1) $CH_3CH_2OH \xrightarrow{PBr_3} CH_3CH_2Br \xrightarrow[\text{dry ether}]{\text{Mg in}} CH_3CH_2MgBr$

or $CH_3CH_2OH \xrightarrow[(2)\text{ Mg in dry ether}]{(1)\ PBr_3} CH_3CH_2MgBr$

---

27. Ethanol is oxidised by potassium permanganate to yield acetic acid. Under carefully controlled conditions the reaction is quantitative and a stoichiometric equation can be written (cf. Frame 23). Under less carefully controlled conditions portion of the acetic acid may be further oxidised and other side reactions may take place. If it is sufficient to indicate the main reaction only,

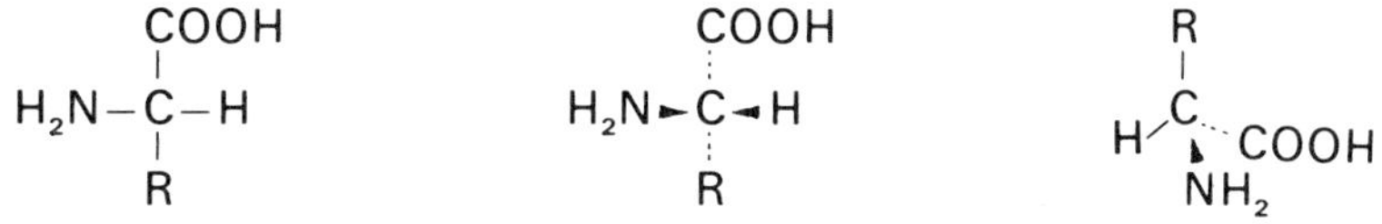

I II III

An amino acid having the configuration shown in I is said to belong to the L-family, and its enantiomer to the D-family. The symbol L- or D- when applied to α-amino acids conventionally represents the absolute configuration in space of the four groups about the α-carbon atom. The symbol L- and D- is quite independent of whether the amino acid happens to show negative or positive rotation. Write structures for L- and D- alanine, $CH_3CH(NH_2)COOH$, using Fischer projection formulae . . . . (1).

---

(1)

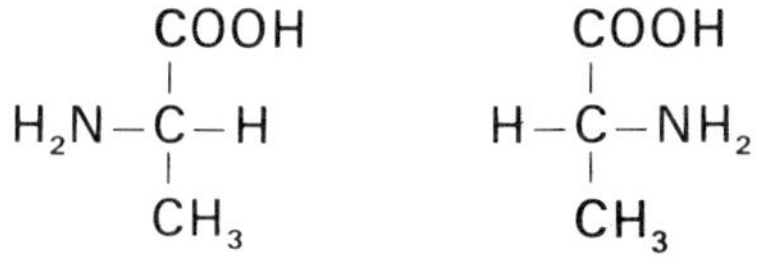

L-alanine D-alanine

An alternative method of defining configuration is given on page 218.

---

31. A D-, L- nomenclature is also applied to sugars (carbohydrates). The simplest sugar is glyceraldehyde, $CH_2(OH)CH(OH)CHO$. This molecule has one asymmetric carbon atom, and the two enantiomers can be shown

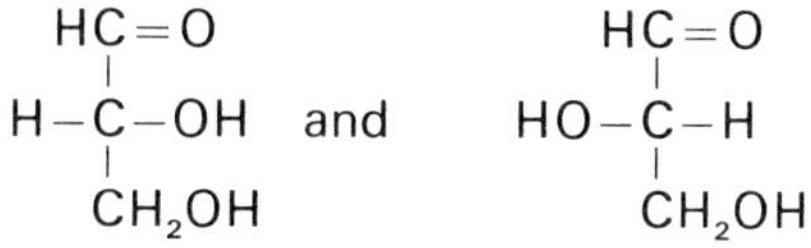

I II

Compound I is known as D-glyceraldehyde, compound II as L-glyceraldehyde. In general, any sugar $RCH(OH)CH_2OH$ is said to belong to the D series if the structure, written in the Fischer convention with the R group at the top, has the secondary –OH group on the right. If the secondary –OH is on the left, the sugar will belong to the L series.

Given that the structure

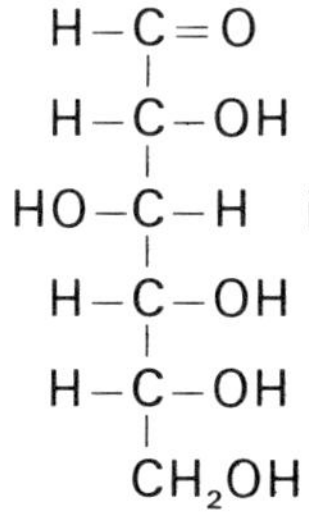

is a Fischer representation of the sugar D-glucose, and that its enantiomeric or mirror image isomer is L-glucose, write down side by side the structures of D- and L-glucose . . . . (1).

---

with emphasis on the nature of the organic product, the following simplified equation would be appropriate

$$CH_3CH_2OH \xrightarrow{KMnO_4} \ldots\ldots\ldots\ldots \ (1).$$

---

(1) $CH_3CH_2OH \xrightarrow{KMnO_4} CH_3COOH$

---

28. The reaction of water with acetamide to give acetic acid and ammonia is an example of *hydrolysis* (*hydro* = water, *lysis* = splitting)

$$CH_3CONH_2 + H_2O \longrightarrow CH_3COOH + NH_3$$

In a hydrolysis reaction, a covalent bond in the starting material is broken, and the elements of water are added. The equation for the hydrolysis of acetamide in boiling dilute hydrochloric acid is

$$CH_3CONH_2 + HCl + H_2O \longrightarrow CH_3COOH + \ldots\ldots\ldots\ldots \ (1).$$

while the equation for alkaline hydrolysis is

$$CH_3CONH_2 + NaOH \longrightarrow CH_3COONa + \ldots\ldots\ldots\ldots \ (2).$$

---

(1) $CH_3CONH_2 + HCl + H_2O \longrightarrow CH_3COOH + NH_4Cl$
(2) $CH_3CONH_2 + NaOH \longrightarrow CH_3COONa + NH_3$

---

29. In both cases, the acetamide is hydrolysed, but in acid solution the products are acetic acid and ammonium chloride, whereas in sodium hydroxide the products are .... (1). It is often convenient to ignore such distinctions and to use general statements such as

$$CH_3CONH_2 \xrightarrow[\text{alkali}]{\text{acid or}} CH_3COOH + NH_3$$

---

(1) sodium acetate and ammonia

---

30. Many organic reactions yield more than one organic product. The chlorination of propane yields a mixture consisting of two monochloropropanes together with minor amounts of more highly chlorinated products, but for every reaction of a chlorine molecule a molecule of hydrogen chloride is produced. When the monochlorinated fraction is separated it is found to consist of 48% of 1-chloropropane and 52% of 2-chloropropane. For each individual product there is a .... (1) equation e.g.

$$CH_3CH_2CH_3 + Cl_2 \longrightarrow CH_3CH_2CH_2Cl + HCl$$
$$CH_3CH_2CH_3 + Cl_2 \longrightarrow CH_3\underset{\displaystyle Cl}{\underset{|}{C}}HCH_3 + HCl$$

but it is difficult to write a single equation to summarise the whole reaction. Further, to start the reaction it must be heated or irradiated and probably the most informative equation is

and bottom project downwards below the plane of the paper. Manipulation of Fischer formulae is restricted to rotations within the plane of the paper, which does not cause a change of configuration. This can be shown by rotating the formula of one enantiomer through 180° within the plane of the paper when it does not become equivalent to the enantiomer.

i.e. 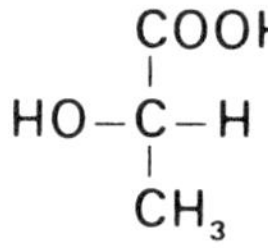 on rotation gives 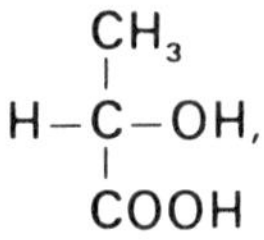 both are equivalent to 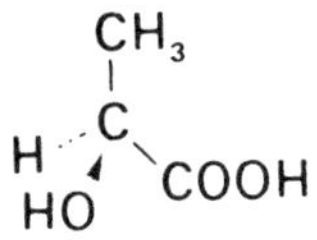

while the enantiomer 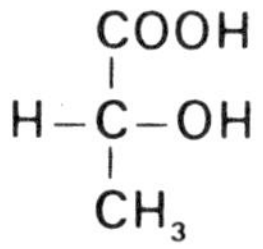 is 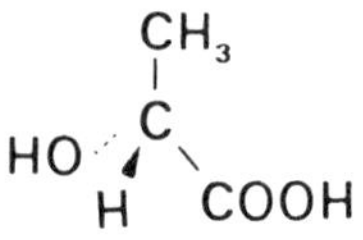 the opposite configuration.

Using this convention the four optical isomers of isoleucine (cf. Frame 24) can be shown:

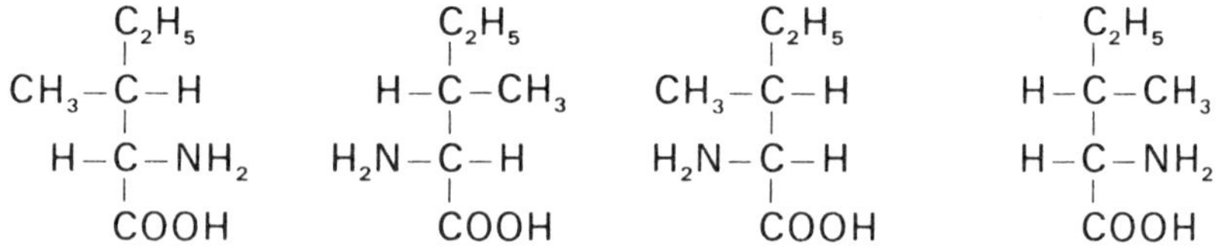

I II III IV

In these four isomers there are two pairs of enantiomers (related as object to non-superimposable mirror image). These pairs are .... (1).

---

(1) I and II; III and IV

---

29. Compounds III and IV above are both optical isomers of I, but neither is the mirror image of I. Compounds I and III are said to be *diastereomers;* so also are compounds I and IV. Diasteromers are optical isomers which are not related as object to .... (1). Other diasteromeric pairs are II and III, and II and IV. An equimolar mixture of forms I and II, or of forms III and IV will be optically inactive. Such a mixture is called the .... (2) form.

---

(1) non-superimposable mirror image (2) racemic

---

30. It is convenient to be able to name enantiomers according to their configuration and not just distinguish them by sign of optical rotation. For any α-amino acid, $RCH(NH_2)COOH$, if the Fischer projection formula is written as in I, this implies that the configuration (or arrangement in space) of the four groups around the asymmetric carbon atom is that given in II, which is equivalent to III.

$$CH_3CH_2CH_3 \xrightarrow[\text{heat or light to start reaction}]{Cl_2} CH_3CH_2CH_2Cl,\ CH_3CHClCH_3 \text{ and } HCl$$

Note that this is . . . . (a balanced/an unbalanced) (2) equation.

---

(1) balanced or stoichiometric (2) an unbalanced

---

31. A possible disadvantage of these abbreviated equations is that it may be important NOT to overlook the nature of the other products of the reaction. Thus the statement

$$CH_3COOH \xrightarrow{SOCl_2} CH_3COCl$$

fails to draw attention to the fact that the two other products are sulphur dioxide and hydrogen chloride:

$$CH_3COOH + SOCl_2 \longrightarrow CH_3COCl + SO_2 + HCl$$

Both these gases are very unpleasant if liberated in quantity in an open laboratory. A warning note could be sounded as follows

$$CH_3COOH \xrightarrow[\text{(fume hood)}]{SOCl_2} CH_3COCl$$

Likewise in the reaction $CH_3COCl \xrightarrow{NH_3} CH_3CONH_2$ it must be remembered that in order to purify the acetamide, it will have to be separated from an equal quantity of . . . . (1).

---

(1) ammonium chloride, NH_4Cl

REACTION MECHANISMS

32. A chemical reaction involves the making and breaking of chemical bonds. In the formation of methyl chloride from methane and chlorine,

$$CH_4 + Cl_2 \longrightarrow CH_3Cl + HCl$$

carbon-hydrogen and . . . . (1) bonds are broken, while carbon-chlorine and . . . . (2) bonds are formed.

---

(1) chlorine-chlorine (2) hydrogen-chlorine

---

33. For the substitution reaction $HO^- + CH_3{-}Cl \longrightarrow HO{-}CH_3 + Cl^-$ in which the chlorine atom attached to the methyl group is substituted by the incoming hydroxide ion, a . . . . (1) bond is broken, and a . . . . (2) bond is formed.

---

(1) carbon-chlorine (2) carbon-oxygen

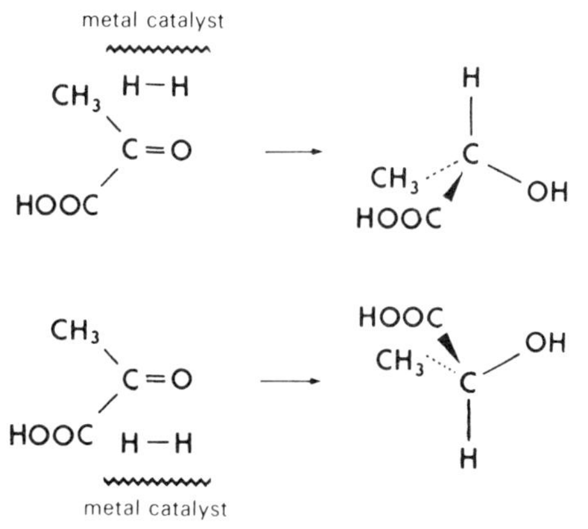

The two enantiomers will therefore be formed in . . . . (1) amounts, so that the mixture will be a . . . . (2).

---

(1) equal or equimolar (2) racemic form or (±)-form

---

26. The addition of hydrogen cyanide to acetaldehyde gives acetaldehyde cyanohydrin, which contains an asymmetric carbon atom.

$$CH_3CHO + HCN \longrightarrow CH_3CH(OH)CN \quad \text{(cf. Program 6)}$$

There is an equal probability that the HCN will add to the . . . . (1) group from one side or the other, so that the product . . . . (will/will not) (2) be optically active.

---

(1) carbonyl or $\rangle C{=}O$ (2) will not

---

27. The racemic form of acetaldehyde cyanohydrin produced in the above way is an equimolar mixture of two optical isomers. The separation of such a racemic form into its two constituent . . . . (1) is known as *resolution*, and is discussed in Frames 38-40.

---

(1) enantiomers

---

28. A convenient way of representing asymmetric molecules on paper is to use Fischer projection formulae. In the Fischer convention the formula

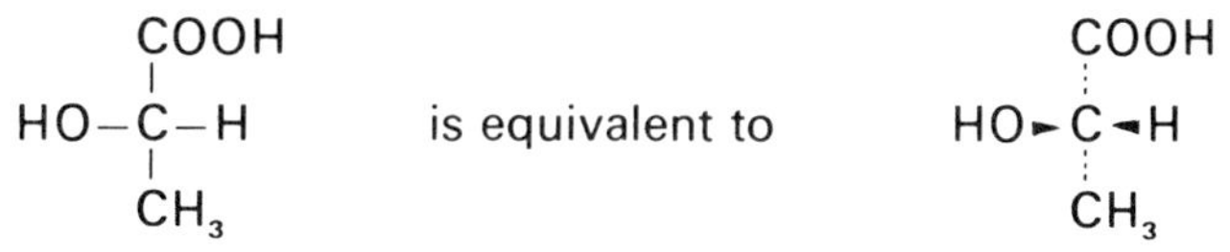

in which the groups written on the left and right of an asymmetric carbon atom project upwards from the plane of the paper, while the groups at the top

34. We can now ask the question: How (and why) does the hydroxide ion react with methyl chloride to give methanol and chloride ion? This is equivalent to asking "What is the mechanism of this reaction?". There are no easy answers to questions of this kind. Some understanding of *why* a reaction occurs can be provided by the science of chemical thermodynamics, which is concerned with the energy states of starting materials and final products, and with the energy changes that occur during the reaction. In the following frames we shall be more concerned with how a reaction occurs, in particular with those factors which, in an organic reaction, favour the formation of one product rather than another, or the formation of several different products rather than a single product. In the discussion we shall draw heavily on the science of electrostatics, which teaches that "like charges .... (1) each other, while unlike charges ...." (2).

---

(1) repel (2) attract

---

35. *Molecular Polarity*. It can be shown experimentally that many neutral molecules behave as small electric dipoles, $\oplus\!\!-\!\!-\!\!-\!\!\ominus$, one region of the molecule having a relative deficiency of electrons (+ve charge) and another region a corresponding excess of electrons (–ve charge). Such a molecule is then said to be *polar* and to exhibit polarity. The molecule H–H consists of two nuclei, each of charge +1, and two negative electrons. The centres of positive and negative charge coincide and consequently the molecule is non-polar. The molecule H–F consists of a nucleus of charge +1, a nucleus of charge +9, and 10 electrons. In this system the centres of positive and negative charge do not coincide and there results a .... (1). The hydrogen fluoride molecule is therefore .... (2).

---

(1) dipole (2) polar

A discussion of why dipoles occur is beyond the scope of this program.

---

36. The degree of molecular polarity can be expressed by a number, derived from experiment, known as the *dipole moment*. Strongly polar molecules, e.g., HF or CH_3Cl, have high dipole moments while non-polar molecules, e.g., H_2 or CH_4, have low or zero dipole moments. Neutral molecules behave as dipoles when .... (1).

---

(1) the centres of positive and negative charge do not coincide.

---

37. The presence of polarity in a molecule can be indicated by the symbols $\delta+$ $\delta-$ written over the structure, or by an arrow $+\!\!\!\longrightarrow$ which shows the direction of charge separation from positive to negative. In the case of H–F,

deoxyribose, 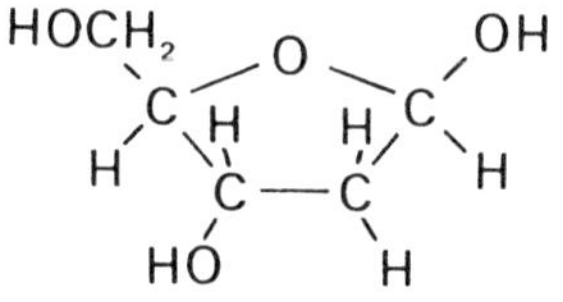 . . . . (4)

For a molecule having *n* asymmetric carbon atoms, a maximum of 2^n optical isomers is possible.

---

(1) two (2) one (3) none (4) three

---

24. Optical isomers which are related as object and non-superimposable mirror image are called *enantiomers*. A given asymmetric molecule, no matter how many asymmetric carbon atoms it may contain, can have only one enantiomer, namely its mirror image form. Draw the enantiomers of alanine $CH_3CH(NH_2)COOH$ in three dimensional projection . . . . (1) (cf. lactic acid in previous frame).

---

(1)

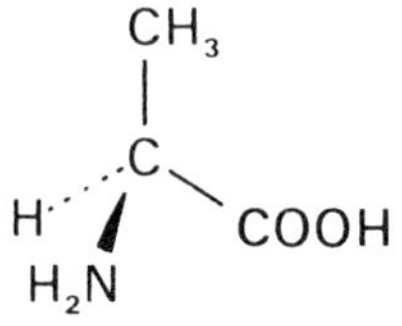

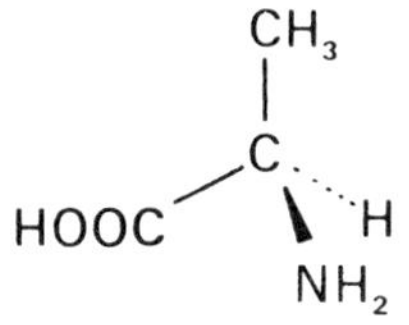

or equivalent representations

---

25. *Racemic forms*. If we have an equimolar mixture of two enantiomers the measured optical rotation will be zero, no matter which solvent or wavelength is used for the measurement. This is because any dextro(clockwise)-rotation produced by one isomer will be cancelled by an equal laevo(anticlockwise)-rotation produced by the other. An equimolar mixture of two enantiomers is called the racemic form of the compound.
Since equal amounts of the (+) and the (−) enantiomers are present, the symbol (±) is used to indicate a racemic form. The symbol dl is also used for the racemic form since equal amounts of the dextro- and laevo- forms are present.

The synthesis of lactic acid in the laboratory, e.g. by the catalytic reduction of 2-oxopropanoic acid, produces the racemic form, because the probability that the activated hydrogen molecule will add to one side of the planar carbonyl group is equal to the probability that it will add to the other side. As shown below, addition from one side or the other gives rise to the two separate enantiomers.

the dipolar nature is commonly indicated by writing $\overset{\delta+}{H}-\overset{\delta-}{F}$ or $\overset{+\!\longrightarrow}{H-F}$. Recent theories, however, suggest that the centre of positive charge is close to the fluorine atom, and the centre of negative charge *beyond the fluorine atom along the line of the H–F bond,* thus $H-\overset{+\!\rightarrow}{F}$. Methyl chloride is likewise often shown $\overset{\delta+}{CH_3}-\overset{\delta-}{Cl}$ or $\overset{+\!\longrightarrow}{CH_3-Cl}$; a closer approximation is probably $CH_3-\overset{+\!\rightarrow}{Cl}$ (cf. Program 4, Frames 32 to 35). Since a dipole is a vector quantity having both magnitude and direction, a molecule having two or more dipoles may yet have a zero dipole moment when placed in an electric field if the individual dipoles oppose each other. Indicate by writing "yes" or "no", whether or not the following molecules will have an observable dipole moment: chloroform ($CHCl_3$) . . . . (1), carbon tetrachloride (CCl_4) . . . . (2), chlorobenzene . . . . (3), *p*-dichlorobenzene . . . . (4).

---

(1) yes (2) no—the vector sum of the four individual C–Cl dipoles is zero

(3) yes (4) no—the vector sum of the opposing C–Cl dipoles is zero

---

38. *Electrophiles and Nucleophiles.* The chemical bonds in stable organic molecules can be regarded as made up of pairs of electrons of opposite spin and the separation or uncoupling of such a pair of electrons requires a large amount of energy. Most of the reactions discussed in this set of programs involve the interactions of ionic and/or polar species in which the electrons remain paired. Polar reagents can be classified as nucleophiles or electrophiles depending on the role they play in a given reaction. If a reagent donates a pair of electrons to an atom in some other molecule or ion so that a new bond is established by sharing of the electron pair, the reagent is said to be a nucleophile, or nucleophilic in behaviour. If a chemical species reacts by accepting an electron pair from an atom in some other molecule or ion, the species is said to be an electrophile, or electrophilic in behaviour. Atomic nuclei are positively charged and the term nucleophilic (nucleus loving) derives simply from the fact that the electron pair of the nucleophile is . . . . (1) by a region of positive charge.

---

(1) attracted

---

39. Nucleophiles include negatively charged ions and molecules possessing atoms with unshared electron pairs. Electrophiles include positively charged ions and molecules containing atoms capable of accepting an electron pair into the valence shell. In the reaction of trimethylamine with boron trifluoride

$$(CH_3)_3N| + BF_3 \longrightarrow (CH_3)_3\overset{+}{N}-\overset{-}{B}F_3$$

the trimethylamine molecule provides a lone pair of electrons to form a bond to the boron atom, thereby increasing the number of valence electrons around

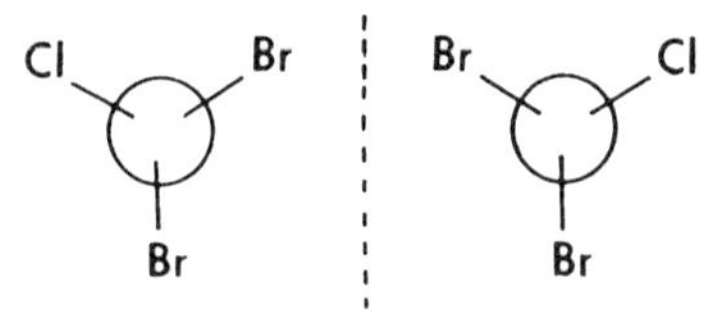

I   II

Formula II is the mirror image of I but is superimposable on it—rotation of II around the H-C axis brings it into coincidence with I. Thus I and II are identical. Write "yes" or "no" to indicate whether or not the following molecules are capable of existing as two distinct optical isomers:

$CH_3CHBrCl$ . . . . (1) $CH_3CH(OH)CH_3$ . . . . (2)

$CH_3CH(OH)CN$ . . . . (3) $(CH_3)_2CHCH_2CH(NH_2)COOH$ . . (4)

$C_6H_5CH(CH_3)N(CH_3)_2$ . . . . (5) $C_6H_5CH(CHO)C_6H_5$ . . . . (6)

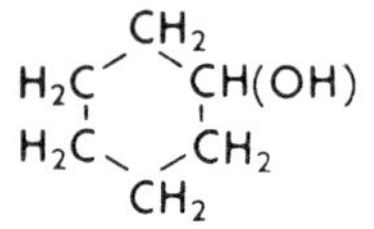 . . . . (7) 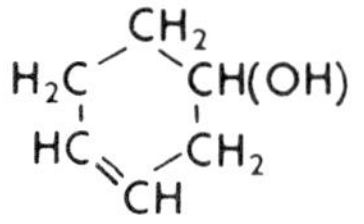 . . . . (8)

---

(1) yes	(2) no	(3) yes	(4) yes
(5) yes	(6) no	(7) no	(8) yes

---

23. In all the asymmetric molecules discussed above there are four different atoms or groups attached to one of the carbon atoms in the molecule. Such a carbon atom can be loosely called an asymmetric carbon atom The two optical isomers of lactic acid, $CH_3CH(OH)COOH$ (2-hydroxypropanoic acid) can be shown in stereochemical projection as

Lactic acid has only one asymmetric carbon atom. Write down the number of asymmetric carbon atoms present in each of the following compounds: the amino acid isoleucine, $CH_3CH_2CH(CH_3)CH(NH_2)COOH$ . . . . (1); the compound $CH_2(OH)CH(OH)CH_2OCOCH_3$ . . . . (2); the compound $CH_2(OH)CH(OCOCH_3)CH_2OH$ . . . . (3); the sugar

the boron atom from six to eight. The nucleophile is . . . . (1) and the electrophile is . . . . (2).

---

(1) $(CH_3)_3N$ (2) BF_3

---

40. In many reactions, the bond-making and bond-breaking processes can be discussed in terms of the nucleophilic behaviour and electrophilic behaviour of atoms or groups of atoms in the reacting species. If we first consider the simple reaction of anhydrous hydrogen chloride with water

the . . . . (1) molecule can be regarded as a nucleophile and the . . . . (2) atom of the hydrogen chloride molecule as an electrophile. The statement "nucleophiles react with electrophiles" is closely analogous to the statement "bases react with . . . . (3)".

---

(1) water (2) hydrogen (3) acids

---

41. The above reaction can be shown in the form

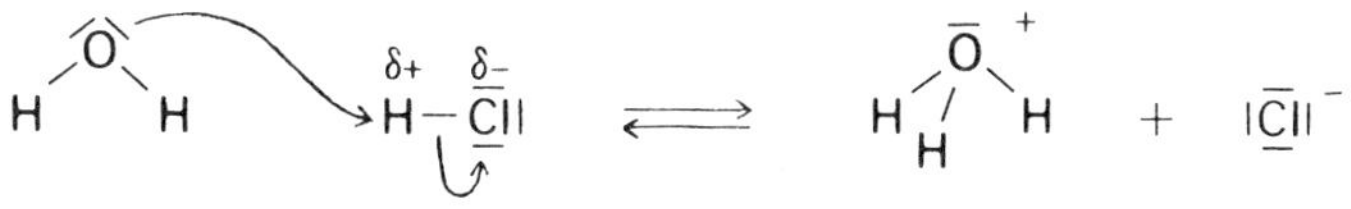

in which we introduce the "curved arrow" symbolism. Expressed in words, this means "A pair of electrons initially located on the nucleophile, H_2O, forms a new covalent bond to the . . . . (1) atom of the molecule HCl. At the same time the hydrogen-chlorine bond is broken, and the pair of electrons forming this bond moves wholly onto the chlorine atom, which departs as an ion Cl^-". Note that the reaction is reversible. Write the reverse reaction showing electron pairs and using the curved arrow symbolism . . . . (2).

---

(1) hydrogen

(2)

---

42. Consider the reaction of sodium hydroxide with methyl chloride. The methyl chloride molecule is neutral but polar, $CH_3-\overset{\leftrightarrow}{Cl}$ (cf. Frame 37). Electrostatic interactions favour a collision of the negative ion with the

20. The stereoisomers I and II in the previous frame are related as object and non-superimposable mirror image, and are examples of *optical isomers.* Such isomers are called optical isomers because they possess the property of rotating the plane of polarisation of a plane-polarised light beam. This optical phenomenon is related to the birefringence shown by many natural crystals (e.g. calcite). Optical activity is measured in an instrument called a polarimeter. If the plane of polarisation of the incident light beam is rotated clockwise on passing through the solution, then the solution is said to have a dextro (d-) or positive (+) rotation, and if the rotation is anticlockwise then the solution is said to have a laevo (l) or negative (−) rotation. Optical activity is expressed quantitatively in the form $[\alpha]_D^t$ where $[\alpha]$ is the specific optical rotation, D indicates that the D line of sodium light (589.3 nm) was used as light source, and t is the temperature of measurement. For a given substance, $[\alpha]$ is defined as the observed rotation, for a light-path length of 10 cm, of a solution having unit concentration (or unit density in the case of a pure liquid). The right- and left-handed forms of an asymmetric substance differ in the *direction* in which they rotate the plane of polarisation of a plane polarised light beam. If a given asymmetric substance has $[\alpha]_D^{20} = +14°$ in ethanol solution, the mirror image form of that substance, again measured in ethanol solution, will have $[\alpha]_D^{20} =$ .... (1).
The subscript D means that .... (2) and the superscript 20 means that .... (3).

---

(1) —14° (2) the D line of sodium light was used to make the measurement
(3) the measurement was made at 20°C

---

21. *Configuration.* The term "configuration" refers to the arrangement in space of the substituents around an atom or rigid part of a molecule. Thus the two optical isomers of CHBrClF (Frame 19) are said to have opposite configurations. It is important to distinguish "configuration" from "conformation". The latter term refers to the different forms which can be adopted by a single molecule as a result of internal .... (1).

---

(1) rotations (cf. Frame 8). Small changes of shape due to a certain flexibility of bond angles are also included in the term "conformational change".

---

22. For a substance CX_4 to be asymmetric, and therefore capable of existing as two distinct optical isomers, it is necessary that all four of the substituents X be different from each other. The molecule $CHBr_2Cl$ is not asymmetric and this is made clear by the formulae below, in which the carbon-hydrogen bond projects downwards into the paper.

positive end of the dipole and, given that the species collide with sufficient kinetic energy, the following reaction occurs

The negative ion HO^- is a nucleophile, while the carbon atom of methyl chloride exhibits electrophilic behaviour. Rewrite the above equation showing all lone pairs and using the curved arrow symbolism . . . . (1).

---

43. The reaction of carbon dioxide with aqueous sodium hydroxide can be shown

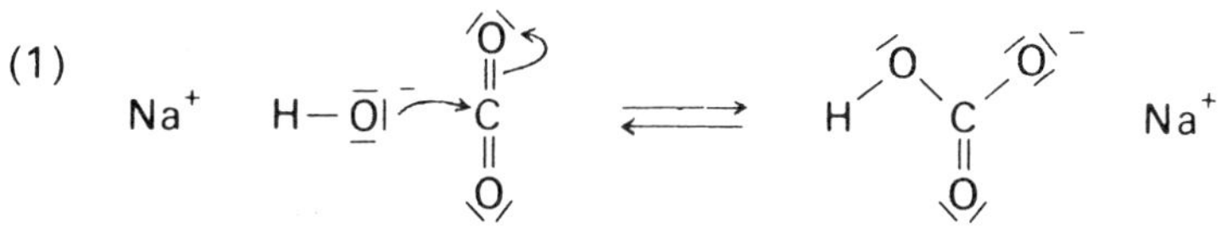

In this reaction, the hydroxide ion is a nucleophile. The carbon atom of the carbon dioxide is electrophilic, since it accepts a pair of electrons from the nucleophile. It can only accept the electrons, however, if a pair of electrons is transferred to one of the two doubly-bound oxygen atoms. It may be noted that the hydrogencarbonate ion is mesomeric

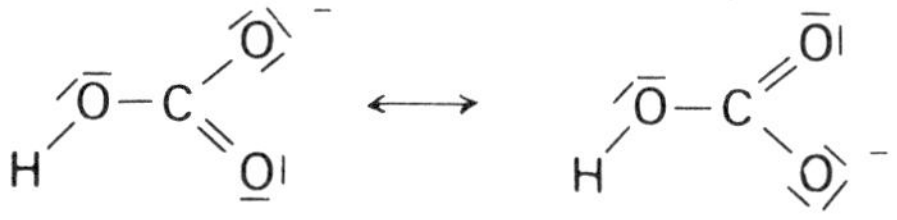

(cf. Frame 10)

The first step in the reaction of formaldehyde with hydrogen cyanide involves addition of the cyanide ion to the >C=O double bond:

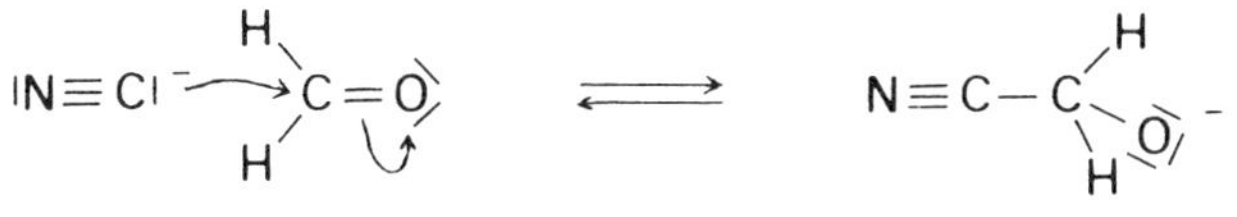

In this process a new carbon-carbon bond is formed, while the carbon-oxygen double bond becomes a . . . . (1) bond, with the oxygen atom becoming . . . . (2) charged.

---

(1) single (2) negatively

---

44. *Carbonium Ions.* Certain organic reactions occurring in polar solvents are thought to proceed via reactive intermediates called carbonium ions (cf.

The asymmetry can be demonstrated with models constructed from plasticine and matches.

Differently marked matches can be thought of as representing different substituents attached to a central carbon atom (the plasticine). The models shown above . . . . (are/are not) (1) mirror images, and . . . . (can/cannot) (2) be superimposed one on the other.

---

(1) are (2) cannot

---

19. A molecule such as CHClBrF can exist in two non-superimposable forms, each of which is the mirror image of the other. The two stereoisomers are represented in figures I and II.

I 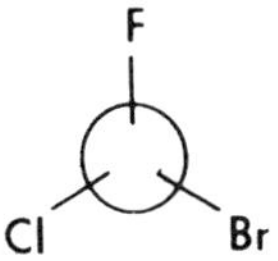 II 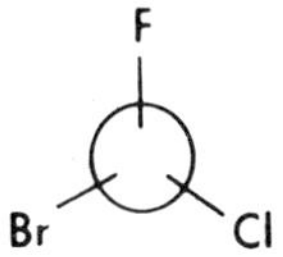

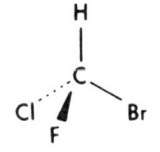

The diagrams show a view of the molecules looking along the C–H bond The large circle represents the carbon atom with the hydrogen atom hidden behind it. Then in one form, (I), the atoms, Br, Cl, F are attached in clockwise fashion to the carbon, while in the other (mirror-image) form, (II), the atoms Br, Cl, F are attached in anti-clockwise fashion. Formula I can be redrawn as

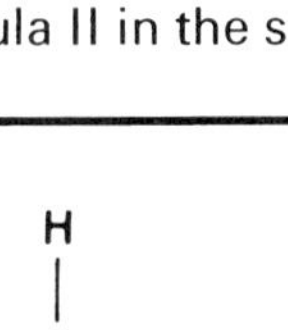

in which the dotted line represents a bond which projects behind the plane of the paper and the thickened line represents a bond which projects in front of the paper. Redraw formula II in the same way . . . . (1).

---

(1)

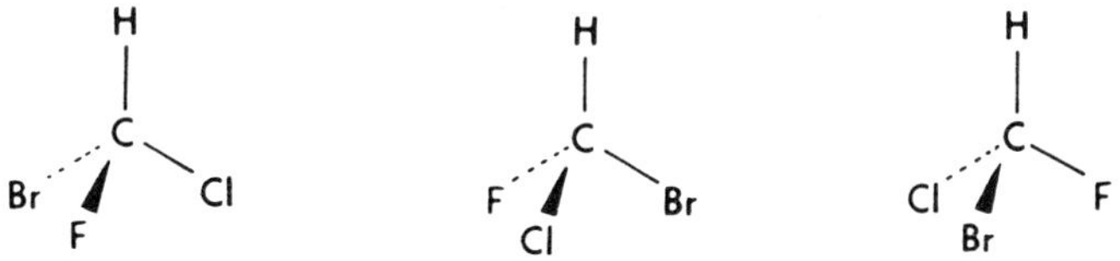

(these three structures are all equivalent)

Program 4, Frame 40). The following equation shows the hydrolysis of 2-chloro-2-methylpropane to 2-methylpropan-2-ol; and the transitory intermediate, $(CH_3)_3C^+$, is a carbonium ion.

$$(CH_3)_3C{-}Cl \rightleftharpoons (CH_3)_3C^+\ Cl^- \xrightarrow{H_2O} (CH_3)_3C{-}OH + H^+ + Cl^-$$

At all times during the reaction the concentration of this ion will be extremely low. In a carbonium ion, the positively-charged carbon atom is bonded to three other groups by single bonds, so that the number of valence electrons around the carbon atom is . . . . (1).

---

(1) six

---

45. *Carbanions.* Organic compounds in which a negative charge is located principally on a carbon atom are called carbanions. Such a species may be produced by removal of a proton from a system $\geq$C–H provided that one or more of the groups attached to the carbon atom confers acidity on the CH bond (e.g. by mesomerism). An example of the formation of a carbanion by this process is shown, where |B represents a base such as OH^-, OEt^-, NR_3 (c.f. Program 7, Frame 20).

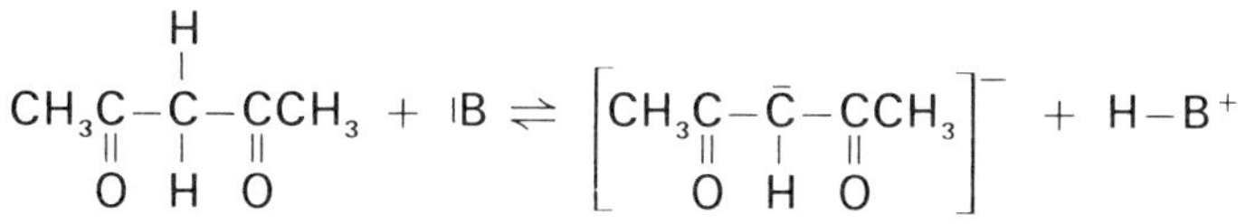

Grignard reagents also behave as though the carbon atom attached to the metal carries a negative charge (c.f. Program 4, Frames 67 and 68). A carbanion may be singly bonded to three other groups, and also carry a lone pair; the total number of electrons around the carbon is then . . . . (1).

---

(1) eight

---

46. *Free Radicals.* In Frame 38 it was stated that most of the reactions in these programs occur between ionic and/or polar reactants in which the electrons remain paired. In certain reactions, usually initiated by light or heat, the reaction mechanism involves the formation of intermediate species containing odd numbers of valence electrons. These intermediates are known as free radicals. A typical free radical is $CH_3\cdot$, the methyl radical, and the "odd" electron is symbolised by the dot. This free radical contains . . . . (1) valence electrons (c.f. Program 4, Frames 18-27).

---

(1) seven

(1) but-*trans*-2-enoic acid | (4) *cis*-1-chloropropene
(2) *cis*-1,2-dichloroethene | (5) 1,1-dibromo-2-methylpropene
(3) 1,1-dichloropropene | (6) 3-methylpent-*trans*-2-ene

A method of designating *cis-trans*-isomers by the use of the prefixes *Z* and *E* is outlined on page 218.

---

16. *Optical isomerism.* An object and its mirror image may or may not be superimposable. An object such as a teacup has a superimposable mirror image. Thus we do not have left-handed and right-handed teacups. If an object and its mirror image are *not* superimposable then right-handed and left-handed forms of that object are possible. For example a bolt may have a left-handed or right-handed spiral thread and bolts of each type are made. Write "yes" or "no" to indicate whether or not the following objects are superimposable on their mirror images:

a nail. . . . . (1), a golf club. . . . . (2), a spiral staircase . . . . (3), a bucket. . . . . (4), a two-bladed propeller . . . . (5), a regular tetrahedron with each apex painted a different colour. . . . . (6), a simple knot. . . . . (7).

---

(1) yes (2) no (3) no (4) yes (5) no
(6) no (7) no

---

17. Some of the above objects which can exist in right- and left-handed forms have their counterpart in organic molecules, especially the propeller, the spiral staircase, and the regular tetrahedron with a different colour at each apex.

Consider a molecule 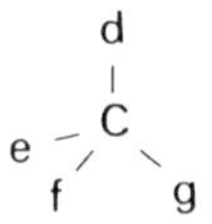

in which the four atoms or groups (d, e, f, g) attached to the central carbon atom are all different. Because of the tetrahedral disposition of the four bonds to the carbon atom, such a molecule. . . . . (is/is not) (1) superimposable on its mirror image.

---

(1) is not

---

18. Asymmetric objects are objects which have non-superimposable mirror images.

Molecules of the type 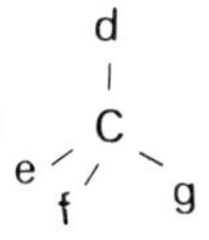

shown above are asymmetric.

Program 3: HYDROCARBONS

CONTENTS

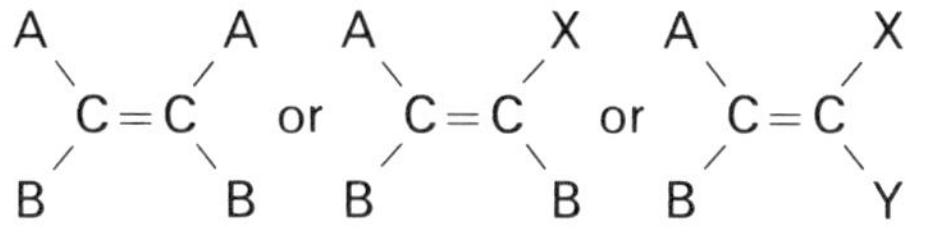

Thus but-2-ene, formula. . . . . (1) can exist as two *cis-trans*-isomers while but-1-ene, formula. . . . . (2) cannot.
Write "yes" or "no" to indicate whether or not the following compounds can exist as *cis-trans*-isomers:
but-2-enoic acid (crotonic acid), $CH_3CH{=}CHCOOH$. . . . . (3),
chloroethene (vinyl chloride), $CHCl{=}CH_2$. . . . . (4),
1-bromo-2-methylpropene, $(CH_3)_2C{=}CHBr$. . . . . (5),
1, 2-dichloroethene $CHCl{=}CHCl$. . . . . (6).

---

(1) $CH_3CH{=}CHCH_3$ (2) $CH_3CH_2CH{=}CH_2$ (3) yes (4) no
(5) no (6) yes

---

15. Consider two *cis-trans*-isomers such as

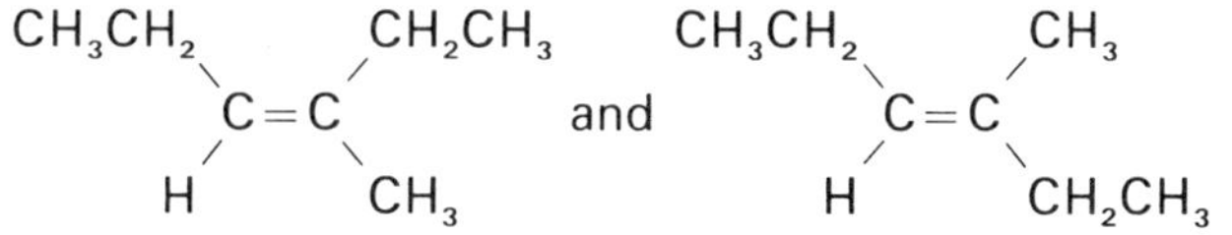

If we examine the longest carbon chain which includes the double bond and find it has the shape __/ then this is termed the *cis* isomer, while the other isomer, of shape /‾‾/ is termed the *trans* isomer.
Where this rule is not applicable we use *cis* to designate the isomer having the same or similar groups on one side of the double bond, and *trans* for the isomer in which these similar groups occur on opposite sides of the double bond. Give the full name of the following compounds:

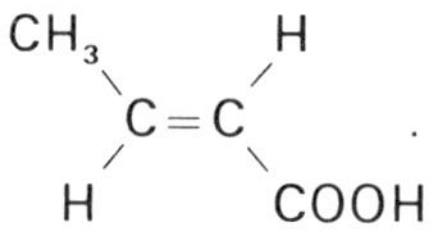 . . . . (1)

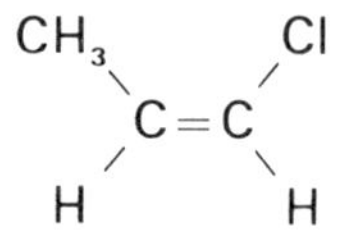 . . . . (4)

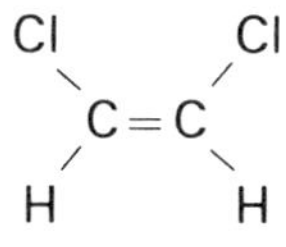 . . . . (2)

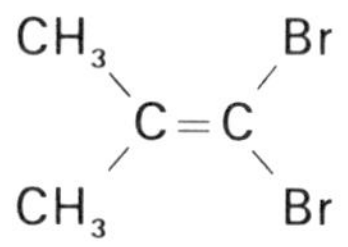 . . . . (5)

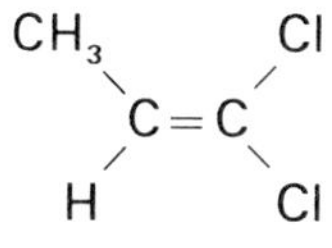 . . . . (3)

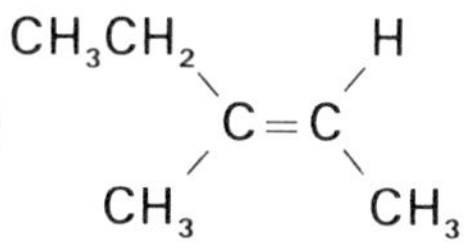 . . . . (6)

HYDROCARBONS

Program 3

INTRODUCTION

In the program on Nomenclature we recognised the following major groups of hydrocarbons:—

1. Alkanes	open chain compounds of general formula	C_nH_{2n+2}
	ring compounds of general formula	C_nH_{2n}
2. Alkenes	open chain compounds with 1 double bond	C_nH_{2n}
	ring compounds with 1 double bond	C_nH_{2n-2}
3. Alkynes	open chain compounds with 1 triple bond	C_nH_{2n-2}
4. Benzene	ring compound with 3 double bonds	C_6H_6

Further variety is added by combinations of these. A compound may have more than one double bond, or a mixture of single, double and triple bonds. Compounds containing benzene rings having any such chain attached are called arenes. We reduce the amount of factual material that must be learned by treating a class as a whole and learning what kinds of reactions can be expected of particular arrangements of atoms called the *functional groups*.

ALKANES

Structure and properties

1. The chemistry of alkanes is probably already familiar and these frames are intended primarily for revision. In the alkanes all carbon atoms are attached by .... (1) bonds to other carbon atoms or hydrogen atoms. The spatial arrangement of the bonds about the carbon atom is .... (2) and the bonds .... (are/are not) (3) polarised.

---

(1) single (2) tetrahedral (3) are not

---

2. The alkanes are chemically inert to many common laboratory reagents such as acids, bases, reducing agents, and mild oxidising agents. They are oxidised by powerful oxidising agents and when ignited react vigorously with oxygen with the evolution of heat. The process of combustion which can be indicated by the equation

$$C_nH_{2n+2} \xrightarrow{\text{excess } O_2} \text{............} \quad (1)$$

can be used to determine the percentage of carbon and hydrogen present in the alkane.

Alkanes serve as raw materials for th^ petrochemical industry where it is possible to control a variety of reactions done under conditions of high temperature and pressure.

---

11. The compound 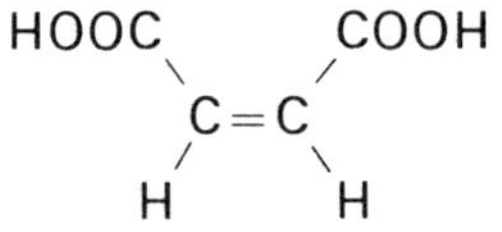 m.p. 130° is known as maleic acid.

It has a *cis-trans*-isomer, fumaric acid, m.p. 287°, of structure. . . . .

. . . . (1). The compound 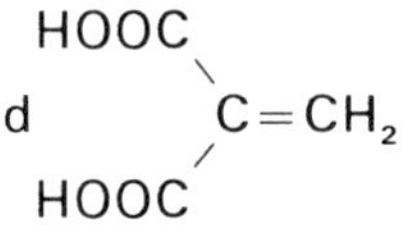 is a. . . . . (2) isomer of both these acids.

---

(1) 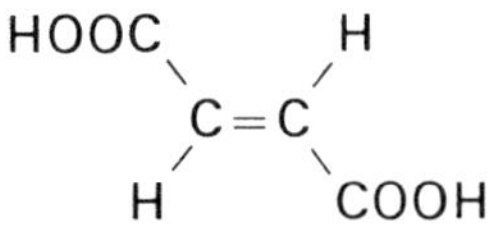 (2) constitutional

---

12. For *cis-trans*-isomers to be possible the molecule must contain, at least in part, a rigid framework to which various other groups are attached. In maleic and fumaric acids this framework is. . . . . (1).

---

(1) a carbon-carbon double bond, 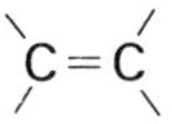

---

13. The two atoms forming such a framework need not be the same—thus in the isomeric aldehyde oximes, 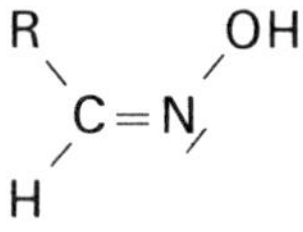 and 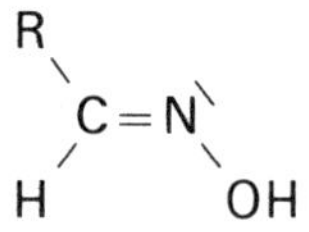 the rigid framework is. . . . . (1).

---

(1) a carbon-nitrogen double bond, 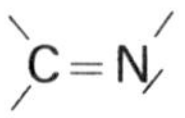

---

14. The mere presence of a carbon-carbon double bond is not a sufficient condition for the appearance of *cis-trans*-isomerism. A further requirement is that there must be two *different* atoms or groups attached to *each* of the double-bond carbons. That is, *cis-trans*-isomers are possible provided the alkene has the structure

(1) $C_nH_{2n+2} \xrightarrow[O_2]{\text{excess}} nCO_2 + (n+1)H_2O$

---

3. Alkanes react with chlorine and bromine when energy is supplied in the form of heat or light. The reaction involves free radicals and is described in more detail in Program 4 (Alkyl Halides). The nett effect of the reaction is the replacement of hydrogen atoms by halogen atoms. The products obtained on replacing each of the hydrogens of methane in turn by chlorine atoms are . . . . (1), ... (2), ... (3), ... (4).

---

(1)

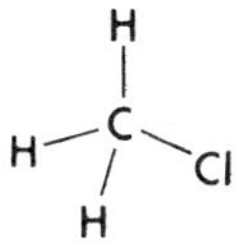

chloromethane
(methyl chloride)

(2)

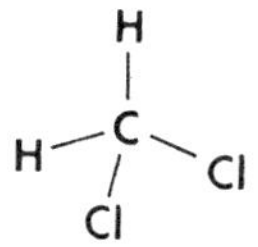

dichloromethane
(methylene chloride)

(3)

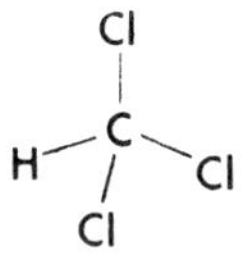

trichloromethane
(chloroform)

(4)

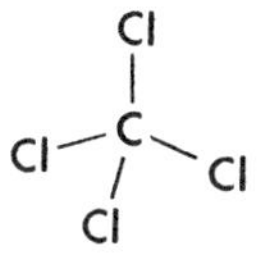

tetrachloromethane
(carbon tetrachloride)

---

4. The importance of such a reaction is that a new functional group is introduced into the molecule. The derivatives are more reactive than the parent compound and have many uses in the laboratory. *Nitro* groups can be introduced into alkanes by vapour phase reaction with nitric acid at about 450°. The nitro compounds, like the chloro and bromo compounds, are reactive and useful compounds.

e.g. $CH_3CH_3 \xrightarrow[450^\circ]{HNO_3} CH_3CH_2NO_2$ (80%) and CH_3NO_2 (20%)

Write the full valence structure of nitromethane CH_3NO_2 . . . . (1).

---

(1)

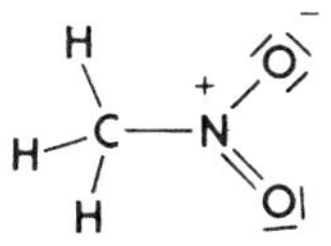

or other mesomeric structures

---

5. Cyclopropane with . . . . (1) carbon atoms and cyclobutane with . . . . (2) carbon atoms are ring structures in which the C–C–C bond angles are 60° and . . . . (3) respectively. These differ considerably from the stable tetrahedral angle of 109° 28′. It is therefore not

The compound CH_2ClCH_2Cl has many conformations, e.g., III, IV, but these are rapidly interconverting at room temperature, so that III and IV represent only a single substance and. . . . . (are/are not) (1) constitutional isomers.

---

(1) are not

STEREOISOMERS

9. Two molecules may have the same structure, in that they contain identical atoms joined together in the same order, and yet may still differ in the manner in which the atoms are arranged in space. A piece of wire bent into the shape  is easily distinguished from a piece of wire bent into the shape , a right-hand spiral is easily distinguished from a left-hand spiral, a right shoe from a left shoe. These examples all have analogies in organic stereochemistry, where we are concerned with the shape and symmetry (or lack of symmetry) of organic molecules.

Molecules which differ *only* with respect to the way in which their constituent atoms are arranged in space are known as. . . . . (1). Molecules which differ in shape in the manner of our bent wire are known as *cis-trans-isomers*, while those which differ because they have opposite symmetries, e.g. as left hand, right hand, are known as *optical isomers.*

---

(1) stereoisomers

---

10. cis-trans-*Isomerism.*

The molecules 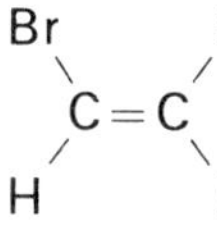 and 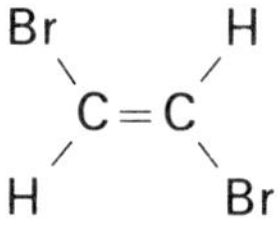 are stereoisomers, being stable, distinct compounds under ordinary conditions, and having identical atoms bonded in similar fashion. They differ only in the way in which the substituent atoms are arranged in space around the carbon-carbon double bond. Such isomers are usually termed geometrical or *cis-trans* isomers.

It is possible to convert a piece of wire of shape 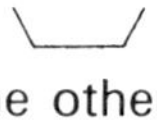 into a piece of shape 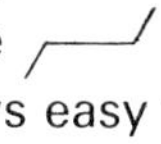 by twisting one end with respect to the other, but this is not always easy to do. The fact that the two different 1,2-dibromoethenes have a separate existence means that the energy barrier to twisting a carbon-carbon double bond through 180° is fairly. . . . . (1). The rigidity of a carbon-carbon double bond stands in contrast to the free rotation usually observed around the carbon-carbon. . . . . (2).

---

(1) large (2) single bond (cf. Frame 8).

surprising that compounds with these ring sizes are often more reactive than the open chain alkanes. Other cyclic alkanes in which the bond angles at carbon are the normal tetrahedral angle are comparable in chemical reactivity to the open chain alkanes.

---

(1) three (2) four (3) 90°

ALKENES

The double bond as a functional group

6. The functional group of the alkenes is the double bond $>C=C<$. This representation tells us that there are . . . . (1) electrons shared between the two carbon atoms and this region of the molecule is therefore, one of . . . . (2) electron density in comparison with the single bond.

---

(1) four (2) high

Addition reactions

7. *Definition*. The most characteristic reactions of alkenes are *addition* reactions. As the name suggests, the product of an addition reaction contains the sum of all the atoms of the reactants

e.g. $C_4H_8 + HCl \longrightarrow C_4H_9Cl$

$C_4H_8 + H_2O \longrightarrow$ . . . . (1)

---

(1) $C_4H_{10}O$

---

8. *Addition of hydrogen*. The general formula of an open chain with one double bond is C_nH_{2n}. Addition of hydrogen takes place in the presence of finely divided platinum or palladium catalyst and the product, C_nH_{2n+2}, is . . . . (1).

$$C_nH_{2n} + H_2 \xrightarrow{\text{Pt or Pd}} C_nH_{2n+2}$$

A cyclic alkene of general formula C_nH_{2n-2} will add hydrogen in the presence of the catalyst to give . . . . (2) of general formula . . . . (3).

---

(1) an open chain alkane (2) a cyclic alkane (3) C_nH_{2n}

---

9. The general formula of an open chain alkene and of a cyclic alkane is the same, namely . . . . (1). In general an alkene can be distinguished from a cyclic alkane by . . . . (2).

---

(1) CH_3COCH_3, $CH_3OCH{=}CH_2$, $H_2C{-}CHCH_3$ (with O bridging the two carbons), $CH_2{=}CHCH_2OH$ and

CH_3CH_2CHO are all isomers of C_3H_6O

$HCOOCH_3$, CH_3COOH and $HOCH_2CHO$ are all isomers of $C_2H_4O_2$.

---

6. It is always possible to distinguish constitutional isomers by differences in physical properties. In addition, many isomers such as e.g., $HCOOCH_3$ (I), and CH_3COOH (II), can be distinguished on the basis of simple chemical tests. Thus compound I is a neutral ester and is not dissolved by sodium hydrogencarbonate solution, whereas its isomer II is an . . . . (1) and dissolves easily in sodium hydrogencarbonate with evolution of . . . . (2).

---

(1) acid (2) carbon dioxide

---

7. The compounds $ClCH_2CHO$, b.p. 85°C, and CH_3COCl, b.p. 52°C, are constitutional isomers. They both have the same molecular formula C_2H_3ClO but clearly possess different structures. Write the structures of the four isomers of formula $C_3H_6Cl_2$ and give their names . . . . (1).

---

(1) $CH_3CH_2CHCl_2$, 1,1-dichloropropane

$CH_3CCl_2CH_3$ 2,2-dichloropropane

$CH_3CHClCH_2Cl$, 1,2-dichloropropane

$CH_2ClCH_2CH_2Cl$, 1,3-dichloropropane

---

8. *Conformation.* Atoms bonded together to form a molecule of a given structure can still undergo certain restricted movements relative to one another, even when they possess only small amounts of thermal energy. These restricted movements can be vibrational or rotational in kind, and those movements which involve relative rotations we speak of as conformational changes. In ethane the two CH_3 groups can rotate relative to each other, with small energy differences between different conformations of the molecule. The conformation of minimum energy can be shown as in Figure I. In this diagram we are looking along the carbon-carbon bond, with the six hydrogen atoms in a staggered relationship. In Figure II the hydrogen atoms are nearly eclipsed.

I 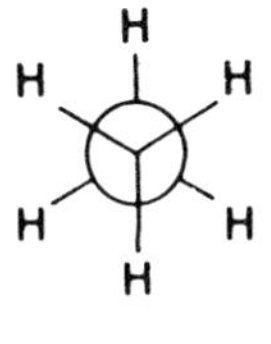

II 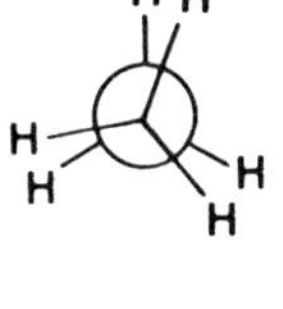

III 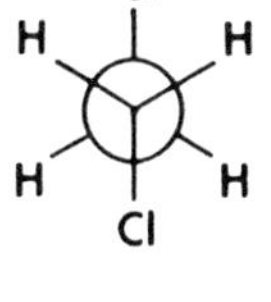

IV 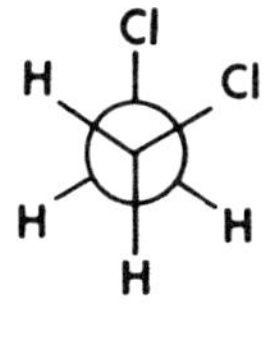

(1) C_nH_{2n}
(2) the addition of hydrogen to the double bond in the presence of catalyst.

---

10. The catalytic addition of hydrogen is known as *hydrogenation*. Hydrogenation can be carried out quantitatively by measuring the volume of . . . . (1) taken up by a known mass of alkene in the presence of a catalyst.

---

(1) hydrogen gas

---

11. *Addition of halogen*. Chlorine and bromine add to alkenes

e.g. $H_2C{=}CH_2 + Cl_2 \longrightarrow H_2C{-}CH_2$ (with Cl on each carbon: Cl Cl)

ethylene or ethene — 1,2-dichloroethane

$CH_3CH{=}CH_2 + Cl_2 \longrightarrow CH_3CHClCH_2Cl$

propylene or propene — 1,2-dichloropropane.

When cyclohexene is shaken with bromine water, the brown colour of the bromine disappears. The equation for this addition reaction is . . . . (1) Decolorisation of bromine water is a simple colour test for distinguishing alkenes from alkanes.

---

(1)

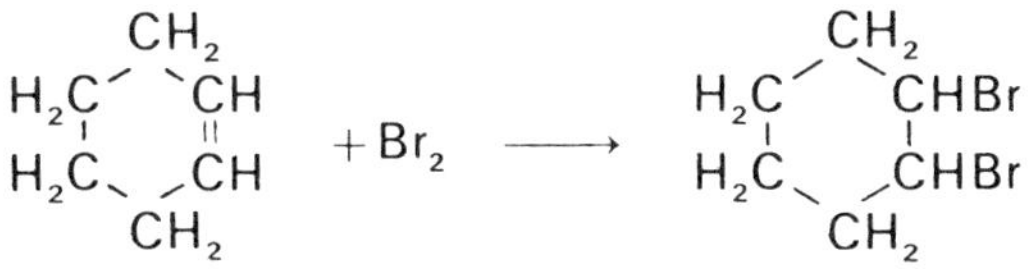

---

12. *Addition of hydrogen halides*. Hydrogen chloride, hydrogen bromide and hydrogen iodide can add to alkenes. One molecule of ethylene adds one mole of hydrogen chloride according to the equation . . . . (1) and the product is called . . . . (2).

---

(1) $H_2C{=}CH_2 + HCl \longrightarrow CH_3CH_2Cl$
(2) chloroethane

---

13. Propylene (propene) $CH_3CH{=}CH_2$ also adds hydrogen chloride or hydrogen bromide but on inspection we see that two different products are possible—$CH_3CHClCH_3$ and $CH_3CH_2CH_2Cl$—depending on the mode of addition. We find, in the addition of hydrogen halides to unsymmetrical alkenes, that there is always predominantly one product. After examining the structures of a number of such adducts Markovnikov found that, *in the addition of an acid HX, the hydrogen atom goes on to the carbon atom with the greater number of hydrogen atoms*. The product of the reaction $CH_3CH{=}CH_2 + HI$ is . . . . (1) called . . . . (2).

---

component atoms are joined together, i.e., they differ in molecular constitution. The compounds CH_3CHCl_2 and CH_2ClCH_2Cl . . . . (are/are not) (2) constitutional isomers.

---

(1) molecular formulae (2) are

---

3. There are three constitutional or structural isomers of formula C_5H_{12}; the structures of these can be shown . . . . (1).

---

(1) $CH_3CH_2CH_2CH_2CH_3$ — pentane

CH_3
$|$
$CH_3CHCH_2CH_3$ — 2-methylbutane (or isopentane)

CH_3
$|$
CH_3CCH_3
$|$
CH_3 — 2,2-dimethylpropane (or neopentane)

---

4. The compounds propan-1-ol, structure . . . . (1). and propan-2-ol, structure . . . . (2), are also constitutional isomers; the molecules differ in the point of attachment of the hydroxyl group to the three-carbon chain.

---

(1) $CH_3CH_2CH_2OH$ (2) CH_3CHCH_3
$|$
OH

---

5. Constitutional isomerism is extremely common in organic chemistry. From the list below collect together those compounds, if any, which are isomeric . . . . (1).

O
$\|$
CH_3CCH_3 2-propanone (acetone)

O
CH_3C, ethanoic acid (acetic acid)
OH

O
HC, methyl methanoate (formate)
OCH_3

$CH_2{=}CHCH_2OH$, prop-2-en-1-ol (allyl alcohol)

$CH_3OCH{=}CH_2$, methyl vinyl ether

H
CH_3CH_2C, propanal (propionaldehyde)
O

$H_2C{-}CHCH_3$, 1,2-epoxypropane
O

H
$HOCH_2C$, 2-hydroxyethanal (glyoxal)
O

---

(1) CH_3CHICH_3

(2) 2-iodopropane. This is known as Markovnikov addition.

---

14. The Markovnikov addition of hydrogen bromide to 2-methylpropene follows the equation . . . . (1).

---

(1) $$CH_3\overset{\displaystyle CH_3}{\overset{|}{C}}=CH_2 + HBr \longrightarrow CH_3\underset{\displaystyle Br}{\underset{|}{\overset{\displaystyle CH_3}{\overset{|}{C}}}}CH_3$$ 2-bromo-2-methylpropane

---

15. In the case of hydrogen bromide *only*, the mode of addition is reversed in the presence of peroxides, which are formed in small amounts by reaction of the alkene with air. Under these conditions 2-methylpropene yields . . . . . . (1). This is called anti-Markovnikov addition.

---

(1) $CH_3\overset{\displaystyle CH_3}{\overset{|}{C}}HCH_2Br$ 1-bromo-2-methylpropane

---

16. *Addition of sulphuric acid.* Markovnikov's rule is valid for acids other than the hydrogen halides. HX may be sulphuric acid, $HOSO_3H$, and according to the rule

$$CH_3CH_2CH=CH_2 + HOSO_3H \longrightarrow \ldots\ldots\ldots\ldots \text{(1)}.$$

---

(1) $CH_3CH_2\underset{\displaystyle OSO_3H}{\underset{|}{C}}HCH_3$ 1-methylpropyl hydrogen sulphate

---

17. *Addition of water.* The addition of water follows Markovnikov's rule, and, in the presence of sulphuric acid, 2-methylpropene adds water according to the equation . . . . (1).

---

(1) $$(CH_3)_2C=CH_2 + H_2O \xrightarrow{\text{acid}} (CH_3)_2\underset{\displaystyle OH}{\underset{|}{C}}CH_3$$ 2-methylpropan-2-ol

---

TEST FRAME

Write the equations for the following additions to but-1-ene:

(1) the addition of hydrogen (hydrogenation)

(2) the addition of chlorine (chlorination)

ISOMERISM

Program 11

It is a basic principle of chemistry that it is possible to assign a molecular formula to every pure, homogeneous substance. Examples are hydrazine, N_2H_4, ethane, C_2H_6, glucose, $C_6H_{12}O_6$. It often happens that quite different compounds, with different physical and chemical properties, have the same molecular formula. This phenomenon is known as *isomerism*, and compounds which are different from each other and yet possess the same molecular formula are said to be isomeric, or to be isomers. Isomerism is common in organic chemistry.

Isomers are possible *either* because the component atoms of the molecules concerned are bonded together in different ways (CONSTITUTIONAL ISOMERS), *or* because the atoms of otherwise identically-bonded molecules are differently arranged in space (STEREOISOMERS).

CONSTITUTIONAL ISOMERISM

1. The structural formula of an organic molecule is a specification of which atoms are joined to which in that molecule. The structural formula of propane, C_3H_8, can be shown . . . . (1). Only one propane is known, and given that carbon is tetravalent and hydrogen univalent, there is only one way in which the eight hydrogen atoms and three carbon atoms of propane can be joined together to form a stable molecule.

---

(1) $CH_3CH_2CH_3$ or

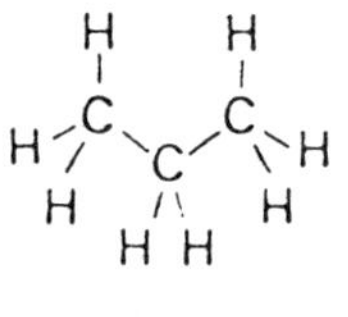

---

2. On the other hand there are two different substances corresponding to the formula C_4H_{10}; in one of these (I) the carbon atoms are joined together in a linear chain, while in the other (II) the carbon atoms are joined in a non-linear or branched chain.

$$\underset{\text{I}}{CH_3CH_2CH_2CH_3} \qquad \underset{\text{II}}{CH_3\overset{\overset{CH_3}{|}}{C}HCH_3}$$

Structures I and II correspond to the substances butane and 2-methyl propane (isobutane) respectively, and these are *constitutional isomers*. Constitutional isomers, also known as structural isomers, are chemical species which have the same . . . . (1) but which differ in the way in which the

(3) the addition of hydrogen bromide (hydrobromination)
(4) the addition of hydrogen bromide in the presence of peroxide
(5) the addition of sulphuric acid
(6) the addition of water.

---

(1) $CH_3CH_2CH{=}CH_2 + H_2 \xrightarrow{\text{Pt or Pd}} CH_3CH_2CH_2CH_3$ butane
(2) $CH_3CH_2CH{=}CH_2 + Cl_2 \longrightarrow CH_3CH_2CHClCH_2Cl$ 1,2-dichlorobutane
(3) $CH_3CH_2CH{=}CH_2 + HBr \longrightarrow CH_3CH_2CHBrCH_3$ 2-bromobutane
(4) $CH_3CH_2CH{=}CH_2 + HBr \xrightarrow{\text{peroxide}} CH_3CH_2CH_2CH_2Br$ 1-bromobutane
(5) $CH_3CH_2CH{=}CH_2 + HOSO_3H \longrightarrow CH_3CH_2\underset{OSO_3H}{\underset{|}{C}}HCH_3$ 1-methylpropyl hydrogensulphate
(6) $CH_3CH_2CH{=}CH_2 + H_2O \xrightarrow{\text{acid}} CH_3CH_2CH(OH)CH_3$ butan-2-ol

---

OXIDATION OF ALKENES

18. *Oxidation by peracids.* Alkenes are oxidised by peracids, $R\overset{\overset{O}{\|}}{C}OOH$, which are formed by the action of hydrogen peroxide on carboxylic acids. The reaction of an alkene with hydrogen peroxide in acetic acid solution yields an epoxide

e.g. $$H_2C{=}CH_2 \xrightarrow{R\overset{\overset{O}{\|}}{C}OOH} \underset{\diagdown\,O\,\diagup}{H_2C-CH_2}$$

epoxyethane (oxiran or ethylene oxide)

In the presence of aqueous acid, the epoxide undergoes hydrolysis to yield a glycol:

$$\underset{\diagdown\,O\,\diagup}{H_2C-CH_2} + H_2O \xrightarrow{\text{acid}} HOCH_2CH_2OH$$

ethane-1,2-diol (ethylene glycol)

Give the reactions for the oxidation of but-2-ene with hydrogen peroxide in acetic acid . . . . (1), followed by acid hydrolysis . . . . (2).

---

(1) $CH_3CH{=}CHCH_3 + CH_3\overset{\overset{O}{\|}}{C}OOH \longrightarrow \underset{\diagdown\,O\,\diagup}{CH_3CH-CHCH_3} + CH_3\overset{\overset{O}{\|}}{C}OH$

(2) $\underset{\diagdown\,O\,\diagup}{CH_3CH-CHCH_3} + H_2O \xrightarrow{\text{acid}} CH_3\underset{OH}{\underset{|}{C}}H-\underset{OH}{\underset{|}{C}}HCH_3$ butane-2,3-diol

---

19. *Oxidation by permanganate.* On reaction with dilute, strongly alkaline solutions of permanganate, alkenes are converted into glycols. Such mild oxidation of 2-methylpropene by permanganate yields . . . . (1).

Program 11: ISOMERISM

CONTENTS

(1) $CH_3C(CH_3)(OH)CH_2OH$ 2-methylpropane-1,2-diol

---

20. Oxidation of alkenes under more vigorous conditions with acid solutions of permanganate breaks the carbon-carbon double bond and the products are acids or ketones according to the structure of the original alkene.

$$CH_3CH{=}CHCH_3 \xrightarrow[KMnO_4]{acid} 2\ CH_3COOH$$

Give the products of oxidation of 4-methylpent-2-ene with cold dilute alkaline $KMnO_4$ . . . . (1) and hot acid $KMnO_4$ . . . . (2)

---

(1) $CH_3CH(CH_3)CH{=}CHCH_3 \xrightarrow[KMnO_4]{\text{cold dil. alkaline}} CH_3CH(CH_3){-}CH(OH){-}CH(OH)CH_3$

(2) $CH_3CH(CH_3)CH{=}CHCH_3 \xrightarrow[KMnO_4]{\text{hot acid}} CH_3CH(CH_3)COOH + CH_3COOH$

---

21. *Oxidation by ozone*. The oxidation of alkenes by ozone involves the preliminary formation of a compound called an ozonide, of structure I.

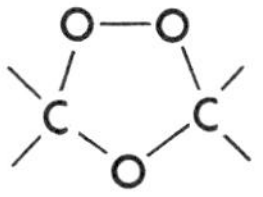

I

Ozonides are very reactive and unstable materials and decompose on addition of water to give aldehydes or ketones and hydrogen peroxide.

Thus: $H_2C{=}CH_2 + O_3 \longrightarrow$

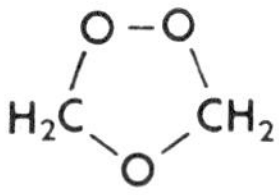

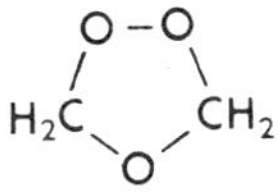

$+ H_2O \longrightarrow 2HCHO + H_2O_2$

Complete the following equations:

$CH_3CH{=}CHCH_3 + O_3 \longrightarrow$ . . . . (1)

(1) $+ H_2O \longrightarrow$ . . . . (2)

---

(1) $CH_3CH{=}CHCH_3 + O_3 \longrightarrow$

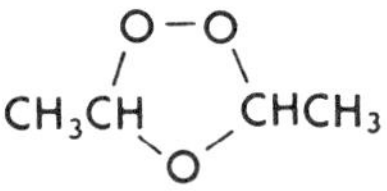

(2)

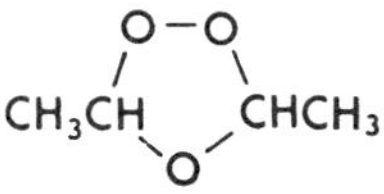

$+ H_2O \longrightarrow 2CH_3CHO + H_2O_2$

---

22. Aldehydes and ketones, which are obtained in the decomposition of an ozonide, are compounds which can be identified readily by the physical properties of some of their crystalline derivatives. Aldehydes, however, are readily oxidised by peroxide so it is essential to destroy this by-product.

$$Ar{-}NH_2 \longrightarrow Ar{-}\overset{+}{N}{\equiv}N \longrightarrow ArH$$

where Ar represents an aromatic ring.
The reagent normally used is hypophosphorous acid, $H_2P(O)OH$, a reducing agent. An example of the replacement of a diazonium group by hydrogen is the conversion of 2,4,6-tribromoaniline into .... (1) by the reaction

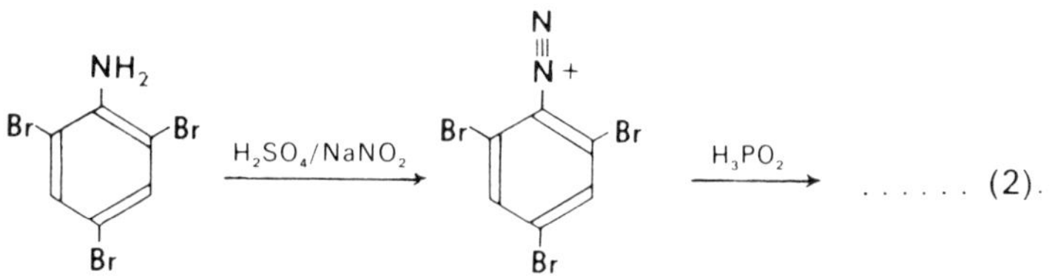

Ethanol can also act as the reducing agent in a similar reaction in which the diazonium salt is converted into the arene when boiled in ethanolic solution. The ethanol is oxidised to acetaldehyde.

---

(1) 1,3,5-tribromobenzene (2)

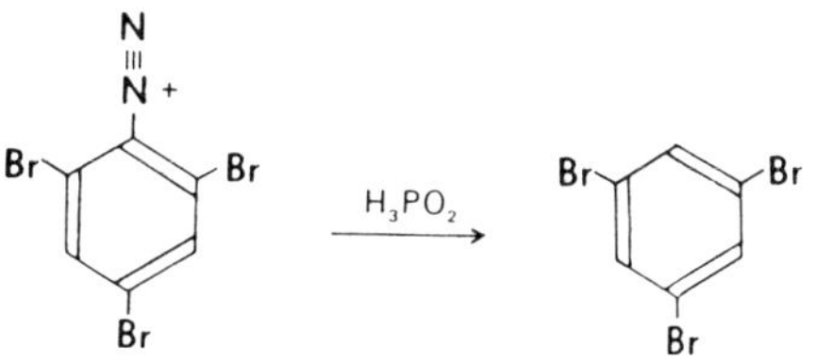

TEST FRAME
Write down the reagents necessary for each substitution shown in the diagram.

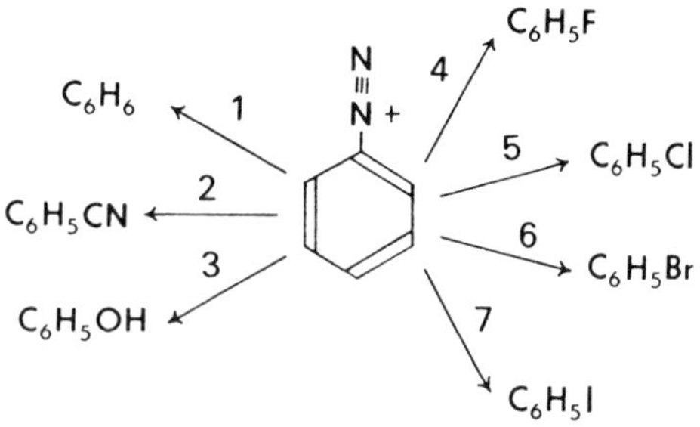

---

(1) H_3PO_2 (2) $Cu(CN)_4^{3-}$ or KCN/CuCN (3) H_2O/H_2SO_4
(4) Heat the dry fluoroborate (BF_4^-) (5) CuCl (6) CuBr (7) KI

The decomposition of the ozonide is therefore carried out in the presence of zinc and dilute acetic acid which reduces the hydrogen peroxide as it forms. Identification of the products, the aldehydes and ketones, is a means of determining the position of the double bond in the original alkene. Write equations for the reaction of 2-methylbut-2-ene with ozone . . . . (1) and the subsequent decomposition of the ozonide in the presence of zinc and acetic acid . . . . (2).

---

(1) 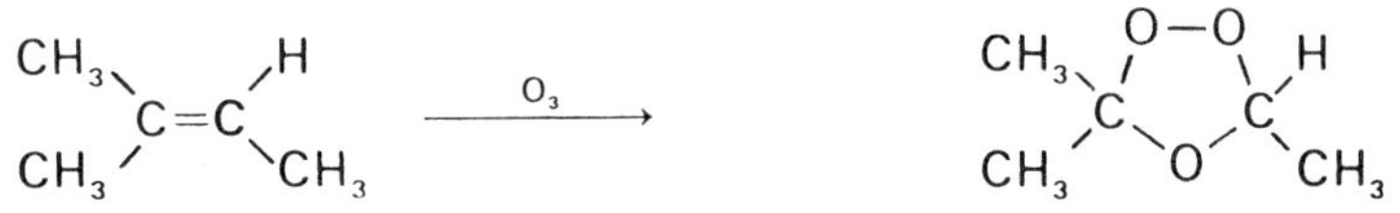

(2) 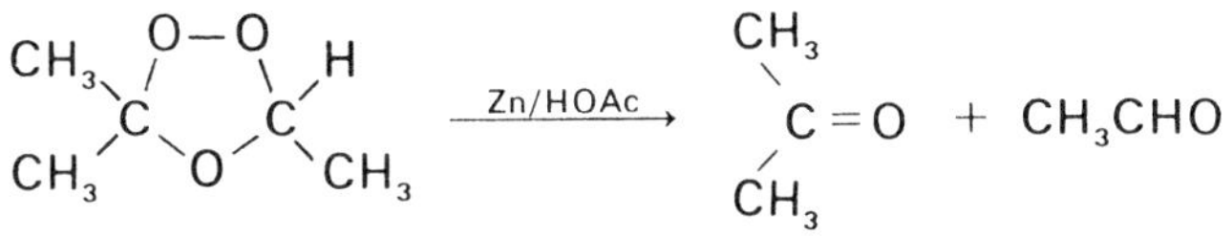

---

23. The process of ozonolysis includes both the preliminary reaction between an alkene and ozone to yield an . . . . (1) and subsequent hydrolysis of the latter to carbonyl compounds. Zinc and acetic acid are present in the second stage to . . . . (2).

---

(1) ozonide (2) destroy H_2O_2

---

24. The application of ozonolysis to cyclopentene gives only one product. This is a dicarbonyl compound. Give the products of ozonolysis of cyclopentene . . . . (1) and 1-methylcyclopentene . . . . (2).

---

(1) $H\underset{\overset{\|}{O}}{C}CH_2CH_2CH_2\underset{\overset{\|}{O}}{C}H$ (2) $CH_3\underset{\overset{\|}{O}}{C}CH_2CH_2CH_2\underset{\overset{\|}{O}}{C}H$

---

25. In summary, the number of double bonds in an alkene may be determined by the process of . . . . (1) and the positions of the double bonds by . . . . (2).

---

(1) hydrogenation (2) ozonolysis

TEST FRAMES

Give equations, with structures shown, for the preparation of the following compounds from appropriate alkenes; pentane-1,2-diol . . . . (1), pentanedioic acid . . . . (2), and hexanedial . . . . (3).

(cf. Program 9, Frame 25). Normally the diazonium salt is prepared in sulphuric acid, and the aqueous solution is then strongly acidified and heated to boiling. Nitrogen is evolved and the phenol is formed. For example *m*-nitroaniline can be converted into *m*-nitrophenol by the reaction sequence:

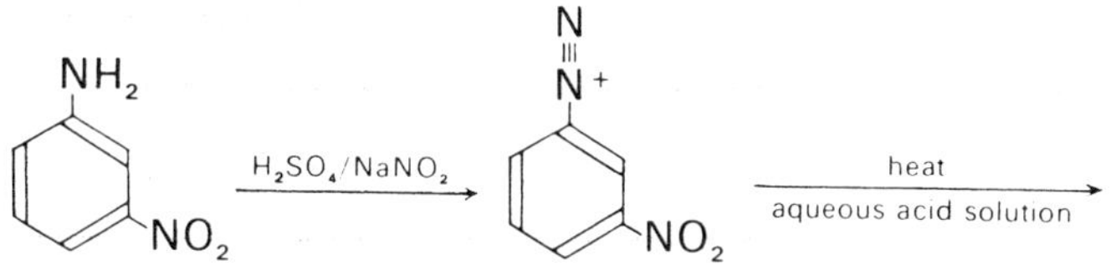

. . . .(1).

---

(1)

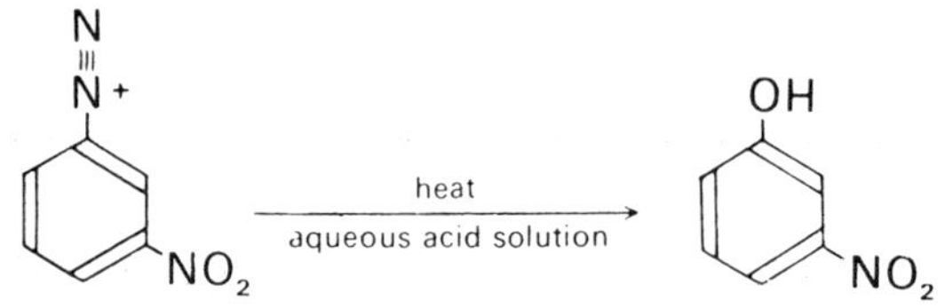

Substitution of $-N_2^+$ *by* $-CN$ *to yield nitriles*

21. The diazonium group can be replaced by a cyano (nitrile) group by reaction with cyanide in the presence of copper(I) complex. This is a further example of a Sandmeyer reaction. The diazotisation is done in sulphuric acid, since, if hydrochloric acid were used, the chloride ions would compete in the reaction. The catalyst is tetracyanocuprate(I) ion, $Cu(CN)_4^{3-}$, formed when potassium cyanide is added to a copper(I) salt.

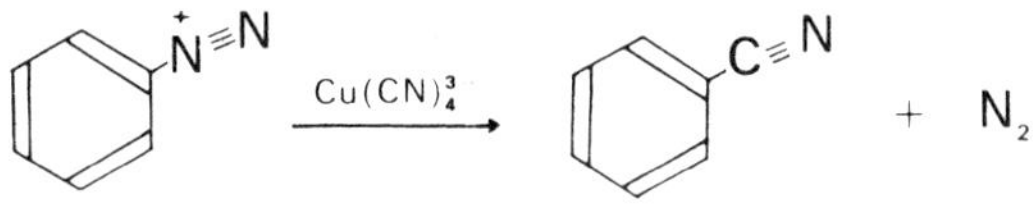

m-Chloroaniline on diazotisation and treatment with $Cu(CN)_4^{3-}$ yields *m*-chlorocyanobenzene (*m*-chlorobenzonitrile) which on hydrolysis with acid yields. . . . . (1) according to the reaction sequence. . . . . (2).

---

(1) *m*-chlorobenzoic acid

(2)

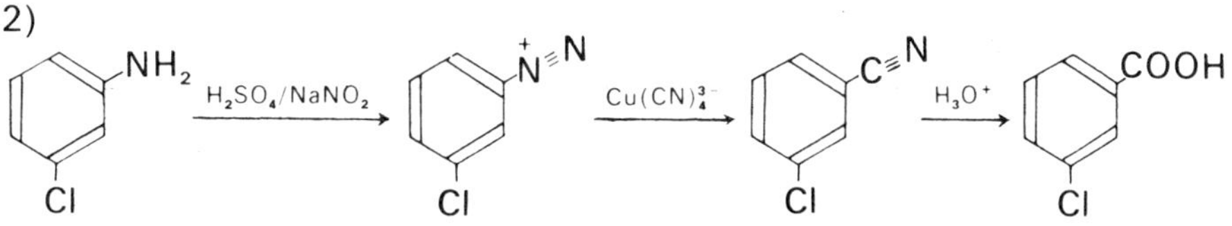

Substitution of $-N_2^+$ *by* $-H$ *to yield arenes*

22. The diazonium ion can be substituted by hydrogen. In other words the group can be removed by reduction. The reaction sequence is given by

(1) $CH_3CH_2CH_2CH{=}CH_2 \xrightarrow{\text{alkaline } KMnO_4} CH_3CH_2CH_2CH(OH)CH_2OH$
Alternatively, oxidation with a peracid followed by acid hydrolysis of the intermediate epoxide would give the diol.

(2) 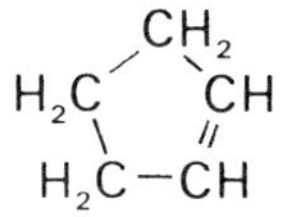 $\xrightarrow{\text{hot acid } KMnO_4}$ $HOOCCH_2CH_2CH_2COOH$

(3) 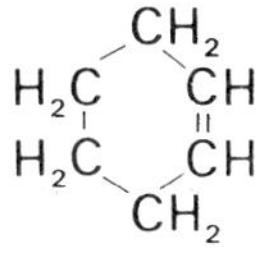 $\xrightarrow{\text{ozonolysis}}$ $HC(=O)CH_2CH_2CH_2CH_2CH(=O)$

Ozonolysis involves reaction with ozone and subsequent hydrolysis with zinc and aqueous acetic acid.

Oleic acid has the formula $C_{17}H_{33}COOH$, which indicates that the molecule has one double bond or a ring in addition to the carboxyl group. Hydrogenation yields stearic acid, $CH_3(CH_2)_{16}COOH$, which indicates that oleic acid contains . . . . (1). Ozonolysis of oleic acid yields nonanal, $C_9H_{18}O$ and a compound which has the properties of both a carboxylic acid and an aldehyde. Write the structure of this latter compound . . . . (2) and of oleic acid . . . . (3).

---

(1) one double bond (2) $O{=}C(H)(CH_2)_7COOH$

(3) $CH_3(CH_2)_7CH{=}CH(CH_2)_7COOH$

POLYMERISATION OF ALKENES

26. Plastics are important products of the organic chemical industry. Everyone is familiar with materials such as polythene, teflon and polyvinyl chloride (PVC). These products are formed by the linking together of a large number of alkene molecules (polymerisation). For example, $H_2C{=}CH_2$ (ethylene) polymerises to polythene,

$$(-CH_2-CH_2-CH_2-CH_2-CH_2-CH_2-)_n.$$

The reaction may be brought about at high temperature and pressure or under milder conditions by the use of catalysts.
Polyvinyl chloride (PVC) has the structure $(-CH_2CH(Cl)CH_2CH(Cl)CH_2CH(Cl)-)_n$.
This polymer is prepared from . . . . (1) and the polymerisation requires the presence of peroxides.

---

(1) $CH_2{=}CHCl$ chloroethene or vinyl chloride

17. In the above substitution no catalyst was required. However, substitution of the diazonium group by a bromine atom requires the addition of a copper(I) salt as a catalyst (Sandmeyer reaction). *o*-Bromobenzoic acid can be prepared from *o*-aminobenzoic acid (anthranilic acid) by such a sequence of reactions according to the equation. . . . . (1).

---

(1)

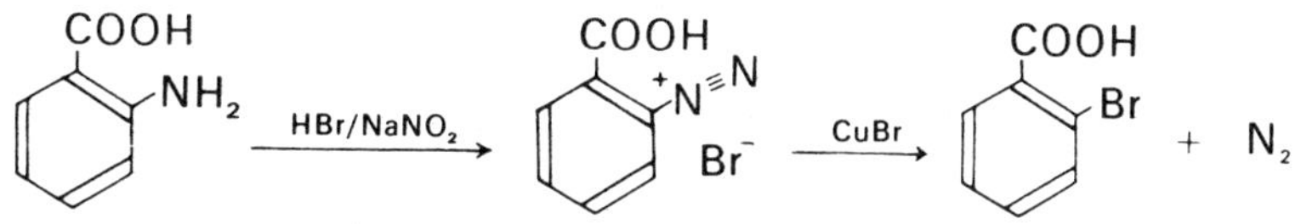

---

18. Introduction of chlorine in the Sandmeyer reaction similarly requires the use of copper(I) chloride. Write the equation for the conversion of *p*-nitroaniline into *p*-chloronitrobenzene. . . . . (1). Note that in this case the diazotisation is done in hydrochloric acid.

---

(1)

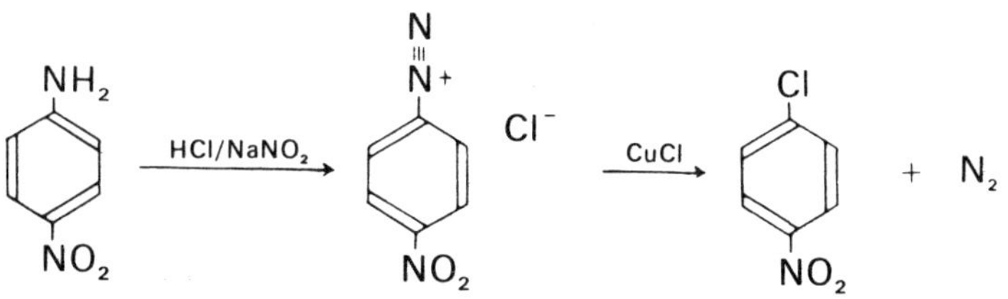

---

19. Substitution of the diazonium group by fluorine is achieved by dry distillation of the diazonium fluoroborate.

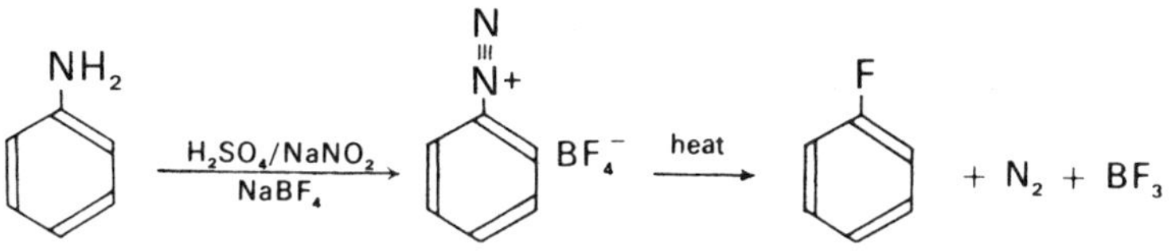

p-Chloronitrobenzene (prepared in the previous frame) can be reduced to *p*-chloroaniline and converted into *p*-chlorofluorobenzene according to the equation. . . . . (1).

---

(1)

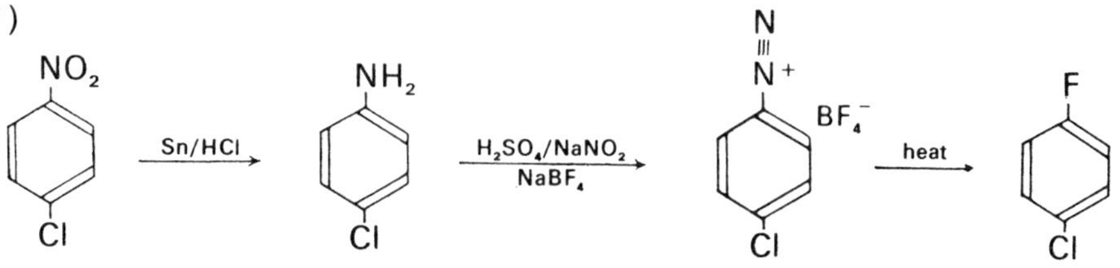

Substitution of $-N_2^+$ by $-OH$ to yield phenols

20. Phenols can be prepared from aromatic amines via diazonium compounds

PREPARATION OF ALKENES

The preparation of alkenes depends on the removal of appropriate atoms or groups from other molecules. Such reactions are called *elimination reactions* and these stand in contrast to the characteristic addition reactions of the alkenes themselves. Industrially, a number of alkenes are prepared by treatment of petroleum fractions (alkanes) at high temperatures. In this "cracking" process, hydrogen gas is eliminated:

$$-CH_2-CH_2- \rightarrow -CH{=}CH- + H_2$$

In the laboratory, alkenes are commonly prepared by elimination of halogens, hydrogen halide or water from suitable starting compounds.

Elimination reactions

27. *Elimination of halogen*. Two adjacent carbon atoms each carrying a halogen substituent must be present for this reaction to occur, e.g.,

$$CH_3\underset{\substack{|\\ Cl}}{C}H\underset{\substack{|\\ Cl}}{C}H_2 \xrightarrow{Zn} CH_3CH{=}CH_2 + ZnCl_2$$

The preparation of but-2-ene by this method would require as starting material . . . . (1).

---

(1) $CH_3\underset{\substack{|\\ X}}{C}H\underset{\substack{|\\ X}}{C}HCH_3$ (X=halogen)

---

28. *Elimination of hydrogen halide.* The removal of a hydrogen atom from one carbon atom and a halogen from the adjacent carbon results in the formation of a double bond. Elimination of hydrogen halide from an alkyl halide is achieved in appropriate cases by treatment with alcoholic solutions of potassium hydroxide

$$CH_3\underset{\substack{|\\ Cl}}{C}HCH_3 \xrightarrow[\text{ethanol}]{\text{KOH in}} \ldots\ldots (1) \qquad (CH_3)_3CCl \xrightarrow[\text{ethanol}]{\text{KOH in}} \ldots\ldots (2)$$

---

(1) $CH_3CH{=}CH_2$ propene
(2) $(CH_3)_2C{=}CH_2$ 2-methylpropene or isobutylene

---

29. A mixture of isomers is often formed in an elimination reaction. Show formulae for the products of dehydrohalogenation of 2-bromobutane . . . . (1).

---

(1) $CH_3CH_2CHBrCH_3 \xrightarrow[\text{ethanol}]{\text{KOH in}} CH_3CH_2CH{=}CH_2$ and $CH_3CH{=}CHCH_3$

(1) 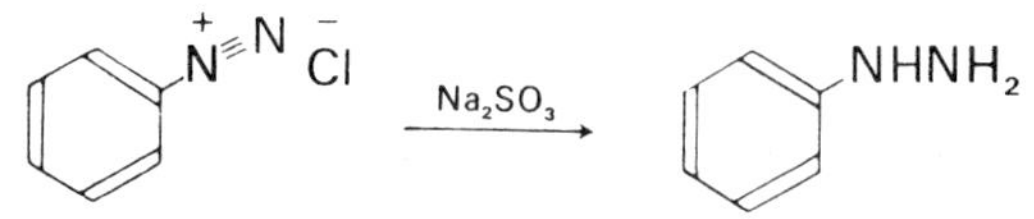

---

14. *p*-Nitroaniline on diazotisation and reduction with sodium sulphite yields *p*-nitrophenylhydrazine according to the equation. . . . . (1).

---

(1)
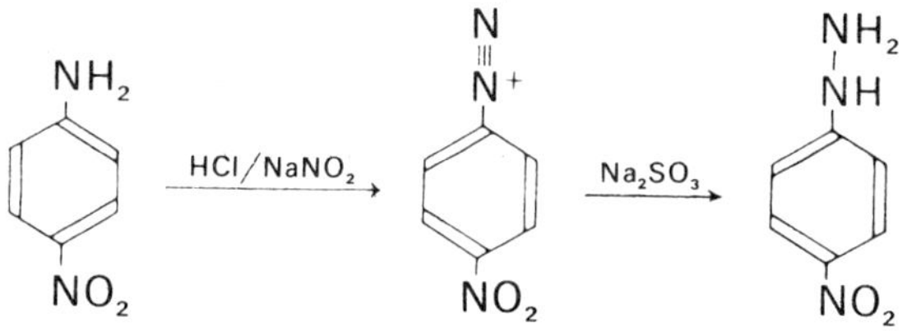

---

15. Phenylhydrazine and *p*-nitrophenylhydrazine are used for making crystalline derivatives of aldehydes and ketones. The structure of the product of reaction of acetaldehyde and phenylhydrazine is. . . . . (1).
(cf. Program 7, Frames 51–57).

---

(1) 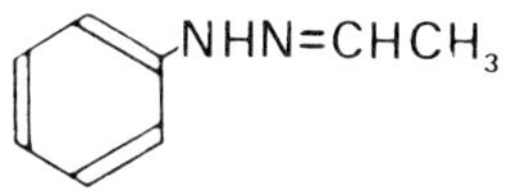

Substitution of $-N_2^+$ by halogen to yield aryl halides

16. Diazonium salts undergo a wide variety of reactions wherein the diazonium group is replaced or substituted by another group, and they are therefore important intermediates in synthetic chemistry. Thus diazonium ions react with iodide ions to form aryl iodides.

e.g., 

A sequence of reactions leading from toluene to *p*-iodotoluene would therefore be. . . . . (1).

---

(1)
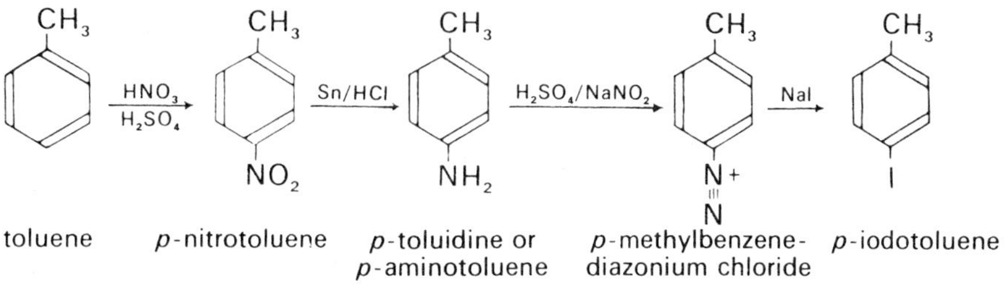

30. *Elimination of water.* Water may be eliminated from alcohols by heating them with sulphuric or phosphoric acid. Cyclopentanol is dehydrated according to the equation . . . . (1). Butan-2-ol, like 2-bromobutane, can give a mixture of isomers of structure . . . . (2).

---

(1)

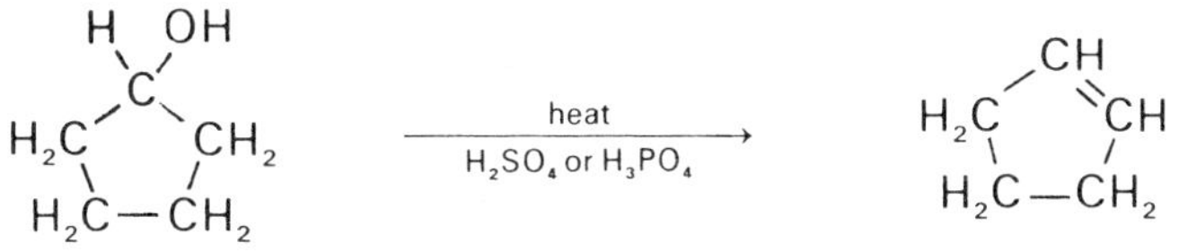

(2) $CH_3CH_2CH(OH)CH_3 \xrightarrow[H_2SO_4 \text{ or } H_3PO_4]{\text{heat}} CH_3CH_2CH{=}CH_2$ and $CH_3CH{=}CHCH_3$

Partial hydrogenation of alkynes

31. Alkenes may also be prepared from alkynes by partial hydrogenation $HC{\equiv}CH + H_2 \xrightarrow{\text{catalyst}} H_2C{=}CH_2$. Further hydrogenation would give . . . . (1).

---

(1) ethane

ALKYNES

Addition reactions—addition of hydrogen, halogens, and hydrogen halides

32. The reactions of the alkynes are, predictably, very similar to those of alkenes. Thus, while an alkene can add one molecule of hydrogen, bromine or hydrogen halide, per double bond, an alkyne can add . . . . (1).

---

(1) 2 molecules per triple bond

---

33. The addition reactions can be controlled to add one molecule of the reactants at a time and hence we can have two stages in the reduction of propyne to propane. The reactions can be shown as . . . . (1).

---

(1) $CH_3C{\equiv}CH \xrightarrow[\text{catalyst}]{H_2} CH_3CH{=}CH_2 \xrightarrow[\text{catalyst}]{H_2} CH_3CH_2CH_3$

---

34. Similarly the addition of halogen to but-2-yne can be shown in two stages . . . . (1).

---

(1) $CH_3C{\equiv}CCH_3 \xrightarrow{X_2} CH_3CX{=}CXCH_3 \xrightarrow{X_2} CH_3CX_2CX_2CH_3$

(1)

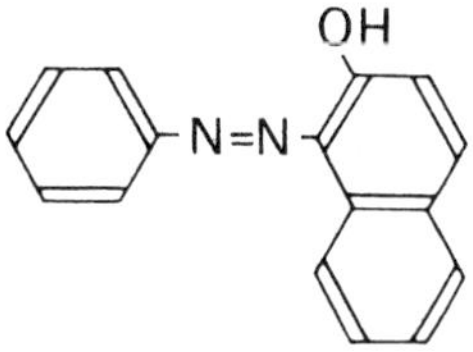

phenylazo-2-naphthol

11. It is easy to determine whether a compound is a primary aromatic amine since on diazotisation and coupling with 2-naphthol, an intense colour, usually orange or red, will be formed. Write the structures of the products obtained by diazotising and coupling the following compounds with 2-naphthol. *o*-chloroaniline . . . . (1), *m*-nitroaniline . . . . (2), and *p*-aminobenzoic acid . . . . (3).

(1)

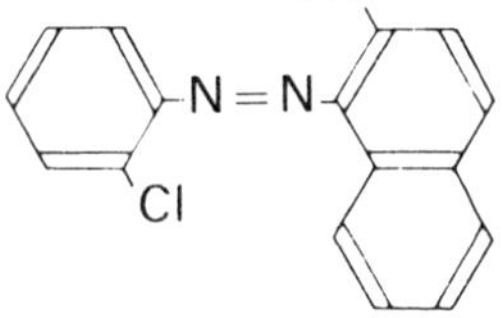

(2)

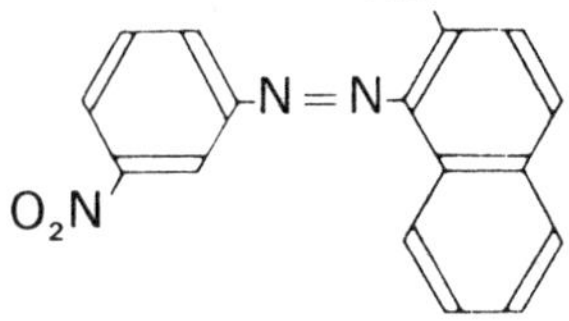

(3)

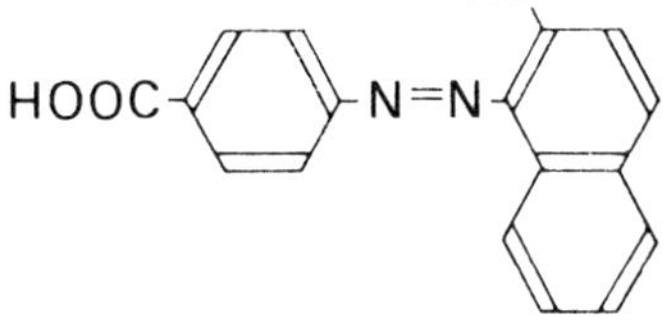

12. Diazonium coupling with aromatic compounds occurs only when highly activating substituents, e.g. $-O^-$ or $-NR_2$, are present on the ring. Coupling takes place in the *para*-position when diazotised sulphanilic acid is added to N,N-dimethylaniline in neutral solution. Complete the equation:

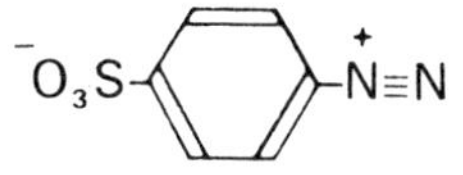

+ . . . . → . . . . (1).

(1)

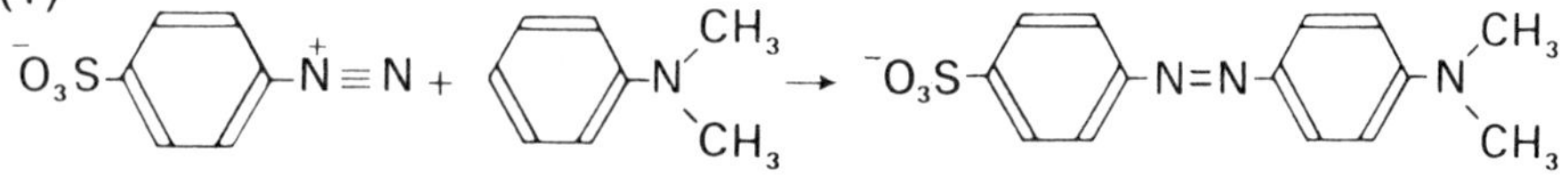

This azo compound is "methyl orange", the dyestuff sometimes used as an indicator in acid-base titrations.

Reduction of diazonium salts to aryl hydrazines

13. Diazonium ions $Ar{-}\overset{+}{N}{\equiv}N$ can be reduced with sodium sulphite to form substituted hydrazines $Ar{-}NHNH_2$. Reduction of benzenediazonium chloride with sodium sulphite yields the useful reagent phenylhydrazine.

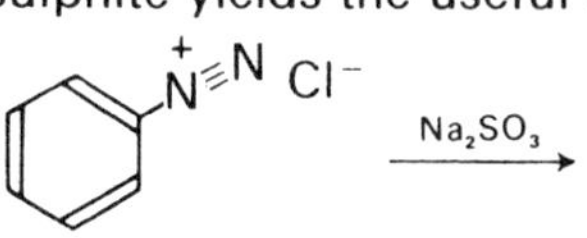

. . . . (1).

35. The addition of one mole of anhydrous hydrogen halide to an unsymmetrical alkyne gives rise to one main product, the structure of which is again predicted by the Markovnikov rule.

e.g. $$RC{\equiv}CH + HCl \longrightarrow \underset{\displaystyle Cl}{\overset{}{RC}}{=}CH_2$$

With excess of hydrogen chloride a second Markovnikov addition occurs. The reaction of pent-1-yne with excess HCl can be shown . . . . (1).

---

(1) $$CH_3CH_2CH_2C{\equiv}CH + HCl \longrightarrow CH_3CH_2CH_2\underset{\displaystyle Cl}{C}{=}CH_2$$

$$CH_3CH_2CH_2\underset{\displaystyle Cl}{C}{=}CH_2 + HCl \longrightarrow CH_3CH_2CH_2\overset{\displaystyle Cl}{\underset{\displaystyle Cl}{C}}CH_3$$

---

36. Chloroethene (vinyl chloride), required for the preparation of polyvinyl chloride, could be prepared from . . . . (1) by . . . . (2), according to the equation . . . . (3).

---

(1) ethyne, acetylene (2) by the addition of one mole of hydrogen chloride
(3) $HC{\equiv}CH + HCl \longrightarrow CH_2{=}CHCl$

Addition of water to give carbonyl compounds

37. One molecule of water adds to acetylene in the presence of H_2SO_4 and catalytic amounts of $HgSO_4$. By analogy with the hydration of an alkene the product which would be expected is . . . . (1).

---

(1) $\underset{\displaystyle OH}{CH}{=}CH_2$

---

38. The structure $\underset{\displaystyle OH}{CH}{=}CH_2$ is the simplest example of an *enol*. The *en*- indicates the presence of a double bond and -*ol*, the attached hydroxyl group. Enols usually rearrange to the more stable isomeric carbonyl compounds. Write the structure of the carbonyl compound arising from $CH_2{=}\underset{\displaystyle OH}{CH}$ . . . . (1).

---

(1) CH_3CHO

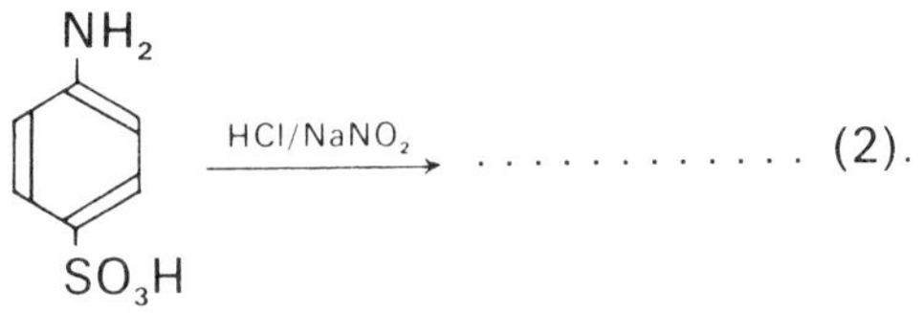

sulphanilic acid
(*p*-aminobenzenesulphonic acid)

---

(1)

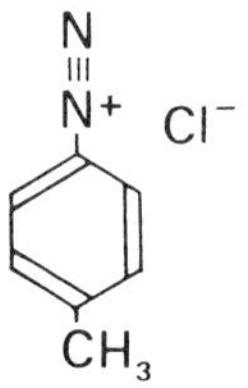

(2)

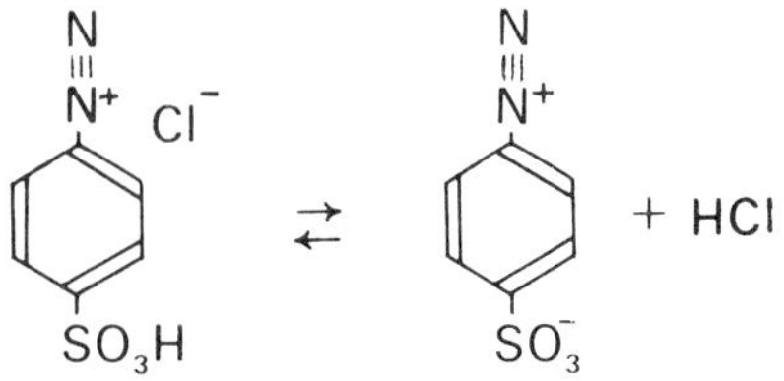

---

9. Simple generalised equations of the type $Ar{-}NH_2 \xrightarrow{HX/NaNO_2}$ . . . . can therefore be written.

---

(1) $Ar{-}NH_2 \xrightarrow{HX/NaNO_2} Ar{-}\overset{+}{N}{\equiv}N\ X^-$

Coupling of diazonium salts with phenols and arylamines

10. Diazonium ions react readily with phenols in alkaline solution to give highly coloured azo compounds. When a cold solution of benzenediazonium chloride is added to an alkaline solution of phenol a deep orange colour is developed and on neutralisation the major product is found to be *p*-hydroxyazobenzene.

$$C_6H_5OH + OH^- \rightleftharpoons C_6H_5O^- + H_2O$$

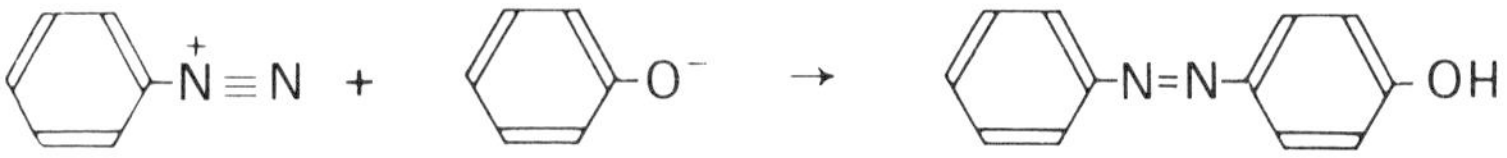

orange

As a general test for the presence of a diazonium ion the coupling is done with 2-naphthol (β-naphthol) as this gives a much deeper colour than does phenol. The cold diazonium solution is added to an alkaline solution of 2-naphthol to give a deep red dyestuff. The coupling takes place in the 1-position.

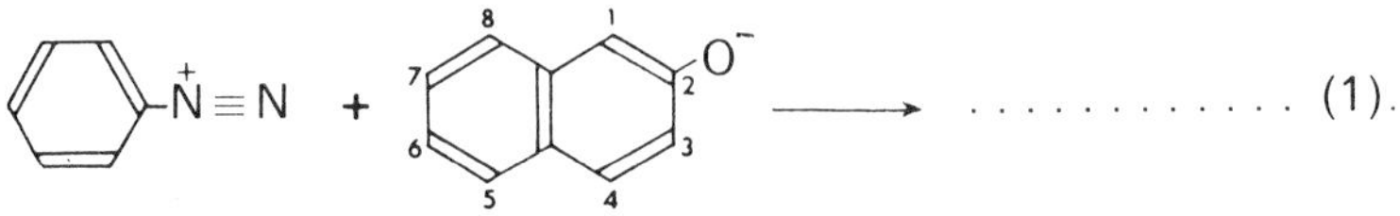

---

39. The addition of water to acetylene may be summarised

$$HC{\equiv}CH \xrightarrow[HgSO_4]{H_2SO_4} H_2C{=}\underset{\displaystyle OH}{\underset{|}{CH}} \rightleftharpoons CH_3\underset{\displaystyle O}{\underset{\|}{C}}H \quad \text{ethanal (acetaldehyde)}$$

A similar process is found on addition of water to propyne . . . . (1).

---

(1) $$CH_3C{\equiv}CH \xrightarrow[HgSO_4]{H_2SO_4} CH_3{-}\underset{\displaystyle OH}{\underset{|}{C}}{=}CH_2 \rightleftharpoons CH_3{-}\underset{\displaystyle O}{\underset{\|}{C}}{-}CH_3 \quad \text{propan-2-one (acetone)}$$

(Markovnikov's rule)

Oxidation to carboxylic acids

40. Alkynes can be oxidised to carboxylic acids with cleavage of the $C{\equiv}C$ bond. The products of oxidation of pent-2-yne are . . . . (1).

---

(1) CH_3CH_2COOH and CH_3COOH

Acidity of terminal alkynes

41. Terminal alkynes, $RC{\equiv}CH$, behave as very weak acids. They give the corresponding anion, $RC{\equiv}C^-$ on reaction with a strong base, e.g. sodium amide, and form heavy metal derivatives, e.g. with ammoniacal silver nitrate or copper(I) chloride:

$$RC{\equiv}CH + Na^+NH_2^- \longrightarrow RC{\equiv}C^-Na^+ + NH_3$$

$$RC{\equiv}CH \xrightarrow[NH_4OH]{AgNO_3} RC{\equiv}CAg$$

$$RC{\equiv}CH \xrightarrow[NH_4OH]{CuCl} RC{\equiv}CCu$$

(These heavy metal "salts" actually have a more complex structure than is indicated by the above formulae).

The ion $HC{\equiv}C^-$ produced from ethyne (acetylene) is known as the acetylide ion; the ion $RC{\equiv}C^-$ is a substituted acetylide ion, and both ions are members of the general class of carbanions.

An alkyne C_5H_8 reacted as an acid towards strong base, and gave a precipitate with ammoniacal silver nitrate. Give possible structures for the hydrocarbon . . . . (1). An isomeric alkyne gave neither of these reactions. What is its structure? . . . . (2).

---

(1) $CH_3CH_2CH_2C{\equiv}CH$ or $(CH_3)_2CHC{\equiv}CH$

(2) $CH_3CH_2C{\equiv}CCH_3$

PREPARATION OF ALKYNES

42. Alkynes of longer carbon chains are prepared by the reactions of acetylide

(cf. Program 5, Frame 20). By contrast, primary *aryl*amines yield fairly stable diazonium salts. For example, when a solution of anilinium chloride in excess hydrochloric acid is cooled in an ice bath to 0–5° and a solution of sodium nitrite is added, benzenediazonium chloride is formed. (di = two, azo = nitrogen, onium = cationic). The sodium nitrite is converted by the excess acid into nitrous acid and this reacts according to the equation:

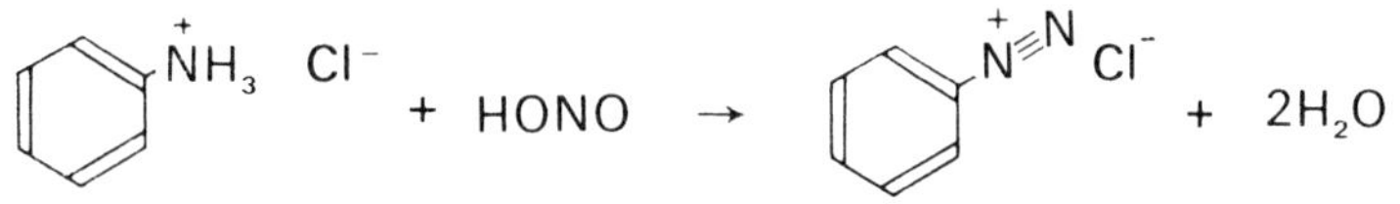

The equation of the overall reaction can therefore be written

$C_6H_5NH_2 + 2HCl + NaNO_2 \rightarrow$ .... (1). The process is called diazotisation, and diazonium salts have great utility in synthetic chemistry.

---

(1) $C_6H_5NH_2 + 2HCl + NaNO_2 \rightarrow C_6H_5N_2^+Cl^- + NaCl + 2H_2O$

---

7. The reaction product, $C_6H_5N_2^+Cl^-$, can be isolated as a crystalline salt which is called............ (1). Diazonium salts when dry are often very sensitive to shock and detonate violently on being heated. In general, reactions are carried out using the easily prepared aqueous solutions.

---

(1) benzenediazonium chloride

---

8. Nearly all compounds which have a primary amino group ($-NH_2$) attached to an aromatic ring undergo this reaction. For instance *p*-nitroaniline reacts with nitrous acid in hydrochloric acid solution according to the equation

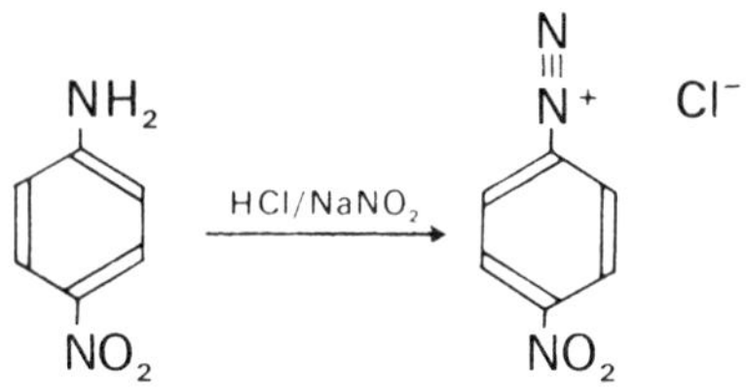

Similarly the following compounds can be diazotised

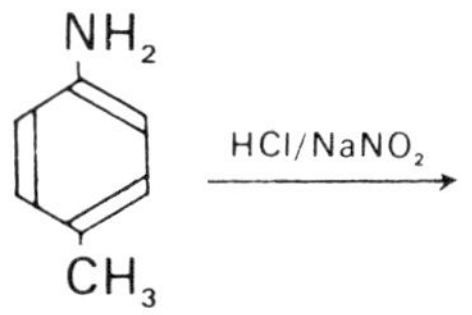

.... (1).

p-toluidine

ions ($R-C\equiv C^-$) with alkyl halides ($R'-X$). The product is $R-C\equiv C-R'$. The acetylide ion is generated by reaction of acetylene with . . . . (1). The substituted acetylide ion from prop-1-yne is . . . . (2). This would react with 1-bromobutane to give the structure . . . . (3) called . . . . (4).

---

(1) a strong base (2) $CH_3C\equiv C^-$ (3) $CH_3C\equiv CCH_2CH_2CH_2CH_3$

(4) hept-2-yne

---

43. Other methods of preparing alkynes are similar to those given for alkenes and are *elimination reactions:*
Dehalogenation with zinc requires a tetrahalogen compound of the type

$$R\underset{X}{\overset{X}{C}}-\underset{X}{\overset{X}{C}}R + 2Zn \longrightarrow R-C\equiv C-R + 2ZnX_2$$

Give the name of the chloro-compound from which pent-2-yne could be made . . . . (1).

---

(1) 2,2,3,3-tetrachloropentane

---

44. *Dehydrohalogenation* can be effected in two stages. The removal of one molecule of hydrogen bromide from 1,2-dibromopropane by treatment with alcoholic potassium hydroxide yields 1-bromopropene according to the equation . . . . (1).

---

(1) $CH_3CHBrCH_2Br \xrightarrow[\text{KOH}]{\text{alcoholic}} CH_3CH=CHBr$

---

45. A stronger base is required for the elimination of hydrogen bromide from 1-bromopropene. Sodium amide, $NaNH_2$, can be used, and the equation is . . . . (1).

---

(1) $CH_3CH=CHBr \xrightarrow{NaNH_2} CH_3C\equiv CH$

BENZENE

46. Benzene, C_6H_6, has a planar, hexagonal structure in which each carbon atom carries one hydrogen atom. The structure of benzene proved a puzzle to early chemists. In spite of the unsaturation implied by the formula it fails to react in simple tests for the presence of $>C=C<$. On catalytic hydro-

3. Primary arylamines can also be made by the Hofmann oxidation of an arylamide (cf. Program 8, Frame 51), for which the general equation is

$$RCONH_2 + 2NaOH + NaOBr \longrightarrow RNH_2 + Na_2CO_3 + NaBr + H_2O.$$

Write the structure of the amide which by this method would yield 3,5-dichloroaniline . . . . (1), 2-chloro-4-methylaniline . . . . (2).

---

(1)

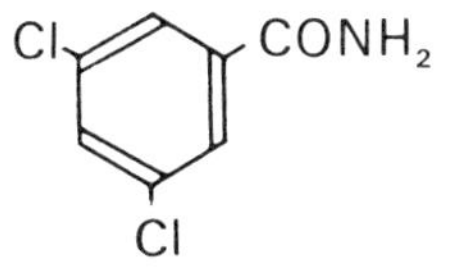

(2)

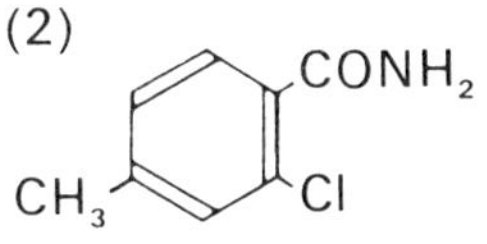

---

4. Aniline can be prepared on an industrial scale by heating chlorobenzene with ammonia at 200° under pressure in the presence of cuprous chloride as a catalyst. Complete the following equation:

$$C_6H_5Cl + 2NH_3 \xrightarrow[CuCl]{200^\circ\ 900\ psi} \ldots\ldots\ldots\ldots (1)$$

---

(1) $C_6H_5Cl + 2NH_3 \xrightarrow[CuCl]{200^\circ\ 900\ psi} C_6H_5NH_2 + NH_4Cl$

Nucleophilic properties of arylamines

5. Like the alkylamines (Program 5), arylamines are basic substances. They form salts with acids and react as nucleophiles towards alkyl halides (N-alkylation) and towards various acid derivatives such as acid chlorides and anhydrides (N-acylation). Write equations for the dissolution of aniline in aqueous acid . . . . (1), and the reaction of aniline with acetic anhydride (ethanoic anhydride) . . . . (2).

---

(1) $C_6H_5NH_2 + H_3O^+ \rightleftharpoons C_6H_5\overset{+}{N}H_3 + H_2O$

(2)

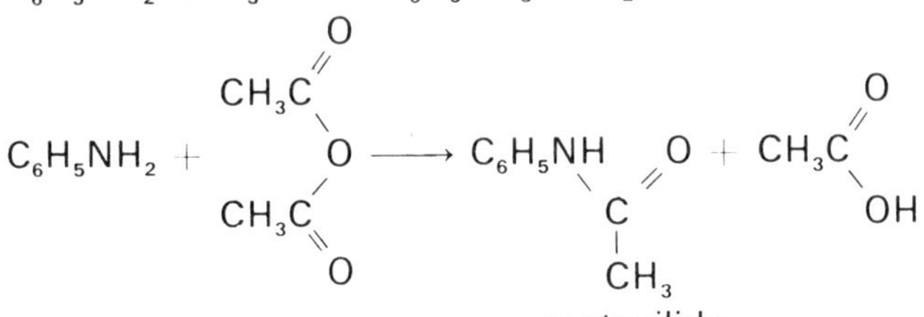

acetanilide

cf. Program 5, Frames 16, 17, and Program 8, Frames 45, 46.

DIAZONIUM SALTS

Preparation from primary arylamines

6. Primary *alkyl*amines react with nitrous acid to give very unstable products

genation, benzene is slowly converted into cyclohexane of structure . . . . (1), with the uptake of three moles of hydrogen. One might therefore deduce that the structure of benzene is I and the problem is to explain the non-reactivity in addition reactions, despite the seeming presence of three double bonds.

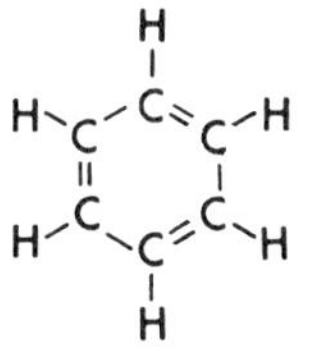

I

---

(1) 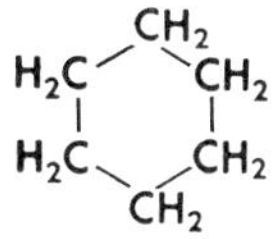

---

47. The characteristic reactions of benzene are *substitution* reactions discussed in Program 9. Theoretically, we can account for the stability of benzene by postulating resonance (or mesomeric) structures of which the chief contributing forms are shown below.

Benzene is frequently represented by one only of these mesomeric structures. Apart from the singly-bonded framework the number of electrons which have to be accommodated is . . . . (1).

---

(1) 6. i.e.

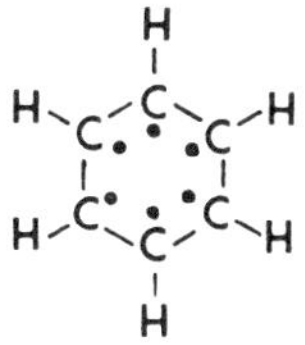

---

48. An alternative theory to account for the structure of benzene suggests that six electrons are accommodated in molecular orbitals above and below the plane of the carbon atoms. This theory has led to a common representation of

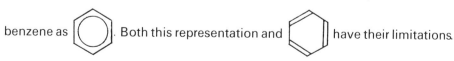

benzene as . Both this representation and  have their limitations.

Aromatic compounds (arenes) can contain more than one ring. For example naphthalene, $C_{10}H_8$, contains two fused six-membered rings. The mesomeric structures for naphthalene are . . . . (1).

ARYLAMINES AND DIAZONIUM SALTS

Program 10

ARYLAMINES

Introduction

1. Primary arylamines contain an amino group, $-NH_2$, attached directly to a carbon atom of the aryl ring, e.g. $C_6H_5NH_2$, aniline; *p*-$CH_3C_6H_4NH_2$, *p*-toluidine or 4-methylaniline. Secondary and tertiary arylamines are obtained when one or both of the hydrogen atoms attached to nitrogen in the primary arylamine are replaced by alkyl or aryl groups.
Write structures for diphenylamine .... (1), N-ethylaniline .... (2), N,N-dimethylaniline .... (3).

---

(1)

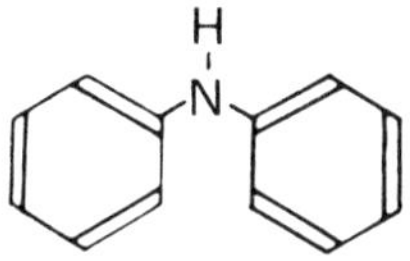

(2)

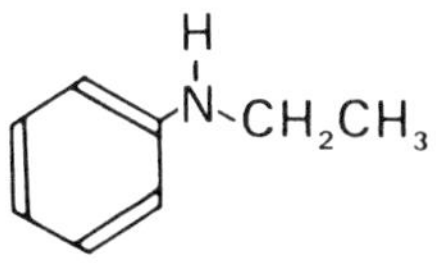

(3)

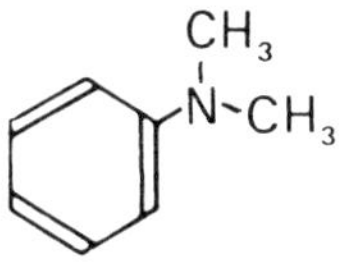

Preparation of primary arylamines

2. Primary arylamines are usually prepared by reduction of the corresponding nitroarene, e.g. by use of a metal/acid combination, or by hydrogenation over a catalyst.

Thus $$C_6H_5NO_2 \xrightarrow[\text{or } H_2/Ni]{Sn/HCl} C_6H_5NH_2$$

Nitroarenes can be made in turn by nitration of the aromatic hydrocarbon (cf. Program 9, Frames 2, 3). Outline a synthesis of *p*-toluidine from toluene .... (1).

---

(1)

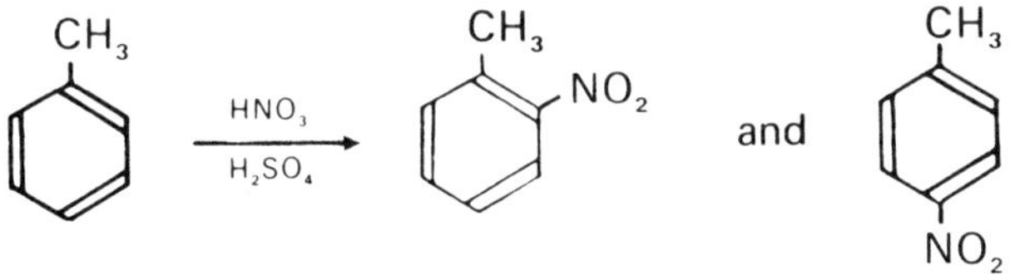

Separate the *o*- and *p*-isomers, and reduce the *p*-isomer:

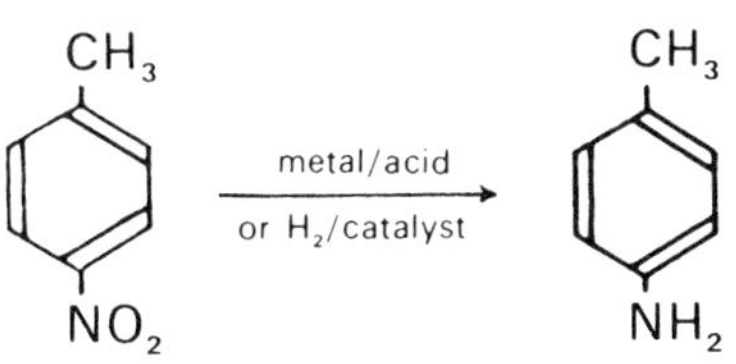

(1)

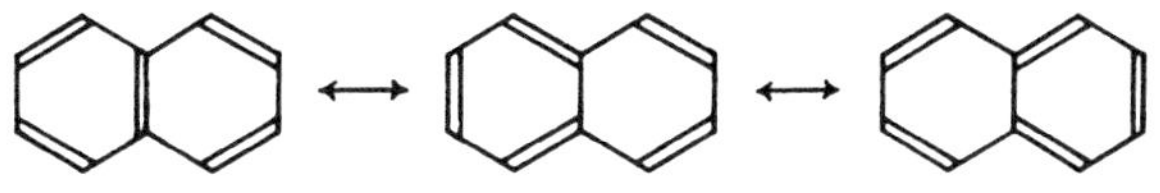

TEST FRAMES

Give equations for reactions 1-12 in the following diagram, using 1-butene as the alkene. Name the compounds on their first appearance.

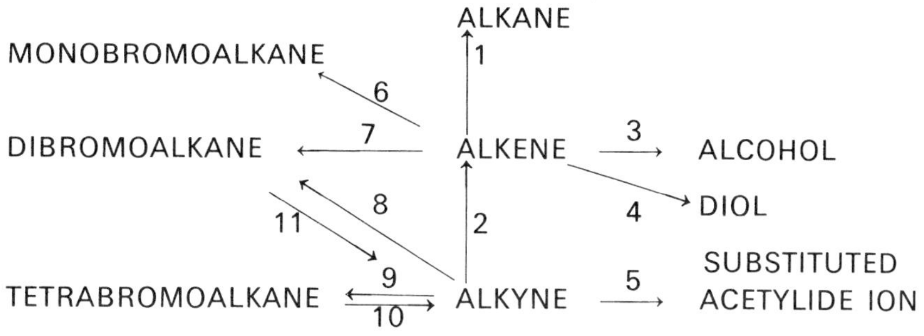

1. $CH_3CH_2CH{=}CH_2 + H_2 \xrightarrow[\text{catalyst}]{\text{Pt}} CH_3CH_2CH_2CH_3$
   but-1-ene — butane

2. $CH_3CH_2C{\equiv}CH + H_2 \xrightarrow[\text{catalyst}]{\text{Pt}} CH_3CH_2CH{=}CH_2$
   but-1-yne

3. $CH_3CH_2CH{=}CH_2 + H_2O \xrightarrow{\text{acid}} CH_3CH_2CH(OH)CH_3$ butan-2-ol

4. $CH_3CH_2CH{=}CH_2 \xrightarrow[\text{or peracid, then acid hydrolysis}]{\text{alk. }KMnO_4} CH_3CH_2CH(OH)CH_2OH$ butan-1,2-diol

5. $CH_3CH_2C{\equiv}CH \xrightarrow[\text{e.g. }NaNH_2]{\text{strong base}} CH_3CH_2C{\equiv}C^-$

6. $CH_3CH_2CH{=}CH_2 + HBr \longrightarrow CH_3CH_2CHBrCH_3$ 2-bromobutane
   or
   $CH_3CH_2CH{=}CH_2 + HBr \xrightarrow[\text{peroxides}]{\text{air or}} CH_3CH_2CH_2CH_2Br$ 1-bromobutane

7. $CH_3CH_2CH{=}CH_2 + Br_2 \longrightarrow CH_3CH_2CHBrCH_2Br$ 1,2-dibromobutane

8. $CH_3CH_2C{\equiv}CH + 2HBr \longrightarrow CH_3CH_2CBr_2CH_3$ 2,2-dibromobutane

9. $CH_3CH_2C{\equiv}CH + 2Br_2 \longrightarrow CH_3CH_2CBr_2CHBr_2$ 1,1,2,2-tetra-bromobutane

10. $CH_3CH_2CBr_2CHBr_2 + 2Zn \longrightarrow CH_3CH_2C{\equiv}CH + 2ZnBr_2$

11. $CH_3CH_2CHBrCH_2Br \xrightarrow[\text{KOH}]{\text{alcoholic}} CH_3CH_2CH{=}CHBr \xrightarrow{NaNH_2} CH_3CH_2C{\equiv}CH$

Program 10: ARYLAMINES AND DIAZONIUM SALTS

CONTENTS

How could you distinguish the following four compounds from each other?

(1) $CH_3CH_2CH{=}CHCH_3$

(2) $CH_3CH_2C{\equiv}CCH_3$

(3) $CH_3CH_2CH_2CH{=}CH_2$

(4) $CH_3CH_2CH_2C{\equiv}CH$

---

(1) and (3) are alkenes and will take up 1 mole H_2 in the presence of catalyst.
(1) on ozonolysis will give $CH_3CH_2CHO + CH_3CHO$
(3) on ozonolysis will give $CH_3CH_2CH_2CHO + HCHO$
(2) and (4) are alkynes and will take up 2 moles H_2 in the presence of catalyst.
(4) is a terminal alkyne and will give a precipitate with Ag^+ in NH_4OH solution.

---

(1)

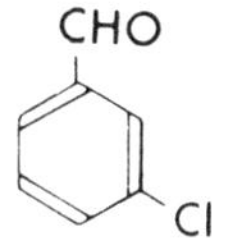
CHO
Cl

(2)

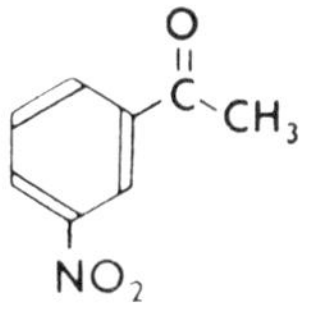
O
C
CH3
NO2

(3)

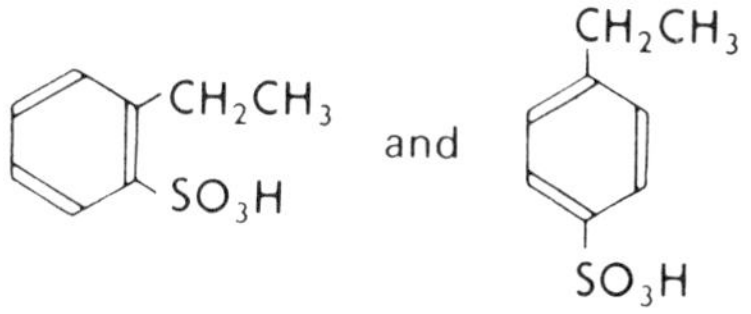
CH2CH3
SO3H
and
CH2CH3
SO3H

(4)

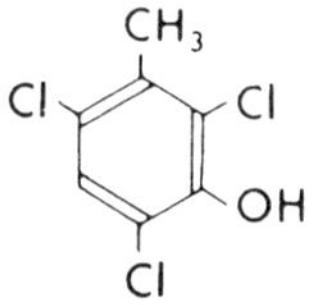
CH3
Cl
Cl
OH
Cl

(5)

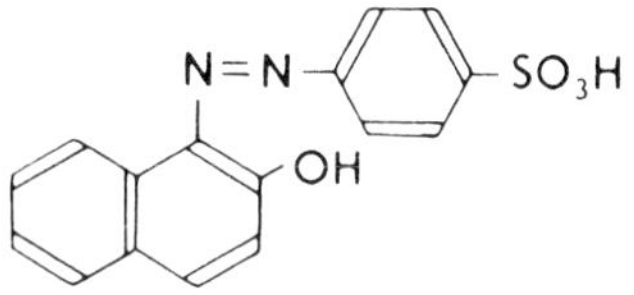
N=N
SO3H
OH

Program 4: ALKYL HALIDES

CONTENTS

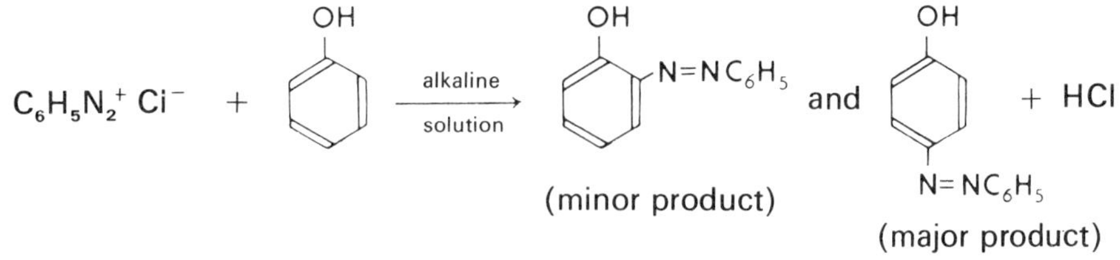

In this substitution the attacking electrophile is. . . . . (1).

---

(1) $C_6H_5N_2^+$, the benzenediazonium cation

---

45. This type of electrophilic substitution is a common means for detecting and identifying aromatic amines, $ArNH_2$. These, with nitrous acid in cold dilute hydrochloric acid, yield diazonium ions, ArN_2^+, which on further reaction with a cold alkaline solution of a phenol give coloured azo compounds.

With 2-naphthol, 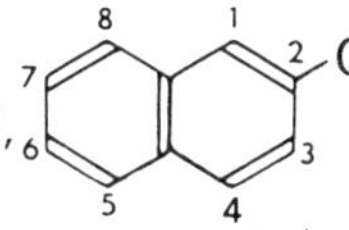

, substitution occurs exclusively at position 1, so that coupling with ArN_2^+ yields an azo compound of structure . . . . (1).

---

(1) 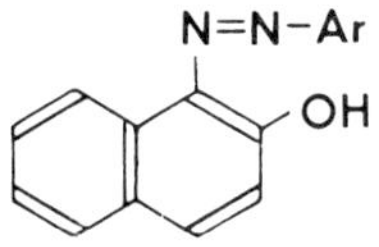

Note. 2-naphthol is a derivative of naphthalene, $C_{10}H_8$, or

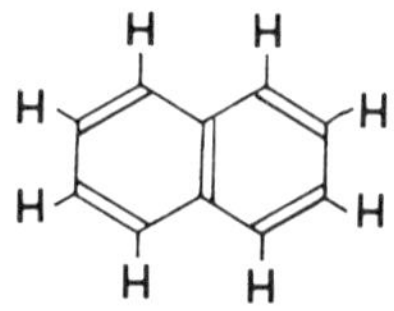

usually abbreviated to
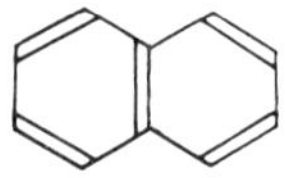

TEST FRAME

Write structures for the product or products of the following reactions: the chlorination of benzaldehyde, C_6H_5CHO, ... (1), the nitration of methyl phenyl ketone, $C_6H_5COCH_3$, ... (2), the sulphonation of ethylbenzene, ... (3), the chlorination of aqueous 3-methylphenol (*m*-cresol) . . . . (4), coupling of the diazonium salt prepared from *p*-aminobenzenesulphonic acid with 2-naphthol . . . . (5).

ALKYL HALIDES

Program 4

INTRODUCTION

1. Alkyl halides or halogenoalkanes have the general formula RX where R is an alkyl group or substituted alkyl group such that the halogen atom, X, is attached to a saturated carbon atom. The chemistry of the alkyl fluorides will be considered only briefly—these compounds differ markedly in their behaviour from the other alkyl halides. The discussion will include compounds such as methyl iodide, ethyl chloride, 2-bromopentane, and 2-chloro-2-methylpropane (t-butyl chloride) whose structures are .... (1), ... (2), ... (3), and .... (4) respectively.

---

(1) CH_3I (2) CH_3CH_2Cl (3) $CH_3CHBrCH_2CH_2CH_3$

(4) $CH_3\overset{\overset{\large CH_3}{|}}{\underset{\underset{\large Cl}{|}}{C}}CH_3$ (t-butyl chloride)

---

2. More complex substances will be considered in particular reactions. These will include 3-chloroprop-1-ene (allyl chloride), $CH_2{=}CHCH_2Cl$, and benzyl bromide, $C_6H_5CH_2Br$. Note that in both these compounds the halogen is attached to a .... (1) carbon atom. Chloro compounds such as chloroethene (vinyl chloride), $CH_2{=}CHCl$, will not be discussed here. In this latter compound the halogen is attached to an .... (2) carbon atom. Aryl halides such as bromobenzene, C_6H_5Br, and chlorobenzene, C_6H_5Cl, are treated in Program 9.

---

(1) saturated (2) unsaturated

PREPARATION OF ALKYL HALIDES

Substitution of the hydroxyl group of an alcohol by a halogen atom

3. A common method for preparing an alkyl halide is by reaction of the corresponding alcohol with a hydrogen halide or phosphorus trihalide.

$$R{-}OH \xrightarrow{HX \text{ or } PX_3} R{-}X$$

Thus, 1-chlorobutane can be prepared from butan-1-ol .... (1).

---

(1) $CH_3CH_2CH_2CH_2OH \xrightarrow{HCl \text{ or } PCl_3} CH_3CH_2CH_2CH_2Cl$

the benzene carbon atoms 1 to 6 around the ring and give the number 1 to the carbon carrying the principal substituent (in this case –OH, since we are naming the compound as a phenol). The systematic name is then . . . . (1).

---

(1) 2,4,6-tribromophenol

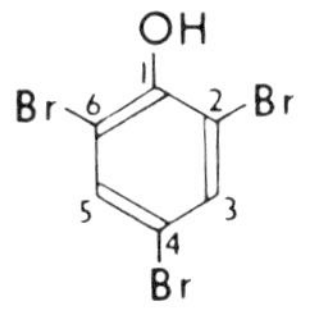

---

42. The amino group, $-NH_2$, like –OH, is an activating group when attached to a benzene ring. The bromination of aqueous aniline, $C_6H_5NH_2$, proceeds in similar fashion to that of phenol and the product is . . . . (1) (give name and structure).

---

(1) 2,4,6-tribromoaniline

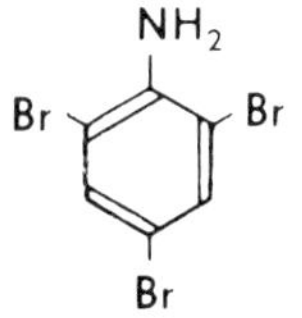

---

43. In the substitution of benzene, the yields of monosubstituted products may be poor if the initial substitution leads to a compound which is more reactive than benzene. Thus in the Friedel-Crafts alkylation of benzene with, say, ethyl chloride and aluminium trichloride (cf. Frames 10-12), the first-formed ethylbenzene will undergo electrophilic substitution more easily than the remaining benzene, so that a good deal of di- and tri- ethylbenzenes will be formed. Write down the structures for these di- and tri-ethylbenzenes . . . . (1), remembering that alkyl groups direct *ortho-*, *para-*.

---

(1)

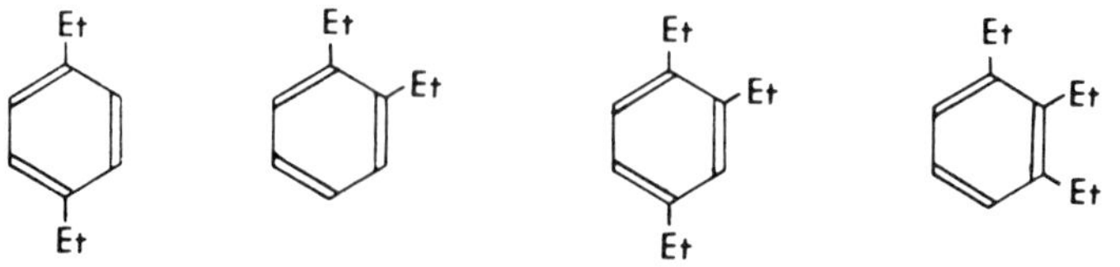

Diazonium Coupling

44. The coupling of a phenol with a diazonium salt, such as benzenediazonium chloride, $C_6H_5N_2^+Cl^-$ (cf. Program 10), to yield an azo compound, ArN=NAr, is another type of electrophilic aromatic substitution.

4. 2-Bromopropane can be prepared from the alcohol . . . . (1) by reaction with hydrogen bromide or phosphorus tribromide.

$$\text{. . . . .} \xrightarrow{\text{HBr or PBr}_3} \text{. . . . .} \quad (2).$$

---

(1) Propan-2-ol or isopropanol (2) $CH_3CH(OH)CH_3 \xrightarrow[\text{PBr}_3]{\text{HBr or}} CH_3CH(Br)CH_3$

---

5. Instead of using hydrogen bromide, a mixture of sulphuric acid and sodium bromide can be used for converting an alcohol into an alkyl bromide. Thus 2-bromopentane can be prepared from . . . . (1) by allowing it to react with sodium bromide and sulphuric acid.

$$\text{. . . . .} \xrightarrow{\text{NaBr + H}_2\text{SO}_4} \text{. . . . .} \quad (2)$$

---

(1) Pentan-2-ol

(2) $CH_3CH_2CH_2CH(OH)CH_3 \xrightarrow{\text{NaBr + H}_2\text{SO}_4} CH_3CH_2CH_2CH(Br)CH_3$

---

6. Alkyl iodides can be prepared by reacting alcohols with hydriodic acid:

$$ROH + HI \longrightarrow RI + H_2O$$

Phosphorus triiodide is relatively unstable and it is more convenient to use a combination of red phosphorus and iodine. For instance methyl iodide is conveniently prepared by reacting methanol with red phosphorus and iodine

$$\text{. . . . .} \xrightarrow{\text{P + I}_2} \text{. . . . .} \quad (1)$$

---

(1) $CH_3OH \xrightarrow{\text{P + I}_2} CH_3I$

---

7. All these reactions proceed by *nucleophilic substitution* and we shall consider one of the reactions in detail. Just as acids (indicated by HA) react reversibly with water according to the equilibrium

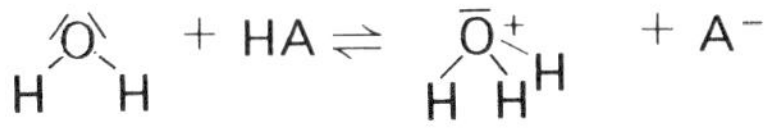

so do they react with alcohols according to the equilibrium

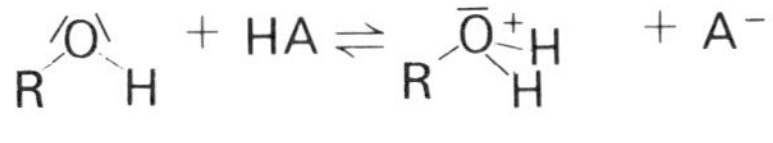

If an alcohol is dissolved in concentrated aqueous acid then both equilibria co-exist as well as the additional equilibrium:

37. Since the nitro group as a primary substituent is . . . . (1) directing, bromination of nitrobenzene gives . . . . (2).

---

(1) *meta*- (2) *m*-bromonitrobenzene (or *m*-nitrobromobenzene)

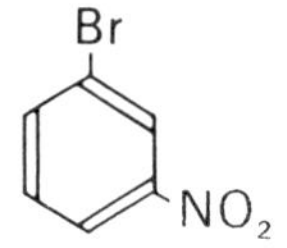

---

38. Methyl phenyl ether has the structure $C_6H_5OCH_3$, and the primary substituent $-OCH_3$, is . . . . (1) directing.

---

(1) *ortho*-, *para*-

---

39. In practice, the acylation of methyl phenyl ether using acetic anhydride in the presence of $AlCl_3$ leads almost entirely to the *para*- isomer, of structure . . . . (1).

---

(1)

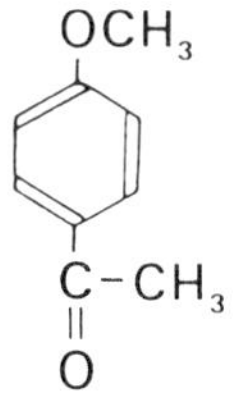

This substance is best named methyl *p*-methoxyphenyl ketone. It is sometimes given the trivial name *p*-methoxyacetophenone.

Polysubstitution

40. With the exception of halogeno substituents, groups which direct *ortho*- and *para*- also *activate* the aromatic ring towards further electrophilic substitution, that is, the substituted compounds react more rapidly than benzene under the same conditions. This means that in some cases, di- and tri- substitution can occur during such a reaction. As an example, the bromination of phenol, C_6H_5OH, in aqueous solution at room temperature (and without the need for any catalyst) gives a compound, $HOC_6H_2Br_3$. Because the $-OH$ substituent directs *ortho*-, *para*-, this compound must have the structure . . . . (1).

---

(1)

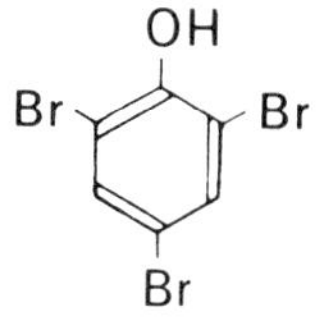

---

41. In a compound such as the above tribromophenol, the *ortho*-, *meta*-, *para*- nomenclature breaks down, and an alternative system is used. We number

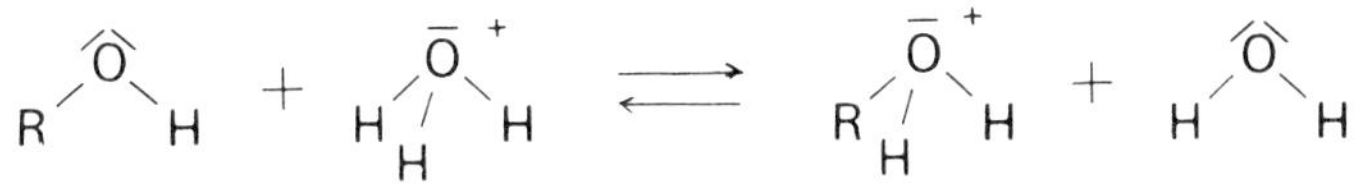

Reactions of anhydrous HBr with water can be shown . . . . (1), and with propan-1-ol the analogous equation is . . . . (2). Protonation of propan-1-ol in strong aqueous acid can be shown . . . . (3).

---

(1) 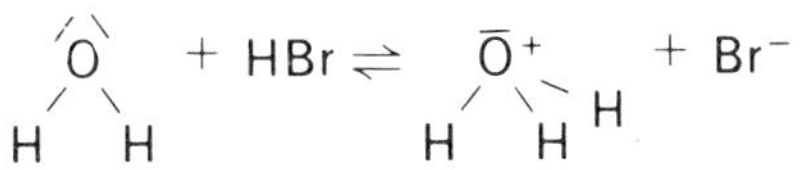

(2) 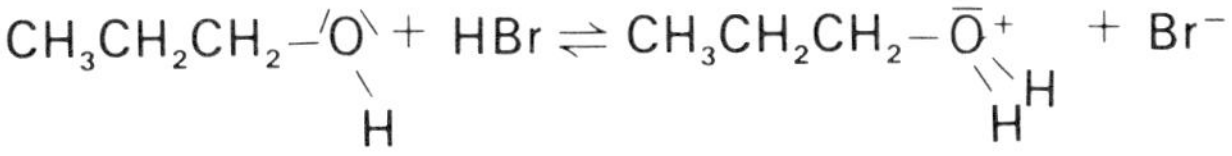

(3) 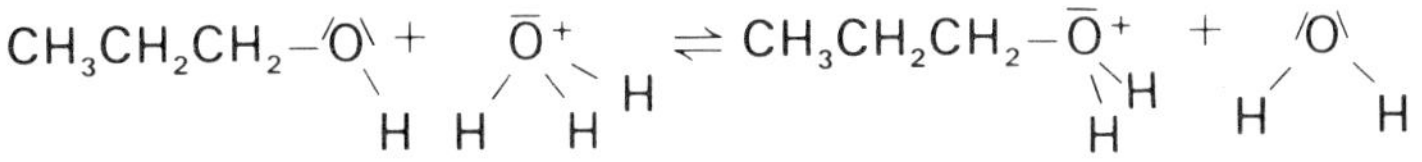

---

8. The carbon-oxygen bond in the protonated alcohol is highly polarised. The powerful attraction of the positive oxygen for the electron pair of the carbon-oxygen bond produces a partial positive charge at the carbon atom.

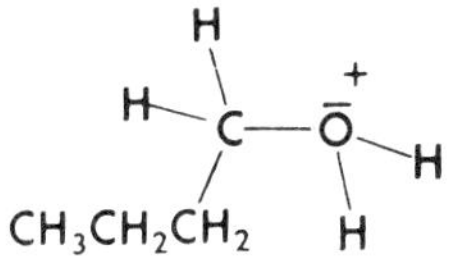

Redraw the above formula, and indicate the carbon atom which acquires a partial positive charge . . . . (1). The nucleophilic bromide ion can now attack the carbon atom and the reaction proceeds by formation of 1-bromobutane and water according to the equation

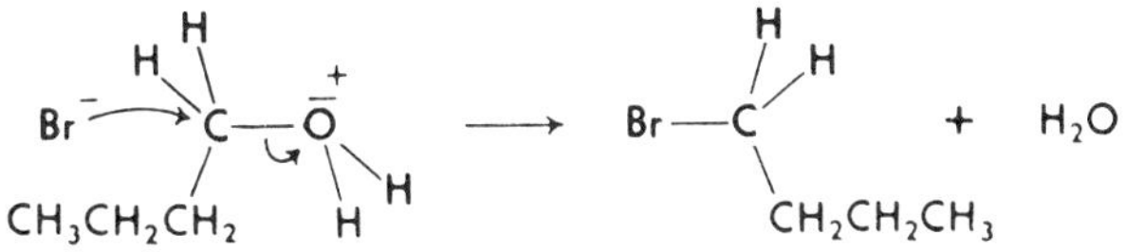

A similar equation for the reaction of hydriodic acid with ethanol would be . . . . . . (2).

---

(1)

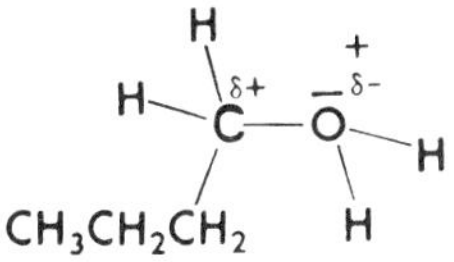

(2)

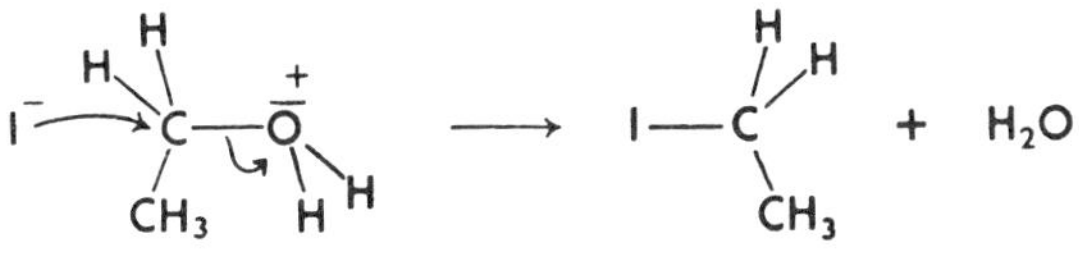

4. ALKYL HALIDES

33. It also happens that *m*-directing groups deactivate the aromatic ring towards further electrophilic attack, so that rather vigorous conditions are usually required to introduce the second substituent in the *meta*-position. Nitration of nitrobenzene is difficult, but can be achieved, and the main product is .... (1) (give name and structure). Similarly, the bromination of benzoic acid leads to .... (2).

---

(1) *m*-dinitrobenzene 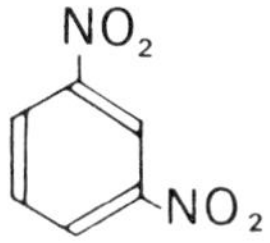

(2) *m*-bromobenzoic acid (3-bromobenzoic acid) 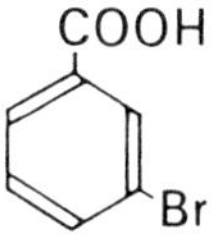

---

34. Five important *o*-,*p*-directing substituents are alkyl, –R; hydroxyl, –OH; alkoxyl, ... (1); amino, ... (2); and halogeno, –Cl, –Br or –I.

---

(1) –OR (2) $-NH_2$

---

35. In the nitration of toluene, the primary substituent, $-CH_3$, is .... (1) directing, so that the nitration reaction (HNO_3/H_2SO_4) leads to .... (2) (give names and structures of products).

---

(1) *ortho*-, *para*-

(2) *o*-nitrotoluene 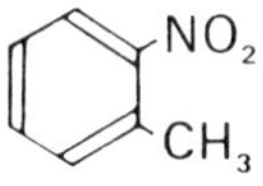 and *p*-nitrotoluene 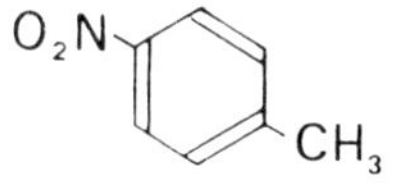

---

36. If we nitrate bromobenzene, where the primary substituent, –Br, is .... (1) directing, the main products will be .... (2).

---

(1) *ortho*-, *para*-

(2) *o*-bromonitrobenzene 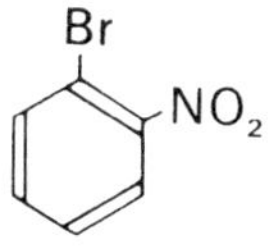 and *p*-bromonitrobenzene 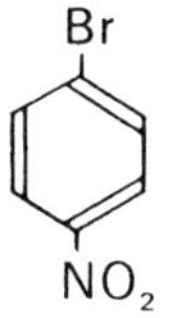

Addition of hydrogen halide to an alkene

9. Alkyl halides can be prepared by the reaction of alkenes with hydrogen halides. We will consider the reaction with hydrogen chloride first since there are complicating factors in the case of hydrogen bromide. The addition of hydrogen chloride to but-2-ene to form 2-chlorobutane proceeds according to the equation:

$$CH_3CH=CHCH_3 + HCl \longrightarrow CH_3CH_2\underset{\underset{Cl}{|}}{C}HCH_3$$

This is called an *addition reaction* since the formula of the product is the sum of the reactants: $C_4H_8 + HCl \longrightarrow C_4H_9Cl$.

In the same way ethyl chloride can be prepared by reaction of . . . . (1) with hydrogen chloride according to the equation . . . . (2), and cyclohexene reacts with hydrogen chloride to yield . . . . (3) according to the equation . . . . (4).

---

(1) ethene   (2) $CH_2=CH_2 + HCl \longrightarrow CH_3CH_2Cl$

(3) chlorocyclohexane (cyclohexyl chloride)

(4)
$$\begin{array}{c}CH_2\\H_2C{\diagup}{\diagdown}CH\\H_2C{\diagdown}{\diagup}\overset{||}{C}H\\CH_2\end{array} + HCl \longrightarrow \begin{array}{c}CH_2\\H_2C{\diagup}{\diagdown}CHCl\\H_2C{\diagdown}{\diagup}CH_2\\CH_2\end{array}$$

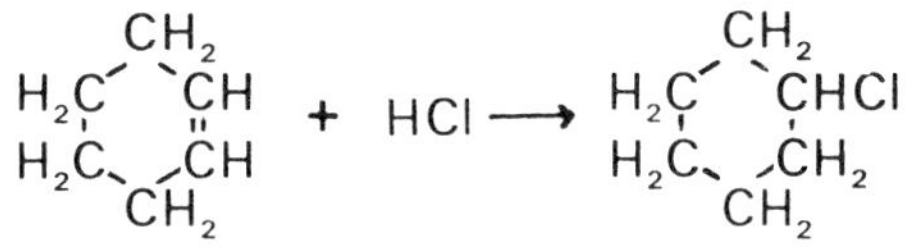

10. In the examples quoted thus far, the alkenes have been symmetrical and the structure of each product could be assigned unambiguously. If we consider the addition of hydrogen chloride to 2-methylpropene $\quad \underset{CH_3C=CH_2}{\overset{CH_3}{|}}\quad$ we see

that there are two possible products. . . . .(1) and. . . . .(2).

---

(1) $CH_3\underset{\underset{Cl}{|}}{\overset{\overset{CH_3}{|}}{C}}CH_3$ (2) $CH_3\underset{}{\overset{\overset{CH_3}{|}}{C}}HCH_2Cl$

---

11. Experimentally it is found that the reaction gives rise to only one of these according to the equation

$$CH_3\overset{\overset{CH_3}{|}}{C}=CH_2 + HCl \longrightarrow CH_3\underset{\underset{Cl}{|}}{\overset{\overset{CH_3}{|}}{C}}CH_3$$

In the product, the *hydrogen atom* of the HCl has added to the carbon atom carrying the . . . . (greater/lesser) (1) number of hydrogen atoms in the starting material.

---

can give rise to three different disubstituted products (the *o*-, *m*- and *p*-isomers). However, in practice, the position of the second substitution is determined by the type of substituent which is already on the ring. That is, in a monosubstituted benzene undergoing substitution the remaining ring-hydrogen atoms are . . . . equally/not equally (1) susceptible to further attack.

---

(1) not equally

---

31. In the further substitution of a compound C_6H_5X to give the disubstituted product C_6H_4XY, the product is usually either the *meta*-isomer or a mixture of the *ortho*-isomer and the *para*-isomer. In consequence we speak of the "directing influence" of a primary substituent already on the ring. Primary substituents fall into two classes, (a) those which direct *meta* and (b) those which direct *ortho* and *para*. A *meta*-directing substituent is one which produces a good yield of *meta*-disubstituted derivative (and very little *ortho*- and *para*-disubstituted products). Conversely, an *ortho*-, *para*-, directing substituent usually gives a mixture of the *ortho*-disubstituted and the *para*-disubstituted derivatives (and very little *meta*-product).

If 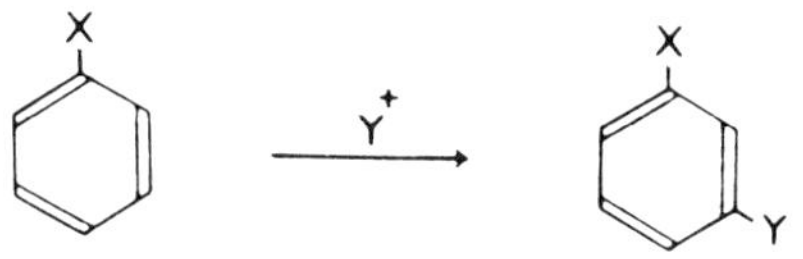 where X is *meta*-directing,

then 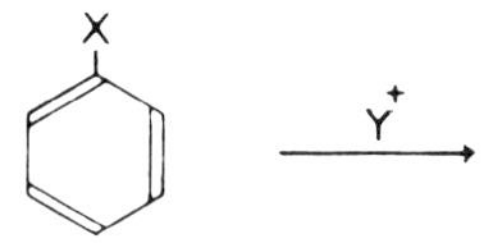 . . . . + ... (1), where X is *o*, *p*-directing.

---

(1)

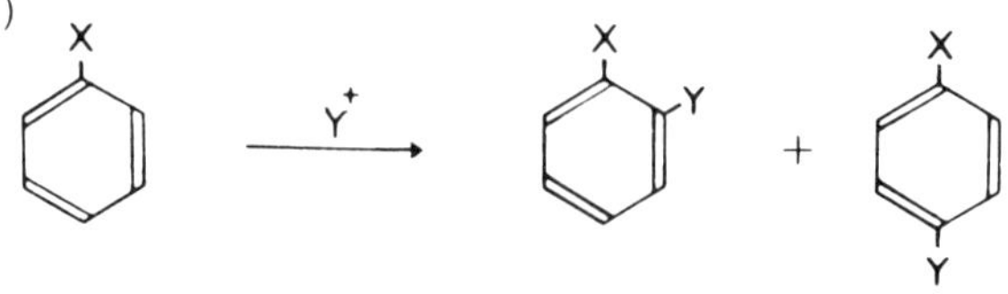

(These two isomers can be separated)

ortho-isomer *para*-isomer

---

32. Five important groups which are *m*-directing are nitro, $-NO_2$; carboxyl, ... (1); ester, –COOR; acyl, –COR, and aldehyde (formyl), ... (2).

---

(1) –COOH (2) –CHO

(1) greater

---

12. From a large number of examples of this type the Russian chemist Markovnikov saw that this was a general rule. In the addition of hydrogen chloride to an alkene, *the hydrogen adds to that carbon carrying the greater number of hydrogens*. Using this rule, write down the equations of the reaction of hydrogen chloride with 2-methylbut-2-ene .... (1) and hex-1-ene .... (2).

---

(1) $CH_3\overset{\overset{CH_3}{|}}{C}=CHCH_3 + HCl \longrightarrow CH_3\underset{\underset{Cl}{|}}{\overset{\overset{CH_3}{|}}{C}}CH_2CH_3$

(2) $CH_3CH_2CH_2CH_2CH=CH_2 + HCl \longrightarrow CH_3CH_2CH_2CH_2\underset{\underset{Cl}{|}}{C}HCH_3$

---

13. There are many *unsymmetrical* alkenes in which the number of hydrogen atoms on each of the two carbon atoms of the double bond is the *same* and application of the rule is not possible. In pent-2-ene, $CH_3CH=CHCH_2CH_3$, there is one hydrogen atom on each carbon atom of the double bond and yet there are two possible products depending on which way the addition proceeds. Experimentally it is found that both products are formed.

$CH_3CH=CHCH_2CH_3 \xrightarrow{HCl}$ .... and .... (1)

---

(1) $CH_3CH=CHCH_2CH_3 \xrightarrow{HCl} CH_3CH_2\underset{\underset{Cl}{|}}{C}HCH_2CH_3$ and $CH_3\underset{\underset{Cl}{|}}{C}HCH_2CH_2CH_3$

---

14. The product formed on the reaction of hydrogen bromide with *symmetrical* alkenes is the simple addition compound. Thus the addition of HBr to but-2-ene and to cyclopentene leads to the expected products according to the equations .... (1) and .... (2).

---

(1) $CH_3CH=CHCH_3 + HBr \longrightarrow CH_3CH_2\underset{\underset{Br}{|}}{C}HCH_3$

(2)

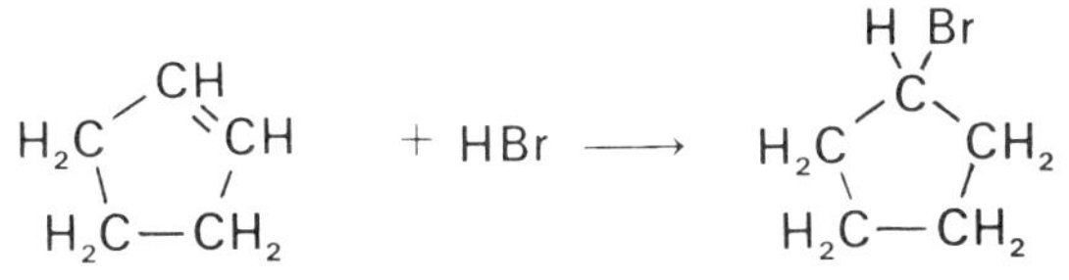

benzene" or a "monosubstituted derivative of benzene". However, when such a derivative has been obtained, and further substitution (not necessarily with the same functional group) is desired, there is more than one position available for the second substituent group. In the case of toluene, $C_6H_5CH_3$, substitution of a second aromatic hydrogen atom by $-CH_3$ could lead, in theory, to the following different, isomeric, products commonly known as xylenes . . . . (1).

---

(1)

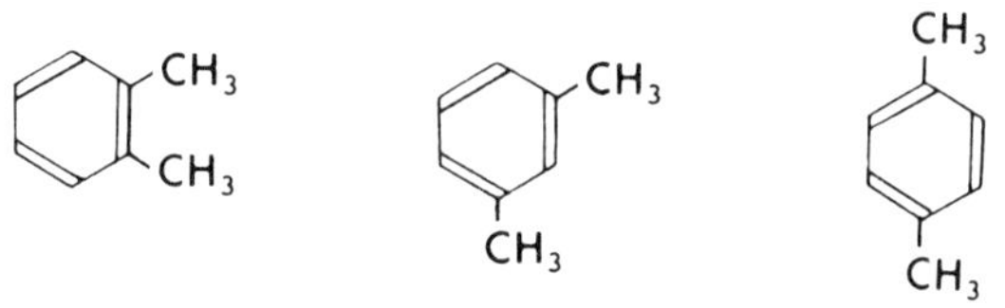

Ortho-, meta-, and para-nomenclature (cf. Program 1, Frames 31-33)

28. In general, a disubstituted benzene derivative C_6H_4XY can exist as . . . . (1) different isomers, and this is true whether the two substituting groups (X, Y) are the same or different.

---

(1) three

---

29. The three possible isomers of a disubstituted benzene are distinguished by the prefixes *ortho*, *meta-*, and *para-*, as shown below for the isomeric monochlorotoluenes:

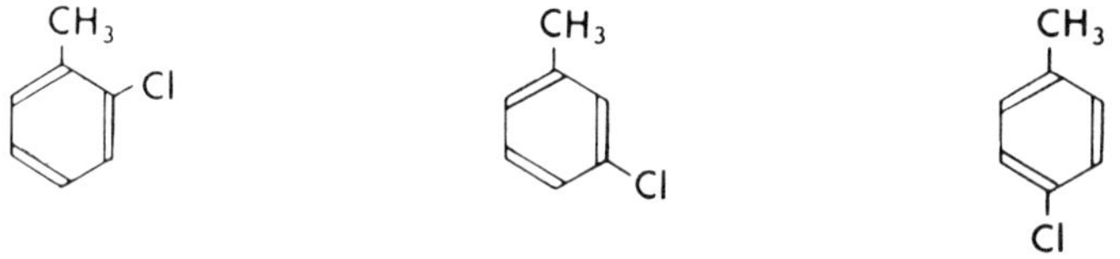

ortho- (o-) *meta- (m-)* *para- (p-)*

Some further examples of disubstituted benzene derivatives are *m*-dibromobenzene . . . . (1), *p*-chloronitrobenzene, ... (2) and *o*-carboxybenzenesulphonic acid, ... (3).

---

(1) 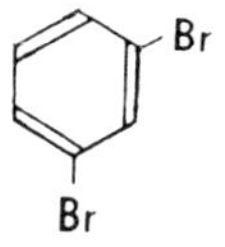
(2) 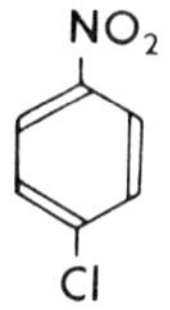
(3) 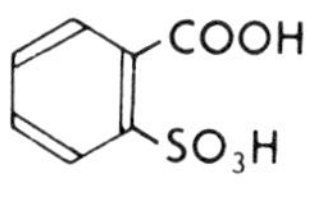

Directing influence of a substituent

30. In principle, the further substitution of a monosubstituted benzene derivative

15. Addition of hydrogen bromide to *unsymmetrical* alkenes in which both carbon atoms of the double bond have the *same* number of hydrogen atoms also leads to a mixture of products. For example, the addition of HBr to pent-2-ene gives a mixture of products according to the equation . . . . (1).

---

(1) $CH_3CH{=}CHCH_2CH_3 \xrightarrow{HBr} CH_3CH_2CH(Br)CH_2CH_3$ and $CH_3CH(Br)CH_2CH_2CH_3$

---

16. In the addition of hydrogen bromide to *unsymmetrical* alkenes having a *different* number of hydrogen atoms on the two carbon atoms of the double bond, two alternative processes are possible, each leading to a different product. Alkenes react with oxygen from the air to form organic peroxides. Most samples of alkenes contain these impurities and unless they are removed they affect the addition of hydrogen bromide. The two kinds of addition are:
 (a) Addition of HBr in the *absence of air and peroxides* (that is after their careful removal) leads to *Markovnikov addition* in which the hydrogen atom adds to the carbon atom carrying the greater number of hydrogens.
 (b) Addition of HBr in the *presence of air and peroxides* leads to *anti-Markovnikov addition* in which the hydrogen atom adds to the carbon atom carrying the lesser number of hydrogens.

 Process (a) can be illustrated by the reaction of hydrogen bromide with 2-methylpropene.

$$CH_3C(CH_3){=}CH_2 + HBr \xrightarrow[\text{and peroxides}]{\text{absence of air}} \text{. . . . . (1)}$$

 This product results from . . . . (2) addition.

---

(1) $CH_3C(CH_3){=}CH_2 + HBr \xrightarrow[\text{and peroxides}]{\text{absence of air}} CH_3C(CH_3)(Br)CH_3$

(2) Markovnikov

---

17. The reaction of 2-methylpropene with hydrogen bromide in the presence of air and peroxides proceeds according to the equation

$$CH_3C(CH_3){=}CH_2 + HBr \xrightarrow[\text{peroxides}]{\text{air and}} \text{. . . . . (1)}$$

 This product results from . . . . (2) addition.

$$ArN_2^+ \xrightarrow[\text{aqueous acid}]{\text{heat in}} ArOH + N_2$$

Balance the following equation:

$$C_6H_5N_2^+\ HSO_4^- + H_2O \xrightarrow{\text{heat}} \ldots\ldots\ldots\ldots\ (1)$$

---

(1) $C_6H_5N_2^+\ HSO_4^- + H_2O \xrightarrow{\text{heat}} C_6H_5OH + N_2 + H_2SO_4$

---

26. By analogy with the hydrolysis of alkyl halides to yield alcohols (cf. Program 4, Frame 31; Program 6, Frames 46-48), it might be thought that aryl halides could be hydrolysed to give phenols. However, monohalogenated benzenes are quite stable to hydrolysis at ordinary temperatures, and forcing conditions are required in order to convert them into phenols. Chlorobenzene reacts with aqueous sodium hydroxide to give sodium phenoxide only at about 300° and 150 atmospheres pressure. In an alkyl halide, the chlorine atom is attached to a tetrahedral carbon atom, whereas in chlorobenzene the chlorine is attached to a . . . . (1) carbon atom.

---

(1) trigonal

TEST FRAME

Nitration, sulphonation and acylation are typical electrophilic substitution reactions of aromatic hydrocarbons, and can be written

$$ArH + X^+ \longrightarrow ArX + H^+$$

In the case of benzene, Ar stands for . . . . (1), for nitration, $X^+ =$ . . . . (2), for sulphonation, $X^+ =$ . . . . (3), and for acylation, $X^+ =$ . . . . (4). Show how diphenyl ketone could be prepared from benzoyl chloride . . . . (5).

---

(1) C_6H_5 (2) NO_2^+ (3) $HOSO_2^+$ (4) $R\overset{+}{C}{=}O$

(5)

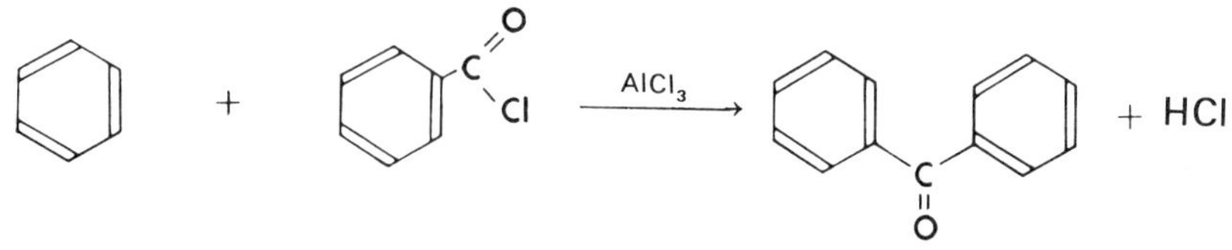

FURTHER SUBSTITUTION OF MONOSUBSTITUTED BENZENES

Introduction

27. In Frames 1-20 we saw that in the electrophilic substitution of benzene, usually only one hydrogen atom is replaced, leading to a "monosubstituted

(1) $$CH_3\overset{\displaystyle CH_3}{\overset{|}{C}}{=}CH_2 + HBr \xrightarrow[\text{peroxides}]{\text{air and}} CH_3\overset{\displaystyle CH_3}{\overset{|}{C}}HCH_2Br$$

(2) anti-Markovnikov

Reaction of alkanes with chlorine or bromine

18. A number of alkyl halides, RX, can be prepared by the reaction of specific alkanes, RH, with chlorine and bromine (but not iodine). In the case of chlorine the reaction is exothermic but heat or light must be supplied in order to initiate the process.

$$CH_3CH_3 + Cl_2 \xrightarrow[\text{to initiate the reaction}]{\text{heat } (\Delta) \text{ or light } (h\nu)} CH_3CH_2Cl + HCl$$

In the reaction of ethane a hydrogen atom on a carbon atom has been substituted by a . . . . (1) atom.

---

(1) chlorine

---

19. The energy required to initiate the reaction is taken up by a small proportion of the chlorine molecules which then undergo dissociation into atomic chlorine.

$$Cl{-}Cl + \text{heat or light energy} \rightleftharpoons Cl\cdot + Cl\cdot$$

The dot represents a single electron. An ethane molecule is then attacked by a chlorine atom according to the equation

$$CH_3\underset{\displaystyle \diagdown H}{\overset{\displaystyle H \diagup}{C}}{-}H + Cl\cdot \longrightarrow CH_3\underset{\displaystyle \diagdown H}{\overset{\displaystyle H \diagup}{C}}\cdot + HCl$$

There is formed an ethyl *free radical* which carries no charge and in which one carbon atom has . . . . (number) (1) electrons in its valency shell.

---

(1) seven

---

20. This ethyl . . . . (1) is an extremely reactive species and can attack an undissociated chlorine molecule to give ethyl chloride and a new chlorine atom which can react with another molecule of ethane and so continue a chain reaction.

$$CH_3\underset{\displaystyle \diagdown H}{\overset{\displaystyle H \diagup}{C}}\cdot + Cl{-}Cl \longrightarrow CH_3\underset{\displaystyle \diagdown H}{\overset{\displaystyle H \diagup}{C}}{-}Cl + Cl\cdot$$

---

phenol is an even weaker acid than carbonic acid, an aqueous solution of the sodium salt of phenol will react with carbonic acid according to the equation

$$C_6H_5ONa + H_2CO_3 \longrightarrow \text{. . . . . (1).}$$

---

(1) $C_6H_5ONa + H_2CO_3 \longrightarrow C_6H_5OH + NaHCO_3$

---

23. The weak acidity of phenols, ArOH, $K_a 10^{-10}$, is in sharp contrast to the virtual neutrality of alcohols, ROH, such as CH_3CH_2OH, $C_6H_5CH_2OH$ etc., which have acid dissociation constants of around 10^{-18}. A mixture of phenol, C_6H_5OH, and benzyl alcohol, $C_6H_5CH_2OH$, could be separated by dissolving the mixture in an inert solvent such as ether or dichloromethane, and shaking this solution with . . . . (1).

---

(1) aqueous sodium hydroxide. This would extract the phenol into the aqueous phase, and after separation of the two phases, the phenol could be recovered by acidification, e.g. with CO_2. The benzyl alcohol could be recovered from the organic phase by distillation.

---

24. Phenols also differ from alcohols in that they usually give blue or violet colours when tested with aqueous ferric chloride. In many other respects, however, they behave like alcohols, especially in forming ethers and esters:

$$ArOH \xrightarrow[NaOH]{CH_3I} ArOCH_3$$ (cf. Program 4, Frames 41, 42: Program 6, Frames 11-16)

$$ArOH \xrightarrow[Na\ OH]{C_6H_5COCl} \text{. . . . . (1)}$$

---

(1) $C_6H_5C(=O)OAr$ an aryl benzoate (cf. Program 8, Frame 31)

Sodium hydroxide is included in the above reaction mixtures in order to convert the phenol into phenoxide ion, $C_6H_5O^-$, which is a more reactive nucleophile: $C_6H_5OH + OH^- \rightleftharpoons C_6H_5O^- + H_2O$

---

25. Phenols can be prepared both from arenesulphonic acids by fusion with sodium hydroxide (cf. Frame 9) and also from aromatic amines by the following process:

$$ArNH_2 \xrightarrow{H_2SO_4/NaNO_2} ArN_2^+\ HSO_4^-$$ (a diazonium salt: cf. Program 10)

(1) free radical

---

21. The overall result of this mechanism is given by the equation:

$$RH + Cl_2 \xrightarrow[\text{initiate reaction}]{\text{heat or light to}} RCl + HCl$$

However the atomic chlorine can attack hydrogen atoms on the alkyl chloride molecules and small amounts of polychlorinated materials are produced. In methane, CH_4, the hydrogens are all equivalent. That is, one cannot distinguish any one hydrogen from any other hydrogen atom. When methane is chlorinated with a limited amount of chlorine, the principal product is chloromethane and this can be separated readily. Small amounts of dichloromethane . . . . (1), trichloromethane . . . . (2) and tetrachloromethane . . . . (3) are formed in the reaction.

---

(1) CH_2Cl_2 (methylene dichloride) (2) $CHCl_3$ (chloroform) (3) CCl_4 (carbon tetrachloride)

---

22. In ethane, H_3C-CH_3, we see that all the hydrogens are . . . . (equivalent/non-equivalent) (1). The principal product of reaction of ethane with one mole of chlorine will be . . . . (2). Small amounts of polysubstituted ethane will also be formed. The main product of the reaction can be separated easily and the reaction can be used preparatively.

---

(1) equivalent (2) chloroethane CH_3CH_2Cl

---

23. Another simple alkane in which all the hydrogens are equivalent is 2,2-dimethylpropane (neopentane) whose structure is . . . . (1). The principal product of its reaction with one mole of chlorine will have the structure . . . . (2) the name of which is . . . . (3).

---

(1)

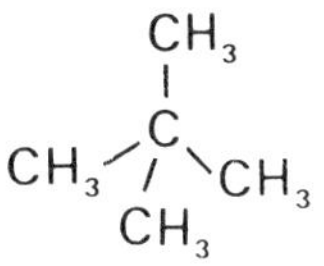

(2)

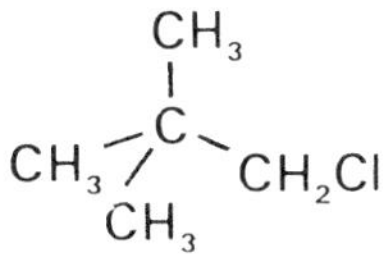

(3) 1-chloro-2,2-dimethylpropane or neopentyl chloride

---

24. In propane,

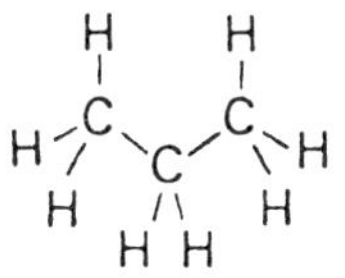

the six hydrogen atoms attached to the terminal carbon atoms are all equivalent but the two attached to the central carbon

(cf. the reaction of benzene with methyl bromide in the presence of aluminium tribromide, Frame 10).
In effect, the attacking electrophilic species is Br^+, formed from bromine by reaction with $FeBr_3$ according to the equation . . . . (1).

---

(1) $Br_2 + FeBr_3 \rightleftharpoons Br^+ + FeBr_4^-$

---

20. Aryl halides are in general much less reactive than alkyl halides, especially towards nucleophilic displacement of the halogen atom. However, both alkyl and aryl halides react readily with metallic magnesium in anhydrous ether to form the corresponding Grignard reagents (cf. Program 4, Frames 66-70). Since Grignard reagents, RMgX or ArMgX, react with water to yield the respective hydrocarbon, a two-step reductive process for the conversion of chlorobenzene into benzene can be shown . . . . (1).

---

(1)

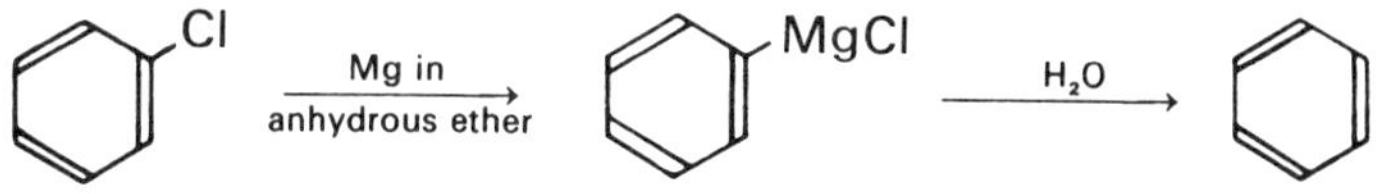

(cf. Program 4, Frame 67).

PHENOLS

Structure

21. The compound C_6H_5OH, in which one of the hydrogen atoms of benzene is substituted by an –OH group, is known as phenol or hydroxybenzene. Phenol, which gives its name to the whole class of such compounds, is an important industrial chemical being used, e.g. in the manufacture of phenol-formaldehyde resins ("Bakelite"). Write down the structural formula of phenol, showing all hydrogen atoms and all lone pairs of electrons . . . . (1).

---

(1)

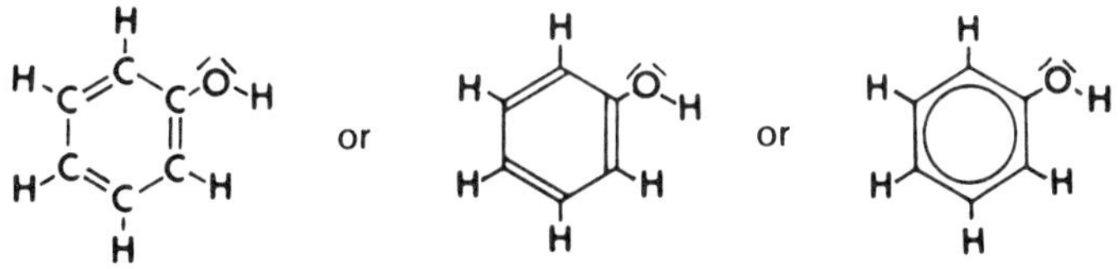

Properties

22. Phenols are very weak acids. Phenol itself has a dissociation constant, K_a, in water of about 1.3×10^{-10}($pK_a = -\log K_a = 9.9$). It can therefore be extracted from solution in an inert solvent with aqueous sodium hydroxide. Because

atom are in a different environment. Of the eight hydrogens present six are in one environment and two are in another. We can say therefore that there are two types of hydrogens present. Reaction of propane with chlorine yields a mixture of isomeric monochloro compounds and hydrogen chloride, together with some di, tri, and polychloro compounds.

$$CH_3CH_2CH_3 \xrightarrow{Cl_2} CH_3CH_2CH_2Cl,\ \underset{\displaystyle Cl}{CH_3\overset{}{C}HCH_3}\ \text{and HCl}$$

For each of the following compounds write down the number of types of hydrogen there are present in the molecule: 2-methylpropane (isobutane) .... (1), pentane .... (2) and 2-methylpentane .... (3).

---

(1) two (2) three (3) five

---

25. If the number of possible isomeric monochloro products is three or more and if, as is generally the case, they have very similar boiling points the reaction cannot be used preparatively. We would conclude therefore that such a reaction would only be used if there were say .... (number) (1) or fewer principal products.

---

(1) three or less. It is difficult to separate more than three principal products using simple distillation methods.

Reaction of chlorine with alkenes and arenes

26. In certain cases the chlorination reaction can be applied to more complex substances to yield monochlorinated compounds.
Propene (propylene) can be chlorinated in the gas phase to give allyl chloride.

$$CH_2{=}CHCH_3 + Cl_2 \xrightarrow{400\text{-}600^\circ} CH_2{=}CHCH_2Cl + HCl$$

The chlorine does not add to the double bond as it would if the reaction were done in solution but atomic chlorine reacts with propene to give a radical according to the equation

$$CH_2{=}CHCH_3 + Cl\cdot \longrightarrow \text{............ (1)}$$

and a chain reaction then ensues by further reaction of this radical with a chlorine molecule.

---

(1) $CH_2{=}CHCH_3 + Cl\cdot \longrightarrow CH_2{=}CHCH_2\cdot + HCl$

(1) ketones or carbonyl compounds (2) ethyl phenyl ketone

---

16. Reaction of acetic anhydride, formula . . . . (1) with benzene in the presence of $AlCl_3$ proceeds according to the equation . . . . (2).

---

(1) $(CH_3CO)_2O$ (2) 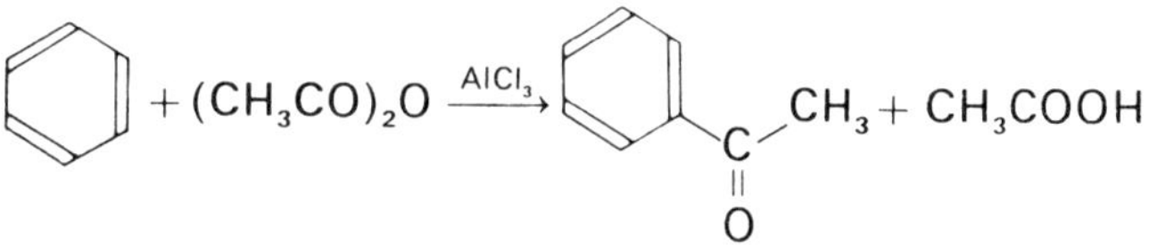

---

17. The substitution product is again a . . . . (1), which on the IUPAC system is named . . . . (2). Its common or trivial name is acetophenone.

---

(1) ketone (2) methyl phenyl ketone

Halogenation

18. Under ordinary conditions (daylight, room temperature) Cl_2 and Br_2 do not react readily with benzene. However, if a catalyst in the form of $FeCl_3$ or $FeBr_3$ is provided, we find that electrophilic substitution occurs. For example

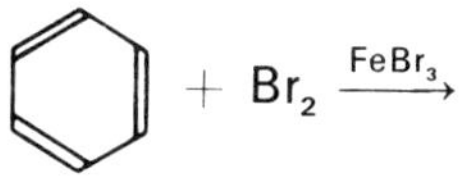

. . . . (1)

---

(1) 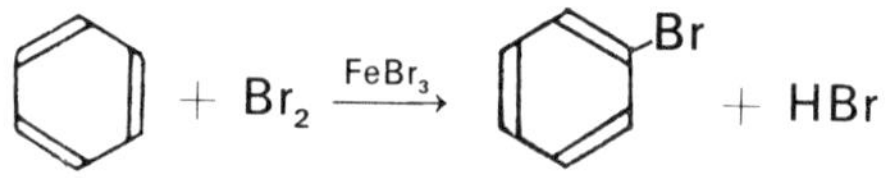

---

19. The reaction of bromine with benzene in the presence of ferric bromide as a catalyst gives rise to bromobenzene and hydrogen bromide. A possible mechanism for this reaction can be written as follows:

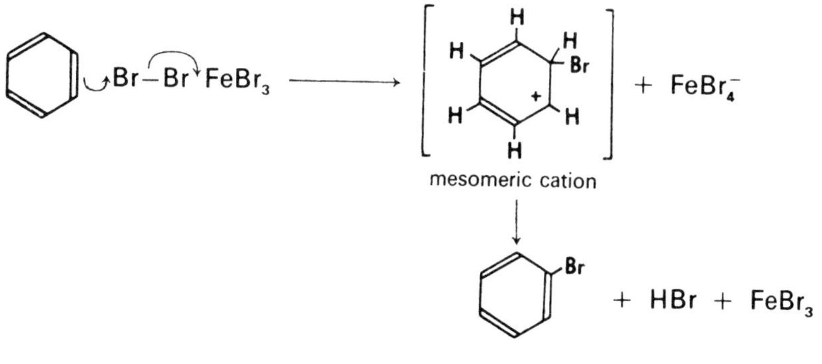

27. Another important reaction of this type is the chlorination of toluene under ultraviolet irradiation. The product of the reaction is benzyl chloride.

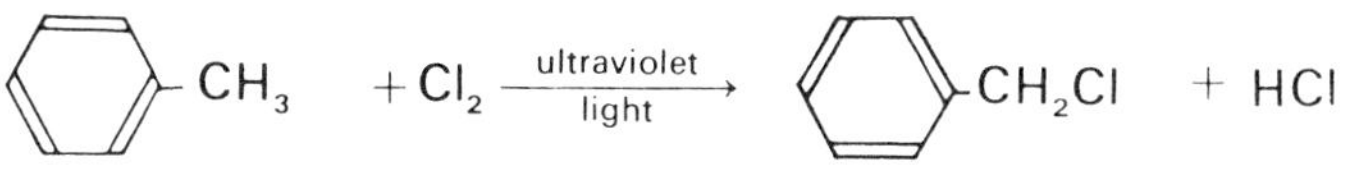

The chlorine does not substitute the aromatic ring as it would in the presence of an ionising catalyst such as iron(III) chloride (Program 9) but atomic chlorine reacts with toluene to form a benzyl radical according to the equation

and the benzyl radical reacts with Cl_2 to form benzyl chloride according to the equation

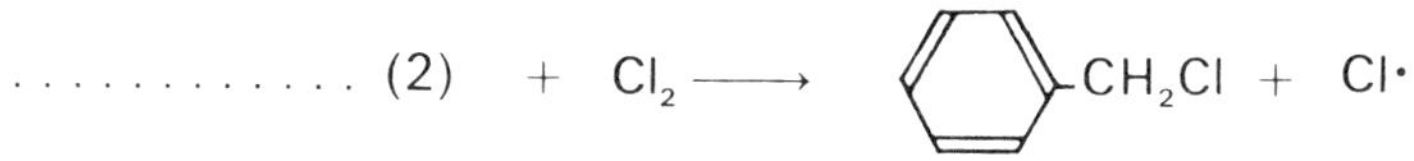

---

(1) 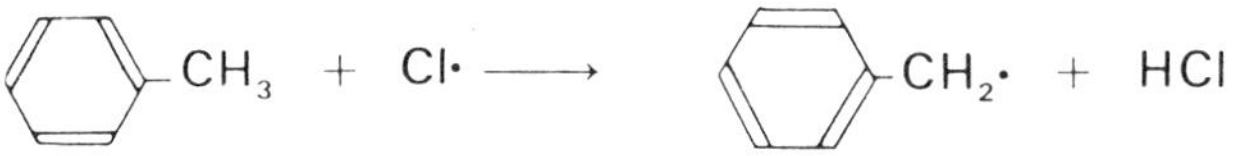

(2) 

REACTIONS OF ALKYL HALIDES

Nucleophilic substitution at the saturated carbon atom

Chemists think in terms of broad classifications in which a great number and variety of reactions are considered as examples of a single fundamental process. One of the most important of these mechanistic classifications is *nucleophilic substitution* at the saturated or tetrahedral carbon atom. It encompasses the reaction of a wide variety of substances with alkyl halides as well as with many other types of compound.

The elements making up the nucleophiles which form the new bond with the carbon atom of the alkyl halide include carbon and the more electronegative elements of the periodic table.

Group	IV	V	VI	VII
	C	N	O	F
		P	S	Cl
				Br
				I

11. Aluminium tribromide has six electrons in the outer shell. On reaction with CH_3Br it acquires a bromide ion and is thereby converted into the complex anion $AlBr_4^-$, which has .... (1) electrons in the outer shell of the aluminium atom.

---

(1) eight

---

12. Isopropylbenzene can be prepared from benzene, aluminium chloride and the alkyl chloride of structure .... (1), according to the equation .... (2).

---

(1) $(CH_3)_2CHCl$ (2) 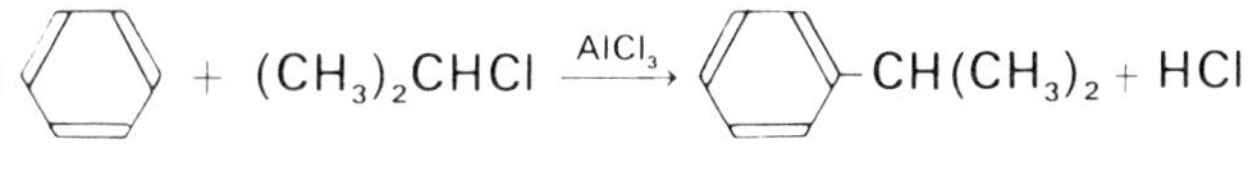

Acylation

13. The above alkylation reactions are examples of a Friedel-Crafts reaction named after the French chemist Friedel and the American Crafts, who discovered the value of $AlCl_3$ as a catalyst in promoting electrophilic substitution reactions of aromatic hydrocarbons. In acylation reactions with the same catalyst, the acyl group, $RC(=O)-$, which can be aliphatic or aromatic, is provided by using an acid chloride, formula .... (1) or acid anhydride, formula .... (2).

---

(1) $RC(=O)Cl$ (2) $RC(=O)-O-C(=O)R$ or $(RCO)_2O$

---

14. When benzene is allowed to react with propanoyl chloride, formula .... (1) in the presence of $AlCl_3$, substitution occurs according to the equation .... (2).

---

(1) CH_3CH_2COCl (2) 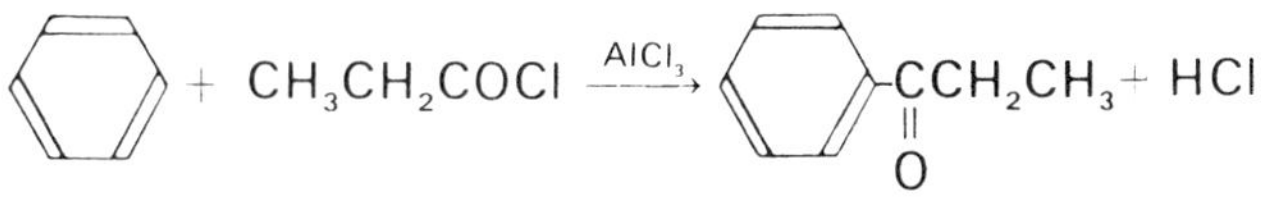

---

15. The product of the above electrophilic substitution belongs to the class of compounds known as .... (1), and its systematic name is .... (2). The electrophilic entity formed from propanoyl chloride by interaction with aluminium chloride is thought to be $CH_3CH_2-\overset{+}{C}=O$, formed as follows: $CH_3CH_2COCl + AlCl_3 \rightleftharpoons CH_3CH_2\overset{+}{C}O + AlCl_4^-$.

Let us look at some of these reactions starting from Group VII and consider the mechanism as we proceed. It should be remembered that *all* the reactions to be discussed take place in solution and the solvent plays an important part in controlling the behaviour of the reactants. A discussion of the details of solvent interaction is outside the scope of these programs.

Group VII nucleophiles—halogen exchange reactions

28. Alkyl chlorides and bromides can be converted into alkyl fluorides by reaction with mercury(II) fluoride.

$$2R\text{–}Br + HgF_2 \longrightarrow 2R\text{–}F + HgBr_2$$

The process can be considered to proceed as shown below, with the fluoride ion acting as a . . . . (1).

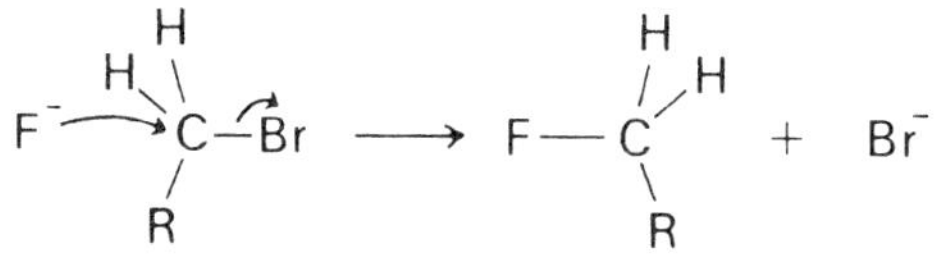

---

(1) nucleophile

---

29. Alkyl chlorides and bromides can be converted into alkyl iodides by reacting them with sodium iodide in acetone solution. Sodium iodide is soluble in acetone but the sodium bromide or sodium chloride formed is insoluble and precipitates from solution as the reaction proceeds.

$$R\text{–}Cl + NaI \xrightarrow{\text{acetone}} R\text{–}I + NaCl$$

Considering this as attack by nucleophilic iodide ion on 1-chloropropane, describe the process by an equation like that in the previous frame . . . . (1). What is the name of the product? . . . . (2).

---

(1)

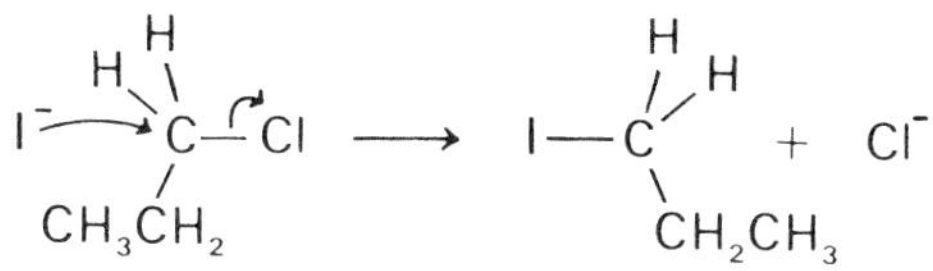

(2) 1-iodopropane or propyl iodide

---

30. To show that chloride ion will react with an alkyl chloride it is necessary to label one of the reactants by using a radioactive tracer. If 2-chloropropane is heated with sodium chloride prepared from radioactive chlorine it is found that the 2-chloropropane becomes radioactive because of the reversible reaction

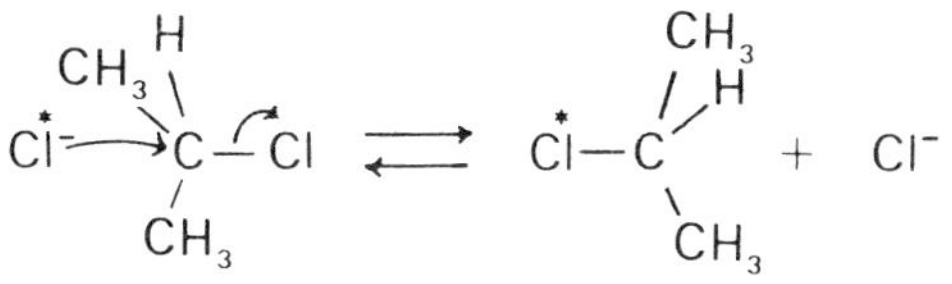

The electrophilic species involved is SO_3H^+, so that an alternative equation is

$$C_6H_6 + SO_3H^+ \rightleftharpoons \ldots\ldots\ldots\ldots (1).$$

---

(1) $C_6H_6 + SO_3H^+ \rightleftharpoons C_6H_5SO_3H + H^+$

---

8. In the above frame the sulphonation of benzene is shown as a reversible reaction, and in general, the arenesulphonic acids can be hydrolysed back to the parent hydrocarbon. Thus $ArSO_3H + H_2O \overset{heat}{\rightleftharpoons}$ .... (1).

---

(1) $ArSO_3H + H_2O \overset{heat}{\rightleftharpoons} ArH + H_2SO_4$

---

9. By contrast, when the sodium salt of an arenesulphonic acid is heated with very strong sodium hydroxide, the $-SO_3Na$ group is replaced by $-ONa$, and subsequent acidification yields a phenol (see Frames 21-24 below)

$$ArSO_3Na + 2NaOH \longrightarrow ArONa + Na_2SO_3$$
$$ArONa \xrightarrow{acid} ArOH$$

In this reaction a carbon- .... (1) bond is broken and a carbon- .... (2) bond is formed.

---

(1) carbon-sulphur (2) carbon-oxygen

Alkylation

10. Methyl bromide can be made to react with benzene to give methylbenzene (toluene) and hydrogen bromide, but only in the presence of a catalyst such as aluminium tribromide.

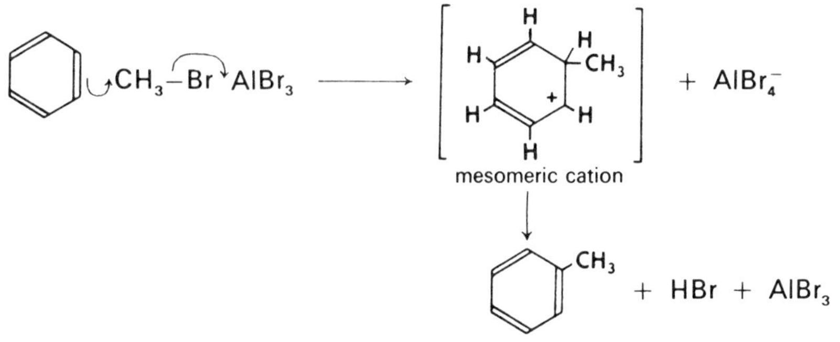

In this reaction, as a result of the complexing of aluminium bromide with the bromine atom of the methyl bromide, the methyl group behaves as an .... (1).

---

(1) electrophile

The same process can be demonstrated in the case of 2-bromopropane using radioactive sodium bromide according to the equation . . . . (1).

---

(1) 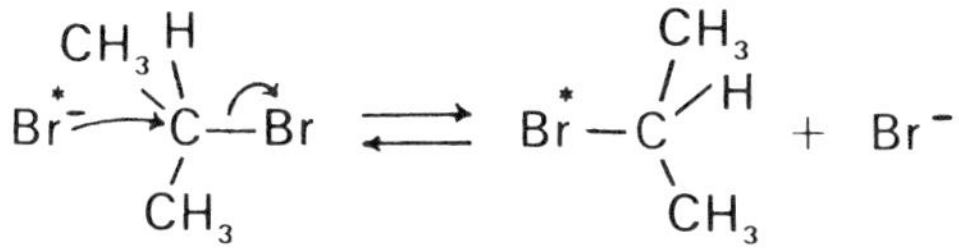

Group VI nucleophiles—synthesis of alcohols, ethers, thiols and thioethers

31. A familiar oxygen nucleophile is the hydroxide ion. Alkyl halides react with sodium hydroxide in aqueous solution to yield alcohols.

$$RX + NaOH \longrightarrow ROH + NaX \quad \text{where } X = Cl,\ Br,\ I.$$

For example, 2-iodobutane (s-butyl iodide) reacts with sodium hydroxide to yield butan-2-ol (s-butanol) according to the equation . . . . (1).

---

(1) $CH_3CH_2CH(I)CH_3 + NaOH \longrightarrow CH_3CH_2CH(OH)CH_3 + NaI$

The simplest such reaction is the conversion of methyl chloride into methanol. Methyl chloride is a dipolar molecule and the magnitude of the dipole can be determined experimentally. But the exact location of the dipole within the molecule cannot be determined and the reason for its existence is a matter of theoretical argument. Frames 32 to 40 give one such theoretical approach to the dipolar nature of methyl chloride together with a theory of nucleophilic substitution.

32. The electronic configuration of the chloride ion, Cl^-, is $1s^2, 2s^2, 2p^6, 3s^2, 3p^6$. The eight outer electrons, $3s^2, 3p^6$, of this ion form four equivalent electron pairs which through mutual repulsion are directed towards the corners of a . . . . (1).

---

(1) tetrahedron

---

33 Since the centre of distribution of each of the electron pairs is at a finite distance from the nucleus each electron pair represents a separation of negative charge from the positive nucleus. This can be represented by an electric dipole $+\!\!\longrightarrow$ going from positive to negative.

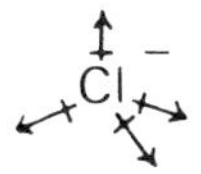

The chloride ion has no overall dipole because the vector sum of four equal dipoles arranged in tetrahedral fashion, is . . . . (1).

That is, a . . . . (1) is displaced from the benzene ring by the incoming nitronium ion.

---

(1) proton or hydrogen ion

Definition of electrophilic substitution

4. Because the nitronium ion carries a positive charge it is electron-seeking and it is therefore said to be an electrophile or an electrophilic reagent. The substitution reaction in which it takes part is described as an aromatic electrophilic substitution.
The nitronium ion is formed in small concentration in an equilibrium reaction. Complete the following equation:

$$HNO_3 + 2H_2SO_4 \rightleftharpoons NO_2^+ + \ldots\ldots\ldots\ldots + \ldots\ldots\ldots\ldots \ (1).$$

---

(1) $HNO_3 + 2H_2SO_4 \rightleftharpoons NO_2^+ + 2HSO_4^- + H_3O^+$

---

5. The other aromatic substitution reactions to be discussed, namely sulphonation, alkylation, acylation, halogenation and diazonium coupling, are also electrophilic substitutions. If the attacking electrophilic species is written X^+, then a general statement for electrophilic substitution of benzene can be written:

$$C_6H_6 + X^+ \longrightarrow \ldots\ldots\ldots\ldots \ (1)$$

---

(1) $C_6H_6 + X^+ \longrightarrow C_6H_5X + H^+$

---

6. The species which attacks benzene in an electrophilic substitution is an electrophile. It follows that benzene in these reactions behaves as a . . . . (1).

---

(1) nucleophile

Sulphonation

7. The sulphonation of benzene is carried out in hot concentrated sulphuric acid, or in cold oleum (a solution of SO_3 in concentrated sulphuric acid).
The product is benzenesulphonic acid, $C_6H_5{-}\overset{\overset{\large O}{\|}}{\underset{\underset{\large O}{\|}}{S}}{-}OH$ and the equation for the reaction is

$$C_6H_6 + H_2SO_4 \rightleftharpoons C_6H_5SO_3H + H_2O$$

(1) zero

---

34. In methyl chloride, CH_3Cl, the electron pair of the carbon-chlorine bond is shared nearly equally between the two atoms. Thus one of the dipoles in the chloride ion is effectively neutralised.

The vector sum of the three remaining dipoles can be considered as a single dipole *in the line of the carbon-chlorine bond* and having the positive end at the . . . . (1) nucleus. This is represented by an arrow ⟼ placed above or below the chlorine atom.

---

(1) chlorine

---

35. The above theory locates the dipole as having the positive end at the chlorine nucleus. Other theories may vary the location of the centres of charge and the general phenomenon is often loosely indicated by the symbol:

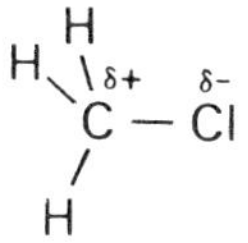

The important feature to remember is the *direction* of the dipole which is from positive to negative in the direction of the . . . . (1) bond, for it is this which controls the direction of approach of any charged or dipolar species.

---

(1) carbon-chlorine

---

36. If a hydroxide ion approaches the methyl chloride molecule along the line of the carbon-chlorine bond as indicated below it will be . . . . (attracted/repelled) (1).

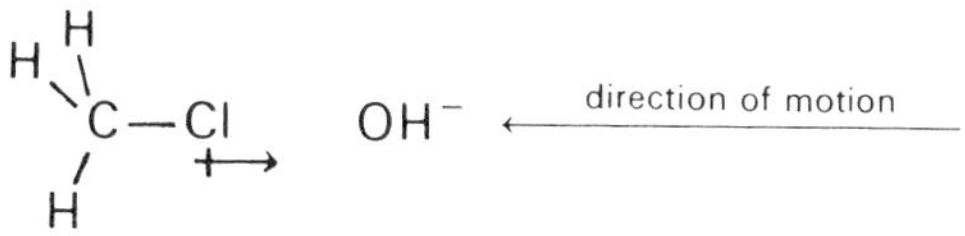

---

(1) repelled, because of electrostatic interaction with the negative end of the dipole.

BENZENE DERIVATIVES

Program 9

1. Benzene, C_6H_6, is a typical aromatic hydrocarbon (cf. Program 1, Frame 29, and Program 3, Frames 46-48). Although highly inflammable, and toxic, it is relatively unreactive to many reagents, and when reactions are carried out to substitute its hydrogen atoms by other atoms or groups, usually only one of these hydrogens is replaced. For this reason it is convenient to use ArH as an abbreviation for an aromatic hydrocarbon, e.g., benzene can be written as C_6H_5H, or as a phenyl group attached to hydrogen, PhH.
A typical benzene derivative is nitrobenzene, formula . . . . (1). In this program we shall discuss six main types of substitution:

NITRATION where the introduced functional group is					$-NO_2$
SULPHONATION	,,	,,	,,	,,	,, $-SO_3H$
ALKYLATION	,,	,,	,,	,,	,, an alkyl group, $-R$
ACYLATION	,,	,,	,,	,,	,, an acyl group, $-COR$
HALOGENATION	,,	,,	,,	,,	,, a halogen atom
DIAZONIUM COUPLING	,,		,,	,,	,, $-N{=}NAr$

---

(1) $C_6H_5NO_2$ or  $-NO_2$ or $PhNO_2$

MONOSUBSTITUTION OF BENZENE

Nitration

2. The monosubstitution of benzene, that is, the introduction of one functional group in place of one hydrogen atom, can be illustrated by the reaction of benzene with a mixture of concentrated nitric and concentrated sulphuric acids. The product is nitrobenzene, and the overall equation for the process can be written

$$C_6H_6 + HNO_3 \xrightarrow{H_2SO_4} \ldots\ldots\ldots\ldots \quad (1)$$

---

(1) $C_6H_6 + HNO_3 \xrightarrow{H_2SO_4} C_6H_5NO_2 + H_2O$

---

3. Investigations of the mechanism of the nitration reaction have shown that the reactive species which attacks the benzene ring is the nitronium ion, NO_2^+. This entity is formed in the nitrating mixture of concentrated nitric and sulphuric acids, and its reaction with benzene can be written

$$C_6H_6 + NO_2^+ \longrightarrow C_6H_5NO_2 + H^+$$

37. If it approaches along the line of the carbon-chlorine bond as indicated it will be . . . . (1)

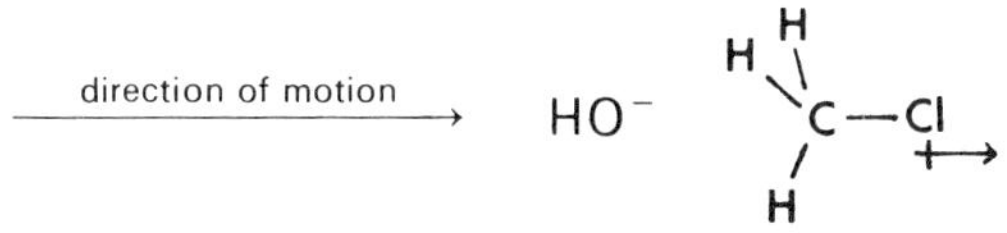

---

(1) attracted

---

38. Because of repulsion by the negative end of the dipole and attraction by the positive end of the dipole, a hydroxide ion will be directed onto the axis of the carbon-chlorine bond as it approaches. That is, the approach to the carbon atom from the side opposite the chlorine will require . . . . (minimum/maximum) (1) energy.

---

(1) minimum

---

39. Despite the attraction of the positive end of the dipole for the hydroxide ion, the near approach to the carbon atom will cause a very rapid build-up of repulsion as the outer electrons of the oxygen and carbon atoms interact. If the kinetic energy of approach is not sufficiently great the ion and the molecule will undergo elastic collision and move apart. Many such collisions may take place between hydroxide ions and methyl chloride molecules without any reaction taking place. But, provided the two reactants approach one another with sufficient velocity then the electrostatic repulsion between the negative hydroxide ion and the valency electrons of the methyl chloride causes a lengthening of the carbon-chlorine bond, and, as the new oxygen-carbon bond is formed, the chlorine departs as a . . . . (1) ion.

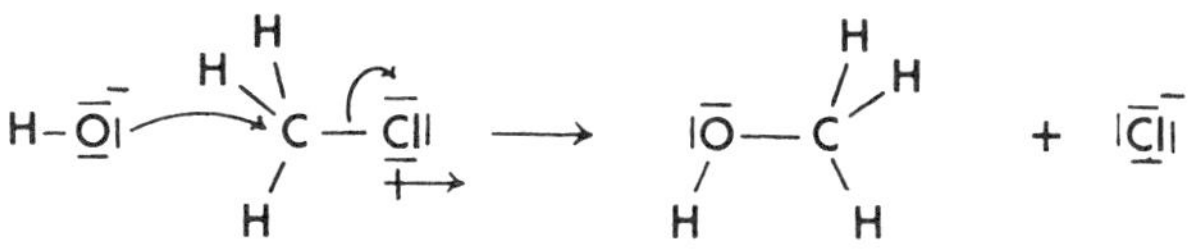

In this reaction the hydroxide ion is a . . . . (2), and the carbon atom is an . . . . (3) centre.

---

(1) chloride (2) nucleophile (3) electrophilic

---

40. The process as a whole represents a *nucleophilic substitution*. All such reactions at primary and secondary carbon atoms can be analysed in terms of the approach of a nucleophile, whether charged or dipolar, along the line of least resistance to the dipolar molecule. This is because the alkyl substituents are non-polar and do not influence the approach of the reagent

Program 9: BENZENE DERIVATIVES

CONTENTS

other than by their size. Further polarisation of the dipolar bonds and substitution by displacement of the electronegative group completes the process. The energy barrier to the substitution is overcome by the . . . . (1) of the collision.

When considering nucleophilic substitution at tertiary carbon atoms, the same arguments apply *only* when the reactions are done in solvent of low polarity. In ionising solvents, a tertiary alkyl halide may undergo substitution by a two stage process involving ionization of the carbon-halogen bond, with formation of halide anion, Cl^-, and a carbonium cation, R_3C^+. The latter species then reacts rapidly with any nucleophile present.

---

(1) kinetic energy

---

41. Alcohols are weaker acids than water but are still sufficiently acidic to react smoothly with metallic sodium to yield hydrogen and the sodium salt of the alcohol, called a sodium alkoxide.

$$R{-}OH + Na \longrightarrow R{-}O^- \ Na^+ + \tfrac{1}{2}H_2$$

An alkoxide ion can react with an alkyl halide to give an ether.

$$R{-}O^- + R'{-}X \longrightarrow R{-}O{-}R' + X^-$$

As an example, sodium dissolves in isopropanol to give sodium prop-2-oxide according to the equation . . . . (1).

---

(1) $$CH_3\overset{\displaystyle H}{\underset{\displaystyle CH_3}{C}}{-}OH + Na \longrightarrow CH_3\overset{\displaystyle H}{\underset{\displaystyle CH_3}{C}}{-}O^- \ Na^+ + \tfrac{1}{2}H_2$$

---

42. Sodium prop-2-oxide reacts with benzyl chloride to give benzyl isopropyl ether according to the equation . . . . (1). (Ethers are named by putting the alkyl radicals in alphabetical order).

---

(1)

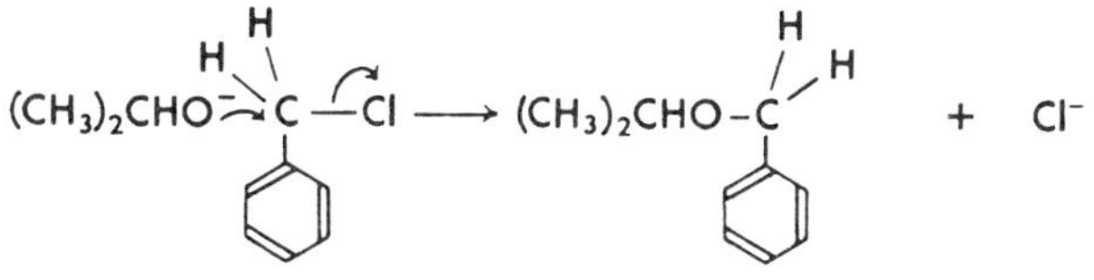

---

43. Another oxygen nucleophile is the anion of a carboxylic acid $RCOO^-$. For example, sodium acetate reacts with 1-chlorobutane to give butyl acetate.

$$CH_3(CH_2)_2CH_2Cl + CH_3COO^-Na^+ \longrightarrow CH_3(CH_2)_2CH_2O\underset{\displaystyle \overset{\|}{O}}{C}CH_3 + NaCl$$

Alternatively

$$C_6H_5COOH + C_6H_5CH_2OH \xrightarrow{\text{acid}} C_6H_5COOCH_2C_6H_5 + H_2O$$

(2) $$C_6H_5CH_3 \xrightarrow{KMnO_4} C_6H_5COOH \xrightarrow{SOCl_2} C_6H_5COCl \xrightarrow{NH_3}$$

$$C_6H_5CONH_2 \xrightarrow{Br_2/NaOH} C_6H_5NH_2$$

In the preparations above, the use of the chart summarising the reactions of acids and their derivatives could have been a help. At the same time, use of the chart might have obscured the actual thought process that was used.

Consider the conversion of toluene into aniline.

Step 1. Write the two structures.

2. Note that the product has *one less carbon atom* than the starting material.
3. Consider the methods for synthesis of an amine. In this program three methods have been given

   (a) $RCN \xrightarrow{LiAlH_4} RCH_2NH_2$

   (b) $RCONH_2 \xrightarrow{LiAlH_4} RCH_2NH_2$

   (c) $RCONH_2 \xrightarrow{Br_2/NaOH} RNH_2$

4. In method (c), the product amine has one less carbon atom than the starting amide, so that method (c) is an obvious choice in solving our problem.
5. The amide required could be prepared by the action of ammonia on an acid chloride, an acid anhydride or an ester. These are all derivatives of a carboxylic acid.
6. We now need to prepare benzoic acid from toluene and this can be done by oxidation.

In summary, a problem of this kind is often best solved by working backwards, from the desired end product to the available starting material.

The mechanism of the attack by acetate ion on 1-chlorobutane can be shown . . . . (1).

---

(1)

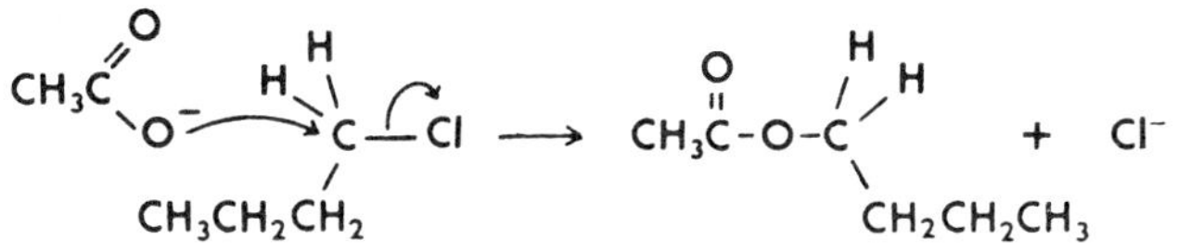

---

44. Thiols and thioethers are formed in reactions analogous to those leading to the corresponding oxygen compounds. Sodium hydrogen sulphide reacts with alkyl halides to give thiols.

$$Na^{+}\ HS^{-} + R{-}X \longrightarrow R{-}SH + Na^{+} + X^{-}$$

For example 2-bromobutane reacts with sodium hydrogen sulphide to give butane-2-thiol according to the equation . . . . (1).

---

(1)

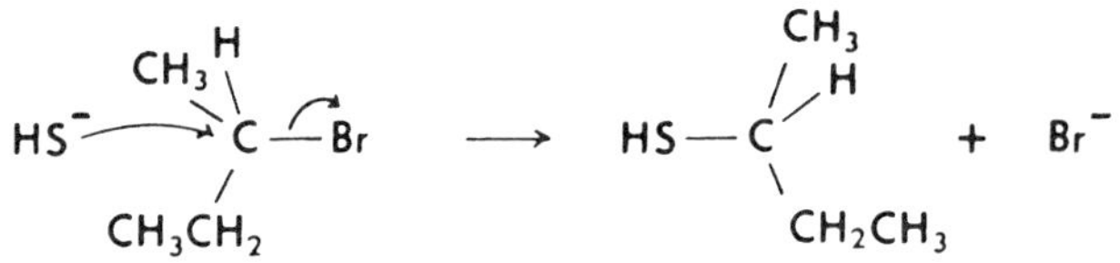

---

45. Thioalcohols or thiols are *more* acidic than water and when dissolved in aqueous sodium hydroxide solution the sodium thioalkoxide is formed.

$$R{-}SH + OH^{-} \longrightarrow R{-}S^{-} + H_2O$$

The thioalkoxide can then react with an alkyl halide to give a thioether.

$$R{-}S^{-} + R'{-}X \longrightarrow R{-}S{-}R' + X^{-}$$

For example, butane-2-thiol dissolved in sodium hydroxide and allowed to react with ethyl bromide yields a sulphide according to the equation . . . . (1).

---

(1)

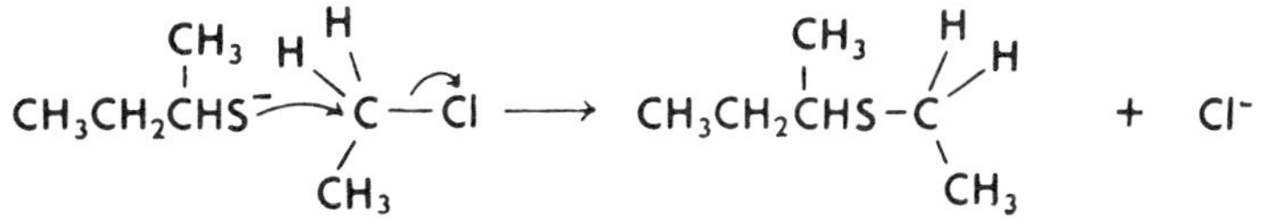

---

46 Symmetrical sulphides or thioethers can be made in one step by reacting two moles of an alkyl halide with one of sodium sulphide. Benzyl chloride is converted in this way into dibenzyl sulphide:

$$2C_6H_5CH_2Cl + Na_2S \longrightarrow \ldots\ldots\ldots\ldots\ (1).$$

---

9. $C_6H_5C(=O)Cl + 2NH_3 \longrightarrow C_6H_5C(=O)NH_2 + NH_4Cl$

10. $C_6H_5C(=O)OR + NH_3 \longrightarrow C_6H_5C(=O)NH_2$ + ROH where R is any alkyl or aryl group

11. $C_6H_5\overset{O}{\overset{\|}{C}}-O-\overset{O}{\overset{\|}{C}}C_6H_5 + 2NH_3 \longrightarrow C_6H_5CONH_2 + C_6H_5COO^-NH_4^+$

12. $C_6H_5COCl + H_2O \longrightarrow C_6H_5COOH + HCl$

13. $C_6H_5COOH + SOCl_2 \longrightarrow C_6H_5COCl + SO_2 + HCl$
or
$C_6H_5COOH + PCl_5 \longrightarrow C_6H_5COCl + POCl_3 + HCl$

14. $C_6H_5COOR + H_2O \xrightarrow{\text{acid}} C_6H_5COOH + ROH$
or
$C_6H_5COOR + HO^- \longrightarrow C_6H_5COO^- + ROH$

15. $C_6H_5COOH + ROH \xrightarrow{\text{acid}} C_6H_5COOR + H_2O$

16. $C_6H_5\overset{O}{\overset{\|}{C}}-O-\overset{O}{\overset{\|}{C}}C_6H_5 + H_2O \longrightarrow 2C_6H_5COOH$

17. $C_6H_5COO^-Na^+ + C_6H_5COCl \longrightarrow C_6H_5\overset{O}{\overset{\|}{C}}-O-\overset{O}{\overset{\|}{C}}C_6H_5 + NaCl$

18. $C_6H_5\overset{O}{\overset{\|}{C}}-O-\overset{O}{\overset{\|}{C}}C_6H_5 + ROH \longrightarrow C_6H_5COOR + C_6H_5COOH$

19. $C_6H_5COCl + ROH \longrightarrow C_6H_5COOR + HCl$

20. $C_6H_5CONH_2 \xrightarrow[\text{HO}^-/\text{BrO}^-]{\text{oxidation with}} C_6H_5NH_2 + CO_2$

Syntheses often require that several different compounds be prepared in sequence. The starting material may be determined by the chemicals which are available in the laboratory. Suggest a series of reactions which could be used for the following conversions:

1. benzyl alcohol into benzyl benzoate
2. toluene into aniline

---

(1) $C_6H_5CH_2OH \xrightarrow{KMnO_4} C_6H_5COOH \xrightarrow{SOCl_2} C_6H_5COCl$

$C_6H_5COCl + C_6H_5CH_2OH \longrightarrow C_6H_5COOCH_2C_6H_5 + HCl$

(1) $2C_6H_5CH_2Cl + Na_2S \longrightarrow C_6H_5CH_2SCH_2C_6H_5 + 2NaCl$

Group V nucleophiles—synthesis of amines

47. So far we have considered only nucleophiles which are anions. Ammonia carries no charge but has a dipole moment due to the presence of a lone pair of electrons. The molecule is a base and behaves as a nucleophile.

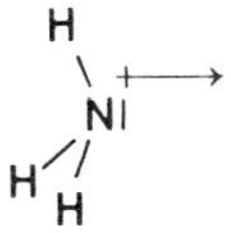

If ethyl bromide is added to aqueous ammonia a complex mixture of products results. There are present the hydrobromide salts of the amines ethylamine, diethylamine and triethylamine whose structures are .... (1) .... (2) and .... (3) respectively, and the quaternary ammonium compound tetraethylammonium bromide whose structure is .... (4) (cf. Program 5).

---

(1) $CH_3CH_2NH_2$ (2) $(CH_3CH_2)_2NH$ (3) $(CH_3CH_2)_3N$

(4) $(CH_3CH_2)_4N^+Br^-$

---

48. To understand why such a mixture results, the familiar ideas of equilibrium and pH need to be invoked. In an aqueous solution of ammonia or of a primary, secondary, or tertiary amine the following equilibria exist

$$NH_3 + H_2O \rightleftharpoons NH_4^+ + OH^-$$
$$RNH_2 + H_2O \rightleftharpoons \ldots\ldots\ldots\ldots \quad (1)$$
$$R_2NH + H_2O \rightleftharpoons \ldots\ldots\ldots\ldots \quad (2)$$
$$R_3N + H_2O \rightleftharpoons \ldots\ldots\ldots\ldots \quad (3).$$

---

(1) $RNH_2 + H_2O \rightleftharpoons RNH_3^+ + OH^-$
(2) $R_2NH + H_2O \rightleftharpoons R_2NH_2^+ + OH^-$
(3) $R_3N + H_2O \rightleftharpoons R_3NH^+ + OH^-$

---

49. Ammonia and the amines are all strong nucleophiles. The approach of the dipolar ammonia molecule to the electrophilic carbon atom of an alkyl halide from the side opposite the halogen leads to a sequence of events similar to that described in the case of an anion

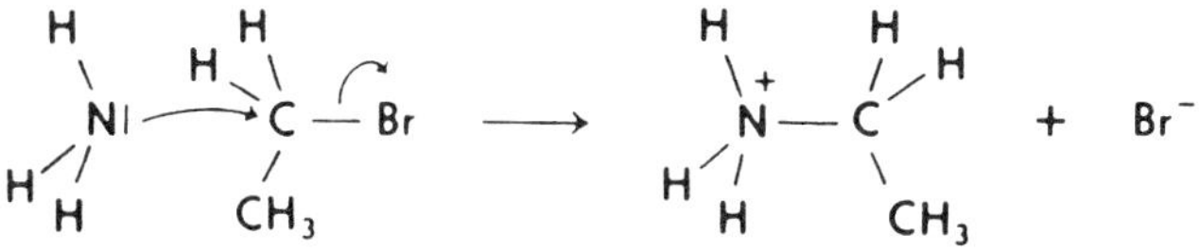

TEST FRAMES

The reactions of the carboxylic acids and their derivatives are summarised in the chart below. Write an equation for each of the numbered reactions using benzoic acid as the central carboxylic acid. Reagents and reaction conditions should be stated as far as possible.

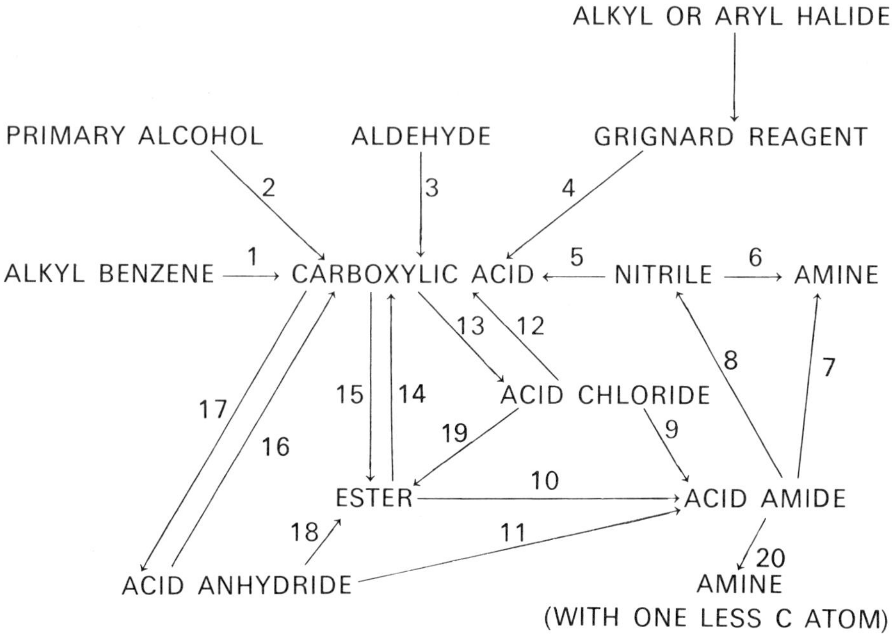

---

1. $C_6H_5R \xrightarrow{\text{alkaline } KMnO_4} C_6H_5COOH$ where R is an alkyl group

2. $C_6H_5CH_2OH \xrightarrow{KMnO_4} C_6H_5COOH$

3. $C_6H_5CHO \xrightarrow[\text{milder oxidants}]{KMnO_4 \text{ or}} C_6H_5COOH$

4. $C_6H_5MgBr \xrightarrow{CO_2 \text{ followed by dilute acid}} C_6H_5COOH$

5. $C_6H_5CN + 2H_2O \xrightarrow{\text{acid or base}} C_6H_5COOH + NH_3$

6. $C_6H_5CN \xrightarrow{LiAlH_4} C_6H_5CH_2NH_2$

7. $C_6H_5C(=O)NH_2 \xrightarrow{LiAlH_4} C_6H_5CH_2NH_2$

8. $C_6H_5C(=O)NH_2 \xrightarrow{P_4O_{10}} C_6H_5CN$

The difference is that the immediate product carries a positive charge but this can readily be lost through an acid-base equilibrium

e.g. $CH_3CH_2NH_3^+ + OH^- \rightleftharpoons$ .... (1).

---

(1) $CH_3CH_2NH_3^+ + OH^- \rightleftharpoons CH_3CH_2NH_2 + H_2O$

---

50. The free ethylamine of the above example is strongly nucleophilic and it can attack a molecule of ethyl bromide by the process outlined in the previous frame.

.... 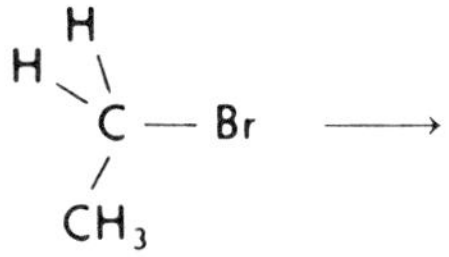 .... (1)

---

(1) 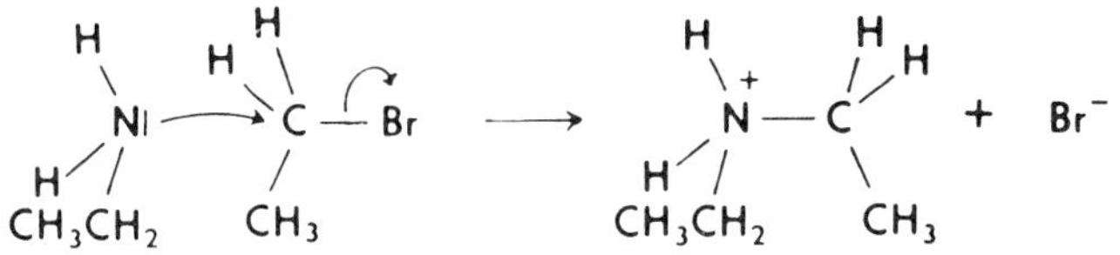

---

51. Again an acid-base equilibrium comes into play and free diethylamine molecules are formed and can attack more of the ethyl bromide according to the reaction .... (1).

---

(1) 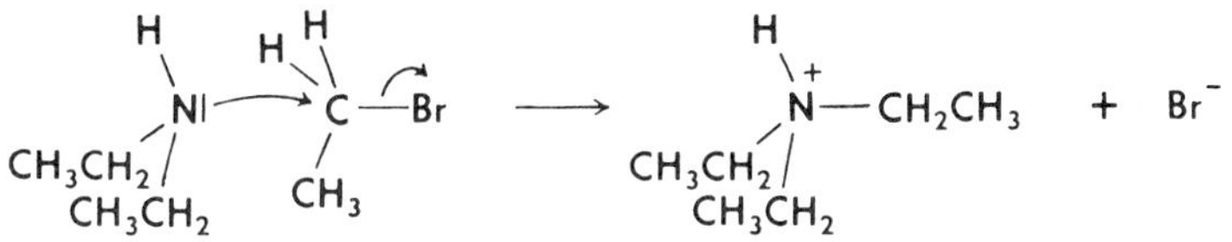

---

52. The final step differs in that the product *cannot* lose the positive charge.

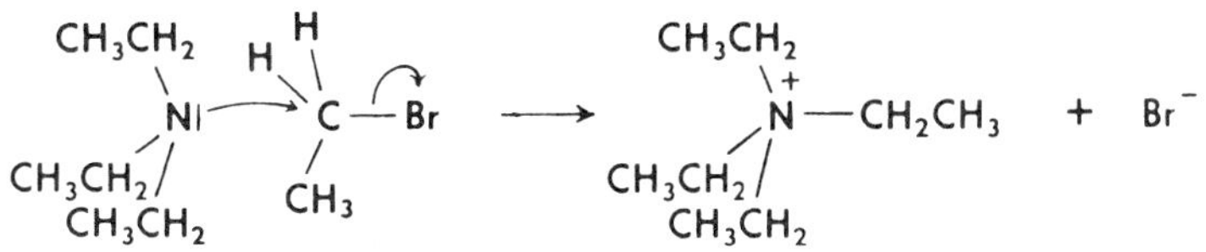

The result is a salt which in this case would be called .... (1).

---

(1) tetraethylammonium bromide

---

53. Thus the reaction of ethyl bromide with aqueous ammonia leads to a complex mixture because of acid-base equilibria and the fact that primary, secondary and tertiary amines are strong .... (1).

---

(1) nucleophiles

(1) $C_6H_5COOCH_3 + NH_3 \longrightarrow C_6H_5CONH_2 + CH_3OH$

(2) $C_6H_5CONH_2 \xrightarrow[\text{or } P_4O_{10}]{POCl_3} C_6H_5CN$

---

57. Nitriles can also be prepared by the dehydration of aldehyde oximes, prepared in turn from aldehydes and hydroxylamine (cf. Program 7, Frames 55-57).

$$RCHO \xrightarrow{H_2NOH} RCH{=}NOH \xrightarrow{\textit{dehydration}} RCN$$

A common dehydrating agent for this purpose is acetic anhydride. The equation for the preparation of benzonitrile (cyanobenzene) from benzaldehyde oxime and acetic anhydride can be written

$$C_6H_5CH{=}NOH + (CH_3CO)_2O \longrightarrow \ldots\ldots\ldots\ldots \quad (1).$$

---

(1) $C_6H_5CH{=}NOH + (CH_3CO)_2O \longrightarrow C_6H_5CN + 2CH_3COOH$

---

58. Nitriles are reduced by $LiAlH_4$ to the corresponding amine (cf. Program 5, Frame 37). Thus $RCN \longrightarrow RCH_2NH_2$. Using this reduction step, outline a series of reactions for the preparation of ethylamine from acetic acid . . . . (1).

---

(1) $CH_3COOH \xrightarrow[SOCl_2]{PCl_5 \text{ or}} CH_3COCl$

$CH_3COCl + 2NH_3 \longrightarrow CH_3CONH_2 + NH_4Cl$

$CH_3CONH_2 \xrightarrow[P_4O_{10}]{POCl_3 \text{ or}} CH_3CN$

$CH_3CN \xrightarrow{LiAlH_4} CH_3CH_2NH_2$

Note: Acetamide could also be reduced directly to ethylamine by $LiAlH_4$.

---

54. For the preparation of a primary amine the reaction of ammonia with an alkyl halide is clearly of little value and the *Gabriel* method is used. This consists of preparing the potassium salt of phthalimide and then reacting it with an alkyl halide.

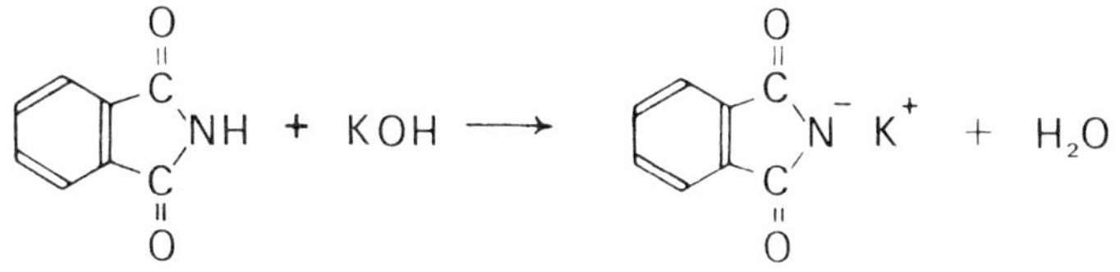

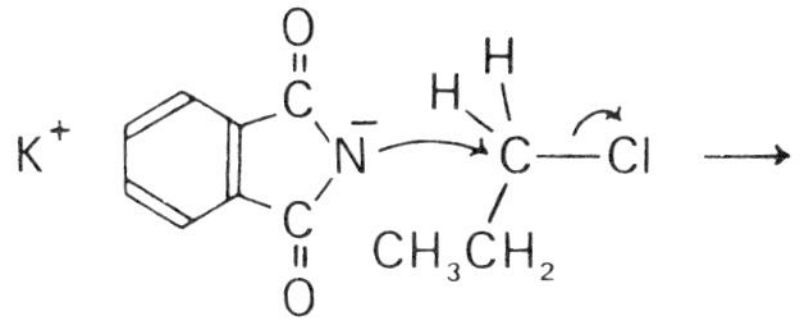

. . . . (1)

---

(1)

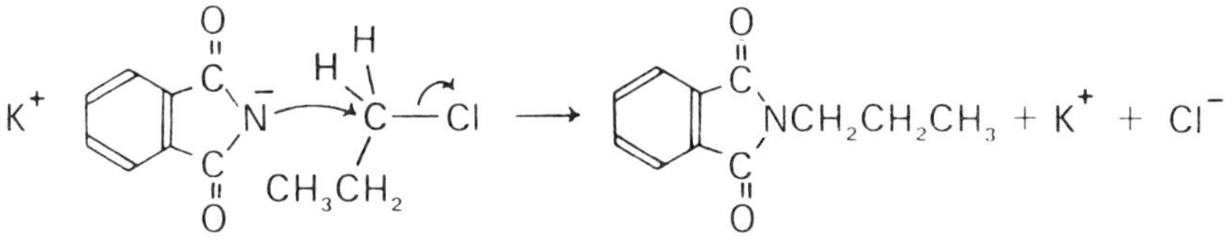

---

55. The N-propylphthalimide formed in the above reaction can be hydrolysed by boiling with strong sodium hydroxide solution to give the sodium salt of phthalic acid and propylamine.

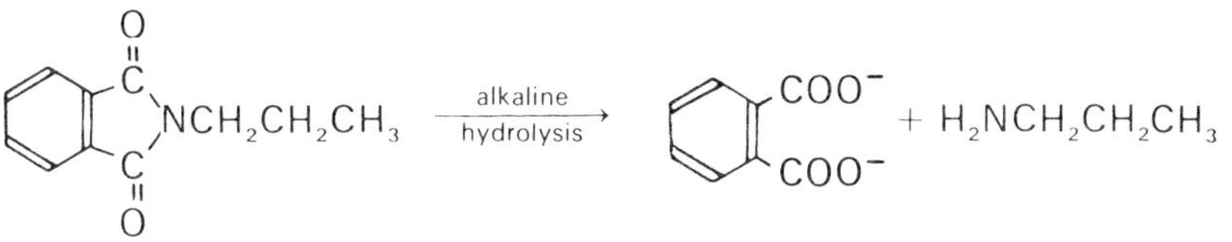

Alternatively the n-propylphthalimide can be treated with hydrazine

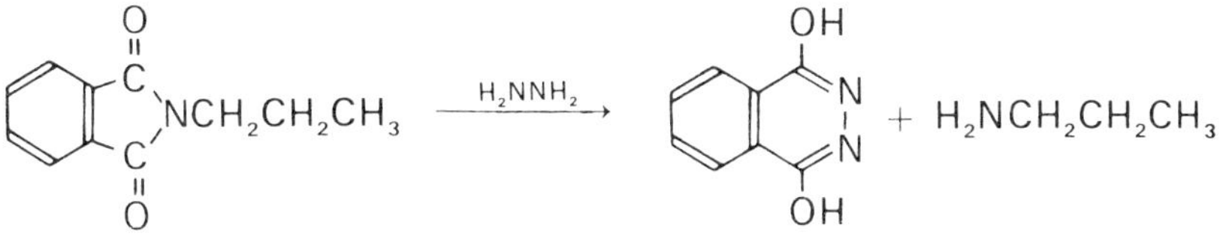

The overall Gabriel synthesis yields a primary amine uncontaminated with . . . . (1) and . . . . (2) amines.

---

(1) secondary (2) tertiary

(1) $(CH_3)_2CHCONH_2 \xrightarrow{POCl_3} (CH_3)_2CHCN$

(2) $(CH_3)_2CHCH_2CONH_2 \xrightarrow{POCl_3} (CH_3)_2CHCH_2CN$

(3) 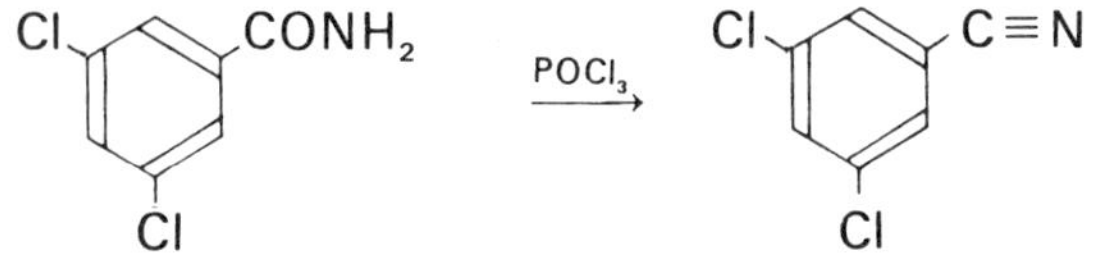

---

54. Nitriles can be hydrolysed by aqueous acid to give carboxylic acids according to the equation . . . . (1) (cf. Frame 21). The synthesis of a nitrile from an acid involves the intermediate formation of a primary amide which is then . . . . (2). The starting acid and the nitrile contain the same number of carbon atoms.

---

(1) $RCN + H_3O^+ + H_2O \longrightarrow RCOOH + NH_4^+$
(2) dehydrated

---

55. Aliphatic nitriles (cyanoalkanes) can be prepared from appropriate alkyl halides by nucleophilic substitution with cyanide ion provided by potassium cyanide. In this reaction, the cyano group replaces the chlorine atom with an increase of one in the number of carbon atoms and the establishment of a new carbon-carbon bond (cf. Program 4, Frame 56).

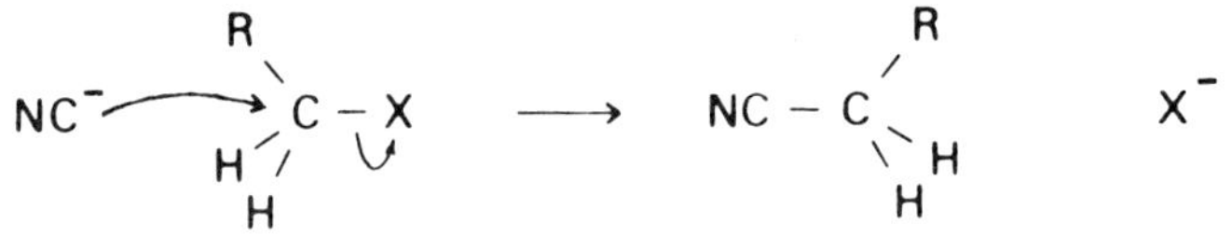

Butanenitrile . . . . (1) is prepared by the action of potassium cyanide on . . . . (2) and vigorous hydrolysis of the nitrile gives . . . . (3).

---

(1) $CH_3CH_2CH_2CN$ (2) $CH_3CH_2CH_2X$ where X=Cl, Br or I.
(3) $CH_3CH_2CH_2COOH$

---

56. Aromatic nitriles are prepared from diazonium salts (cf. Program 10, Frame 21).

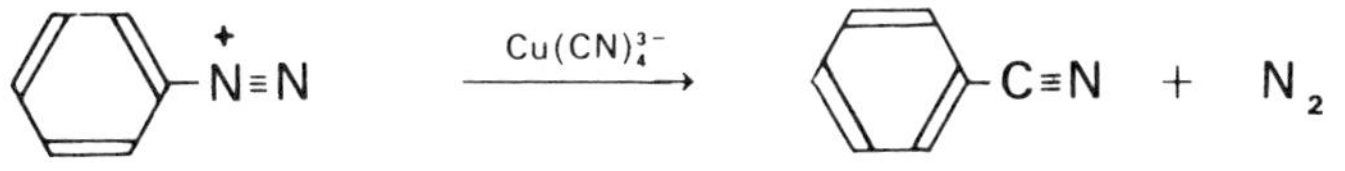

Give the equations for an alternative synthesis of this nitrile from methyl benzoate . . . . (1).

Group IV nucleophiles—formation of carbon-carbon bonds

56. The application of the following reactions (Frames 56 to 61) to organic synthesis is discussed in Program 12. A carbon compound which reacts with an alkyl halide to form a carbon-carbon bond must contain a *nucleophilic* carbon atom. An anionic species in which the charge is wholly or partially carried by a carbon atom is termed a *carbanion* and such an anion can be formed by a normal acid-base reaction if the hydrogen atom bound to the carbon atom in question is sufficiently acidic:

$$\begin{array}{c}\diagdown \\ \diagup C - H \end{array} \quad |B \quad \rightleftharpoons \quad \begin{array}{c}\diagdown \\ \diagup C| \end{array}{}^{-} \quad H-B^{+}$$

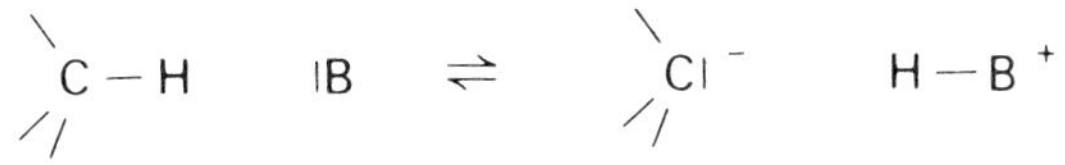

A compound containing an acidic hydrogen atom bound to carbon is hydrogen cyanide, HCN. This will react with a base to give the cyanide ion which contains a nucleophilic carbon atom.

$$HO^{-} + H-C\equiv N| \quad \rightleftharpoons \quad H_2O + |N\equiv C|^{-}$$

Alkyl halides react with sodium cyanide to form nitriles. For example 1-iodobutane (butyl iodide) reacts with sodium cyanide according to the equation . . . . (1).

---

(1)

$$|N\equiv C|^{-} \longrightarrow \begin{array}{c} H\ H \\ \diagdown | \\ C - I \\ \diagup \\ CH_3CH_2CH_2 \end{array} \longrightarrow \begin{array}{c} H \\ |N\equiv C-C\diagup{}^{H} \\ \diagdown \\ CH_2CH_2CH_3 \end{array} + \quad I^{-}$$

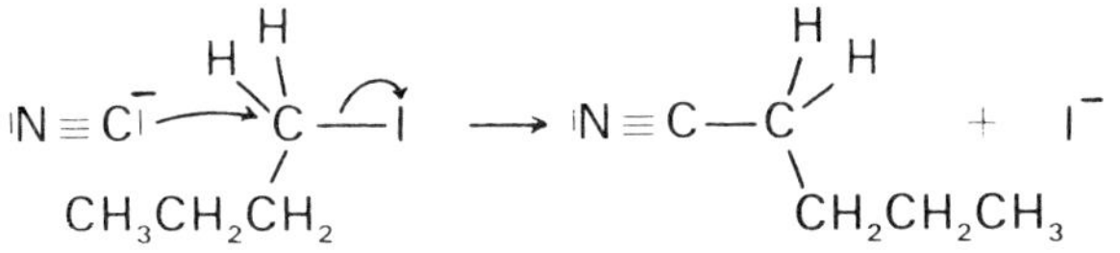

57. Acetylene is very weakly acidic and when added to the strong base sodium amide, $NaNH_2$, in liquid ammonia solution it is converted into sodium acetylide.

$$HC\equiv CH + \begin{array}{c} \overset{-}{N} \\ H\ \ H \end{array} + Na^{+} \rightleftharpoons HC\equiv C|^{-} + Na^{+} + \begin{array}{c} \overset{-}{N}\diagdown{}_{H} \\ H\ \ H \end{array}$$

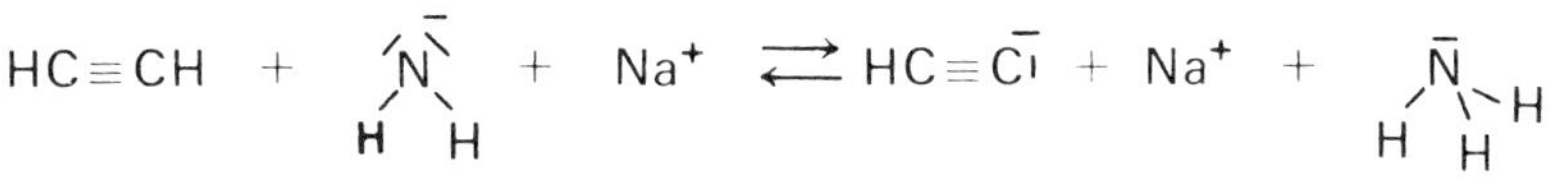

Alkyl halides react with sodium acetylide in liquid ammonia solution to form substituted acetylenes. Write the equation of the reaction of the acetylide ion with benzyl chloride . . . . (1).

---

(1)

$$HC\equiv C|^{-} \longrightarrow \begin{array}{c} H \\ H\diagdown | \\ C - Cl \\ \diagup \\ C_6H_5 \end{array} \longrightarrow \begin{array}{c} H \\ HC\equiv C-C\diagup{}^{H} \\ \diagdown \\ C_6H_5 \end{array} + Cl^{-}$$

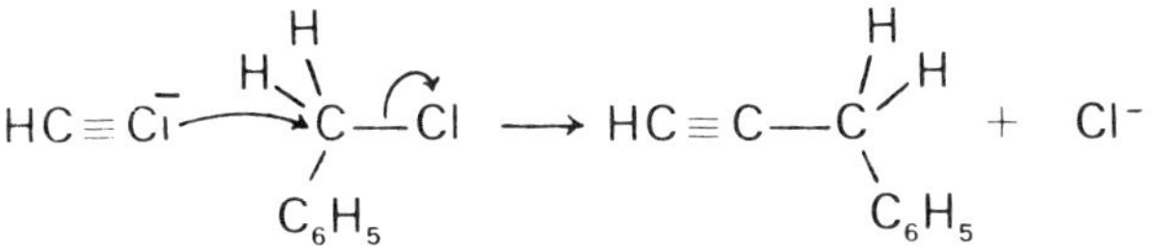

58. Similarly any alkyne having a terminal acetylene group can be converted into a substituted acetylide ion.

$$R-C\equiv C-H + NaNH_2 \rightleftharpoons R-C\equiv C^{-}Na^{+} + NH_3$$

(1) 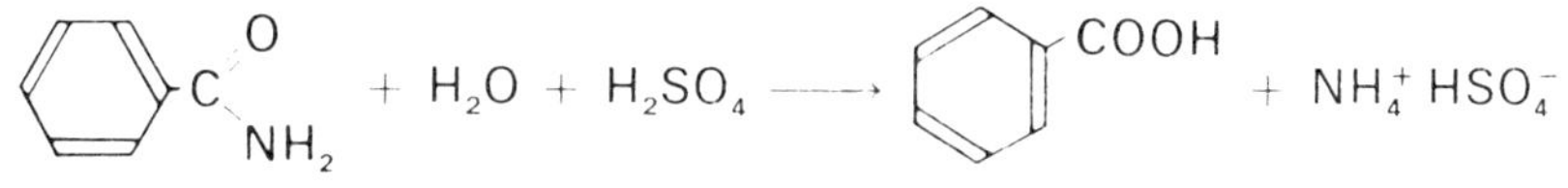

(2) $CH_3CH_2CONHCH_3 + NaOH \longrightarrow CH_3NH_2 + CH_3CH_2COO^- Na^+$

$$CH_3CH_2COO^- Na^+ \xrightarrow{\text{acid}} CH_3CH_2COOH$$

---

51. Amides are reduced by $LiAlH_4$ to give an amine with the same number of carbon atoms as the amides. Benzamide is reduced to .... (1) and N,N-dimethylpropanamide gives .... (2) (cf. Program 5, Frames 33-35).

---

(1) $C_6H_5CH_2NH_2$ benzylamine

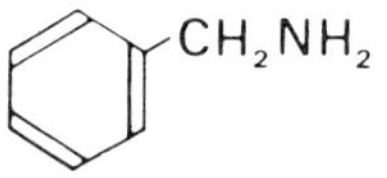

(2) 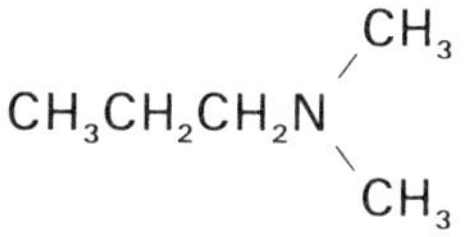

dimethylpropylamine

---

52. *Hofmann oxidation of amide to amine.* When a primary amide is added to a solution of bromine in aqueous sodium hydroxide at 0°, and the mixture is then heated rapidly to 90°, the hypobromite brings about an oxidation of the amide to amine and carbon dioxide:

$$RCONH_2 + 2NaOH + NaOBr \longrightarrow RNH_2 + Na_2CO_3 + NaBr + H_2O$$

or in shorthand form

$$RCONH_2 \xrightarrow[HO^-/BrO^-]{\text{oxidation with}} RNH_2 \text{ (cf. Program 5, Frame 38)}$$

Benzamide oxidised by alkaline sodium hypobromite gives .... (1) while benzamide reduced with lithium tetrahydridoaluminate gives ...... (2).

---

(1) $C_6H_5NH_2$ aniline (2) $C_6H_5CH_2NH_2$ benzylamine

---

53. A primary amide, $RCONH_2$, may be dehydrated with phosphorus pentoxide or phosphoryl chloride to yield a nitrile.

$$RCONH_2 \longrightarrow RCN + H_2O$$

Give the formula of the primary amide which, on dehydration, will give 2-cyanopropane .... (1), 3-methylbutanenitrile .... (2) and 3,5-dichlorocyanobenzene .... (3).

Using this information write a series of reactions for the synthesis of hex-2-yne from acetylene . . . . (1).

---

(1) $H-C\equiv C^-Na^+ + CH_3CH_2CH_2Br \longrightarrow CH_3CH_2CH_2C\equiv CH + NaBr$

$CH_3CH_2CH_2C\equiv C^-Na^+ + CH_3I \longrightarrow CH_3CH_2CH_2C\equiv CCH_3 + NaI$

The addition of the alkyl groups can be done in the reverse order.

$H-C\equiv C^-Na^+ + CH_3I \longrightarrow CH_3C\equiv CH + NaI$

$CH_3C\equiv C^-Na^+ + CH_3CH_2CH_2Br \longrightarrow CH_3CH_2CH_2C\equiv CCH_3 + NaBr$

---

59. Ethyl 3-oxobutanoate (acetoacetate), CH_3COCH_2COOEt, when added to a solution of sodium ethoxide in ethanol is converted into sodium ethyl acetoacetate.

$$CH_3COCH_2COOEt + NaOEt \rightleftharpoons CH_3CO\bar{C}HCOOEt\ Na^+ + EtOH$$

On addition of allyl chloride, $CH_2=CHCH_2Cl$, a nucleophilic substitution takes place according to the equation . . . . (1) (see also Program 12).

---

(1) $CH_3CO\bar{C}HCOOEt\ Na^+ + CH_2=CHCH_2Cl \rightarrow CH_3COCH(CH_2CH=CH_2)COOEt + NaCl$

---

60. Diethyl propanedioate (malonate), $EtOOCCH_2COOEt$ or $CH_2(COOEt)_2$, when added to a solution of sodium ethoxide in ethanol is converted into sodium diethyl malonate.

$$CH_2(COOEt)_2 + Na^+EtO^- \rightleftharpoons \bar{C}H(COOEt)_2\,Na^+ + EtOH$$

Addition of an alkyl halide results in attack by the nucleophilic carbanion. For example, addition of methyl iodide yields diethyl methylmalonate according to the equation . . . . (1) (see also Program 12).

---

(1) $\bar{C}H(COOEt)_2 + Na^+ + CH_3I \longrightarrow CH_3CH(COOEt)_2 + NaI$

Elimination reactions

61. Alkyl halides, when heated with a concentrated solution of potassium hydroxide in ethanol are converted into alkenes. When 3-bromopentane is treated under these conditions it yields mainly *trans*-pent-2-ene according to the equation . . . . (1). (For *cis*- and *trans*- isomers cf. Program 11).

---

(1) $CH_3CH_2CHBrCH_2CH_3 \xrightarrow[\text{in ethanol}]{\text{heat with KOH}} CH_3CH_2CH=CHCH_3$

(1) $R'NH_2$, a primary amine (2) $R'R''NH$, a secondary amine

---

48. Write equations for the synthesis of N-ethylbenzamide from benzoyl chloride .... (1), of N,N-dimethylacetamide from acetic anhydride .... (2) and of 2-methylpropanamide from ethyl 2-methylpropanoate .... (3). (cf. Program 5, Frames 16-18).

---

(1) $C_6H_5COCl + 2CH_3CH_2NH_2 \longrightarrow C_6H_5CONHCH_2CH_3 + CH_3CH_2NH_3^+ \; Cl^-$

(2) $(CH_3CO)_2O + 2(CH_3)_2NH \longrightarrow CH_3CON(CH_3)_2 + (CH_3)_2NH_2^+ CH_3COO^-$

(3) $(CH_3)_2CHCOOCH_2CH_3 + NH_3 \longrightarrow (CH_3)_2CHCONH_2 + CH_3CH_2OH$

---

49. Amides can sometimes be prepared by heating ammonium (or substituted ammonium) salts of the carboxylic acids. The water which forms in the reaction is removed by distillation. The procedure is not normally used since hydrolysis and volatilisation of the ammonium salt leads to considerable losses. The method can be used when both the acid and the base are relatively involatile, e.g. in the production of nylon from 1,6-diaminohexane and hexane-1,6-dioic acid. The equation for the reaction of a simple ammonium salt is

$$RCOO^- NH_4^+ \xrightarrow{heat} RCONH_2 + H_2O$$

N-methylacetamide, of structure .... (1) can be prepared from the .... (2) salt of .... (3) acid, i.e., ... (4).

---

(1) $CH_3-C(=O)-NHCH_3$ (2) methylammonium (3) acetic

(4) $CH_3-C(=O)-O^- \quad CH_3\overset{+}{N}H_3$

---

50. Amides are hydrolysed in the presence of acid to yield the free acid and an ammonium or amine salt

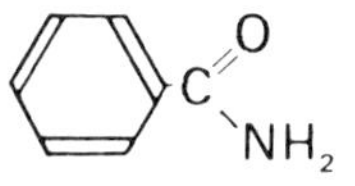

$+ H_2O + H_2SO_4 \longrightarrow$ .... (1).

Carboxylic acids are also obtained by reaction of amides with aqueous alkali. The reaction with base is followed by acidification to give the free acid

$$CH_3CH_2CONHCH_3 + NaOH \longrightarrow \text{............} \; (2).$$

---

62. The overall reaction results in the elimination of the elements of . . . . (1) from the 3-bromopentane.

---

(1) HBr, hydrogen bromide

---

63. A simple way of looking at this so-called β-elimination process is as follows: Rotation about the carbon-carbon bond permits a planar arrangement in which the $H-C_\beta$ bond is parallel to the $C_\alpha-Br$ bond.

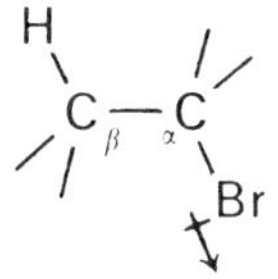

The electric field due to the dipolar carbon-bromine bond will attract the electrons of the $H-C_\beta$ bond and also direct the attacking anionic base into the region of the hydrogen atom. Removal of the proton by the base permits the pair of electrons of the $H-C_\beta$ bond to form a double bond provided that at the same time the bromine atom departs as a . . . . (1) ion. This process can be shown as follows:

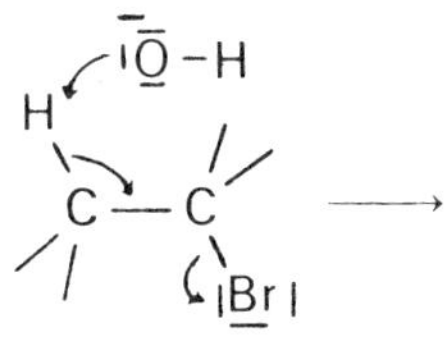

. . . . (2).

---

(1) bromide (2)

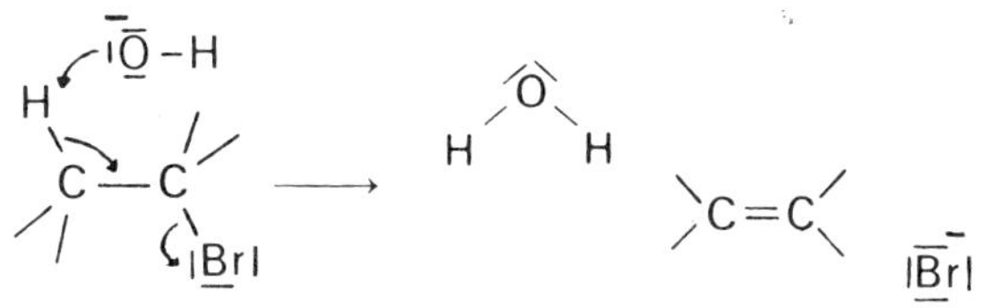

---

64. We could therefore write the equation for the elimination of hydrogen bromide from 3-bromopentane in the form:

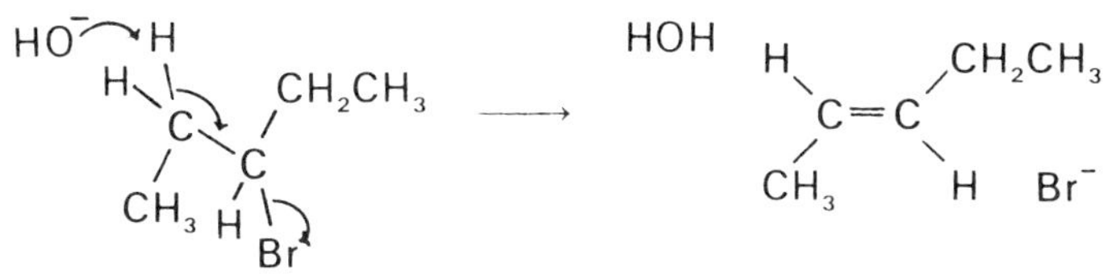

butanoate . . . . (2) and benzyl benzoate . . . . (3).
(cf. Program 6, Frame 44).

---

(1) ethanol and benzyl alcohol (2) methanol and butan-1-ol
(3) benzyl alcohol (2 moles)

---

44. Esters, e.g., RCO_2Me, react with excess of a Grignard reagent to yield tertiary alcohols of structure $R-\underset{R'}{\overset{R'}{C}}-OH$ in which the group R′ is derived from the Grignard reagent R′MgX. Reaction of phenylmagnesium bromide with ethyl propanoate yields a tertiary alcohol of structure . . . . (1).
(For further details of this reaction cf. Program 6, Frame 41).

---

(1) $CH_3CH_2\underset{C_6H_5}{\overset{C_6H_5}{C}}-OH$

---

45. *Amides* (primary). We have already encountered three main methods for preparing primary amides, namely the action of ammonia on . . . . (1), ... (2) and . . . . (3).

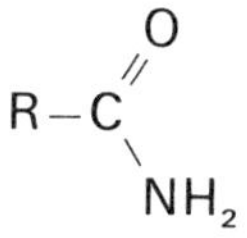

---

(1) acid chlorides (2) acid anhydrides (3) esters

---

46. Write a general equation for the synthesis of an acid amide from the corresponding acid chloride . . . . (1).

---

(1) $RCOCl + 2NH_3 \longrightarrow RCONH_2 + NH_4Cl$

---

47. The synthesis of secondary amides (I) and tertiary amides (II) will require the ammonia to be replaced by . . . . (1) and . . . . (2) respectively.

I
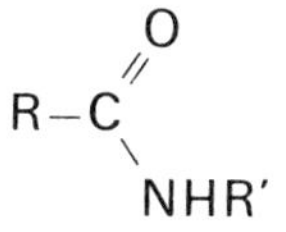
II
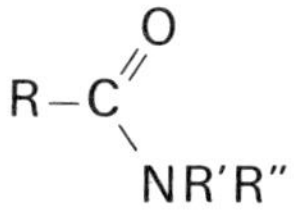

Because of the symmetry of 3-bromopentane the number of hydrogen atoms which can be attacked in this way to give the same product is . . . . (1).

---

(1) four

---

65. In 3-bromopentane and in chlorocyclohexane the symmetry of each molecule makes the hydrogen atoms on the vicinal carbon atoms equivalent and only one elimination product results in each case.
If 2-chlorobutane is allowed to react with potassium hydroxide in ethanol two elimination products are formed.

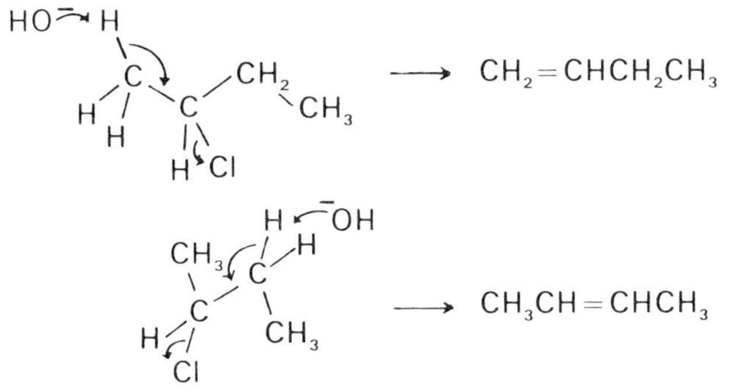

These products are named . . . . (1).

---

(1) but-1-ene and but-2-ene. The Saytzeff and Hofmann rules governing the nature of the principal product from such reactions are beyond the scope of this program.

Formation of Grignard reagents (cf. Program 6, Frames 33-41; Program 7, Frame 49; Program 9, Frame 18; and Program 12, Frame 1).

66. Alkyl halides (and aryl halides, cf. Program 9, Frames 18-20) react with magnesium metal in dry diethyl ether (EtOEt) to form Grignard reagents, the molecular formulae of which are written as RMgX for convenience. The reaction takes place only if the ether has been dried thoroughly, normally over metallic sodium, and the equation of the reaction is

$$RX + Mg \xrightarrow{\text{anhydrous ether}} RMgX$$

For example 1-bromobutane reacts with magnesium in dry ether (EtOEt) to give butylmagnesium bromide according to the equation

$$\ldots\ldots\ldots\ldots + Mg \xrightarrow{\text{EtOEt}} \ldots\ldots\ldots\ldots \quad (1)$$

---

40. Esters can be hydrolysed by acid or base. The acid-catalysed reaction, which has already been given is

$$RCOOR' + H_2O \rightleftharpoons \text{. . . . .} \ (1)$$

The equilibrium can be displaced to the right by the addition of a large amount of water.

---

(1) $RCOOR' + H_2O \rightleftharpoons RCOOH + R'OH$

---

41. The overall hydrolysis of an acid chloride by base includes the subsequent acidification to produce the free acid. Similarly the reaction of an ester with excess base yields the sodium salt of the acid. The free acid is obtained subsequently by acidifying with mineral acid. The reaction of an ester with a base to form a salt and an alcohol is known as saponification since this is the chemical reaction involved in the very old established process of soap-making. In the manufacture of soap, a fat, which is a triester of glycerol $RCOOCH_2CH(OCOR)CH_2OCOR$ is heated with sodium hydroxide solution and the sodium salts of the long chain aliphatic acids are isolated. The process may be represented by the equation

$$RCOOR' + NaOH \longrightarrow RCOO^-Na^+ + R'OH$$

Write the equation for the saponification of benzyl butanoate . . . . (1) followed by acidification of the salt and isolation of the acid . . . . (2).

---

(1) $CH_3CH_2CH_2COOCH_2C_6H_5 + NaOH \longrightarrow$
$CH_3CH_2CH_2COO^- Na^+ + HOCH_2C_6H_5$

(2) $CH_3CH_2CH_2COO^- Na^+ + HCl \longrightarrow CH_3CH_2CH_2COOH + NaCl$

---

42. Esters react with anhydrous ammonia to give the amide and alcohol. The reaction is normally done in alcoholic solution at room temperature.

$$RCOOR' + NH_3 \longrightarrow RCONH_2 + R'OH$$

Ethyl propanoate reacts with ammonia according to the equation . . . . (1).

---

(1) $CH_3CH_2COOEt + NH_3 \longrightarrow CH_3CH_2CONH_2 + EtOH$

---

43. Esters may also be reduced by $LiAlH_4$ and the products are two alcohol molecules, one corresponding to the carboxylic acid part, the other being the alcohol originally used to make the ester. Give the names of the alcohols produced on reduction of ethyl benzoate . . . . (1), methyl

(1) $CH_3CH_2CH_2CH_2Br + Mg \xrightarrow{EtOEt} CH_3CH_2CH_2CH_2MgBr$

---

67. When water or any weak or strong acid is added to a Grignard reagent the alkane is formed. The alkyl group in the reagent behaves as though it carries a pair of electrons to which a proton can add. In many other reactions Grignard reagents behave as powerful nucleophiles, confirming the suggestion that the alkyl group carries an available pair of electrons. That is, the carbon atom attached to the magnesium carries a partial negative charge and behaves as a carbanion. A solution of methyl magnesium iodide, CH_3MgI, in anhydrous ether can be used to detect a weak acid such as ethanol since the gas . . . . (1) is evolved according to the equation:

$$CH_3MgI + C_2H_5OH \longrightarrow \ldots\ldots\ldots + \ldots\ldots\ldots \quad (2)$$

---

(1) Methane (2) $CH_3MgI + C_2H_5OH \longrightarrow CH_4 + C_2H_5OMgI$

---

68. In dilute ethereal solution the Grignard reagent exists principally as RMgX but small concentrations of R_2Mg and MgX_2 are present due to the reversible equilibrium

$$2RMgX \rightleftharpoons R_2Mg + MgX_2$$

The Grignard reagent, RMgX, contains a carbon-magnesium bond and this is polarised in the sense $\overset{\delta-}{>C}-\overset{\delta+}{Mg}X$ which can be explained on the basis of mesomeric structures such as

$$\begin{array}{ccc} CH_3-Mg-I & \longleftrightarrow & CH_3^{-}\ {}^{+}Mg-I \\ \updownarrow & & \updownarrow \\ CH_3-Mg^{+}I^{-} & \longleftrightarrow & CH_3^{-}\ Mg^{2+}I^{-} \end{array}$$

In two of these contributing mesomers the structure CH_3^- is written. An ion containing a carbon atom carrying a negative charge is called a . . . . (1) and it behaves in chemical reactions as a . . . . (2).

---

(1) carbanion (2) nucleophile

---

69. Consider the electron balance in forming a Grignard reagent from metallic magnesium and methyl iodide. Magnesium is a divalent electropositive element and readily loses two electrons:

$$Mg(1s^2, 2s^2, 2p^6, 3s^2) \longrightarrow 2e^- + Mg^{2+}(1s^2, 2s^2, 2p^6)$$

We can now write the partial equation

$$CH_3-I + 2e^- \longrightarrow CH_3^- + I^-$$

and the sum of these equations is . . . . (1).

---

(1) $Mg + CH_3-I \longrightarrow CH_3^- + Mg^{2+} + I^-$ which is one of the mesomeric structures for the Grignard reagent shown in Frame 68.

37. Acid anhydrides react with ammonia to form amides and with alcohols to form esters giving in the case of a linear anhydride one molecule of the acid derivative and one molecule of the acid or its salt. Write the equation for the reaction of acetic anhydride with ammonia . . . . (1). In the case of a cyclic anhydride the liberated carboxyl group is retained in the structure. Write the equation for the reaction of phthalic anhydride with methanol . . . . (2).

---

(1) $(CH_3CO)_2O + 2NH_3 \longrightarrow CH_3CONH_2 + CH_3COO^- NH_4^+$

(2)

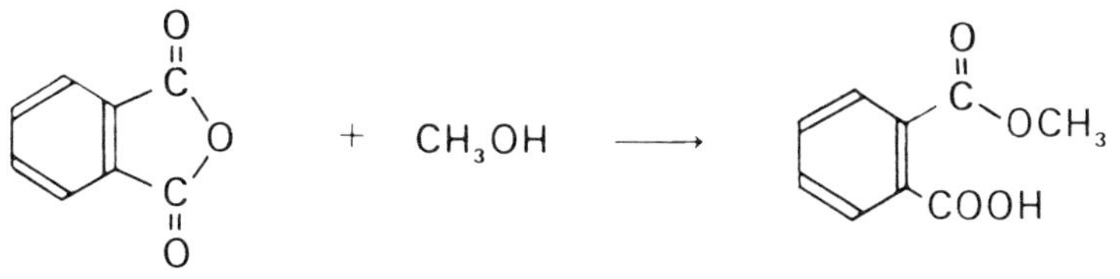

---

38. *Esters.* Esters are formed in the equilibrium reaction

$$RCOOH + R'OH \underset{}{\overset{acid}{\rightleftharpoons}} RCOOR' + H_2O$$

Appreciable amounts of both reactants and products are normally present at equilibrium but the equilibrium can be displaced in favour of ester formation if the concentration of one or other of the reactants i.e. . . . . (1) or . . . . (2) is increased, or if a product, e.g. the water, is continuously removed.

---

(1) RCOOH (2) R'OH

---

39. When it is not possible to obtain good yields of an ester by direct reaction of the carboxylic acid and alcohol, the ester is prepared from an acid chloride or acid anhydride according to the reactions already given. Write an equation for the formation of an ester using acetic anhydride and phenol (hydroxybenzene) . . . . (1), and for the preparation of benzyl benzoate using an acid chloride . . . . (2).

---

(1)

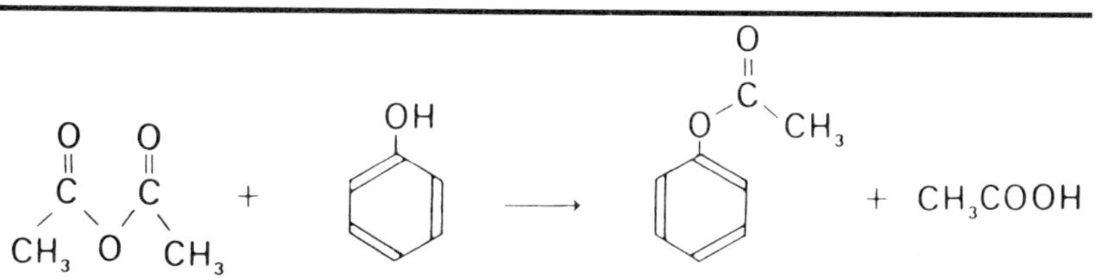

(2) $C_6H_5COCl + C_6H_5CH_2OH \longrightarrow C_6H_5COOCH_2C_6H_5 + HCl$

In this latter reaction a slight excess of base, e.g. NaOH, is usually added to neutralise the hydrochloric acid as it is formed, and the reaction is kept cold to minimise loss of benzoyl chloride by hydrolysis.

70. In Frame 67 the reduction of an alkyl halide to the corresponding alkane by way of the Grignard reagent and an acid was described. Alkyl halides can be reduced to alkanes by other metal/acid combinations, e.g., Zn/acetic acid or Na/ethanol. They can also be reduced catalytically, e.g. H_2/Pt, or by lithium aluminium hydride (lithium tetrahydridoaluminate).
Reduction may be considered as the transfer of electrons from the reducing agent to the reductant. How many electrons are transferred per molecule of RX in the reduction $RX \xrightarrow{\text{Zn/acetic acid}} RH$? . . . . (1).

---

(1) two; thus $RX + 2e^- + H^+ \longrightarrow RH + X^-$

TEST FRAMES
Give the names of the following compounds and write the equations for the formation of each from the corresponding alcohol. Give alternative reagents in each equation if more than one is available.

(1) CH_3I (2) $CH_3CH_2CHBrCH_3$ (3) $C_6H_5CH_2Cl$

---

(1) iodomethane or methyl iodide $CH_3OH \xrightarrow[\text{or P + I}_2]{\text{HI}} CH_3I$

(2) 2-bromobutane or s-butyl bromide

$$CH_3CH_2CHOHCH_3 \xrightarrow[\text{or NaBr + H}_2\text{SO}_4]{\text{HBr or PBr}_3} CH_3CH_2CHBrCH_3$$

(3) benzyl chloride $C_6H_5CH_2OH \xrightarrow{\text{HCl or PCl}_3} C_6H_5CH_2Cl$

Give the structures and names of the expected products in the following reactions:
(1) hydrogen chloride is reacted with hept-3-ene
(2) A sample of 2-methylbut-2-ene is purified until free of peroxides and then allowed to react with hydrogen bromide
(3) Hydrogen bromide is allowed to react with an impure sample of 2-methylbut-2-ene. Give reasons for your answer in (2) and (3) . . . . (4).

---

(1) $CH_3CH_2CHClCH_2CH_2CH_2CH_3$ and $CH_3CH_2CH_2CHClCH_2CH_2CH_3$
3-chloroheptane 4-chloroheptane

(2) 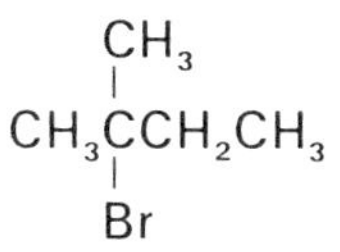

2-bromo-2-methylbutane

(3) 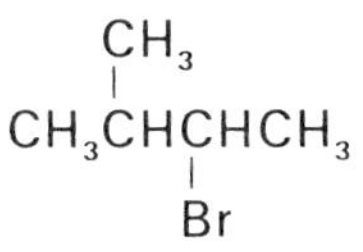

3-bromo-2-methylbutane

(1) $CH_3CH_2CH_2OH$
propan-1-ol

(2) CH_3CH_2CHO
propanal, propionaldehyde

Acid anhydrides

33. Acid anhydrides also react with nucleophiles but are less reactive than the acid chlorides. By far the most frequently encountered anhydride is acetic anhydride . . . . (1) which is prepared commercially by a process which is overall, a dehydration of acetic acid.

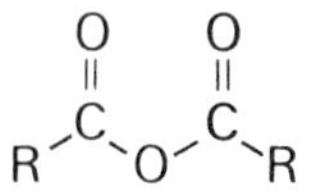

---

(1)

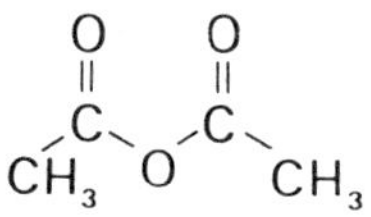

---

34. The anhydrides which can be formed from some dicarboxylic acids are cyclic structures since both the carboxyl groups are present in one molecule. Provided the anhydride ring contains five or six atoms, an anhydride may be formed by heating a dicarboxylic acid. Benzene-1,2-dicarboxylic acid (phthalic acid) is converted into phthalic anhydride of structure . . . . (1).

---

(1)

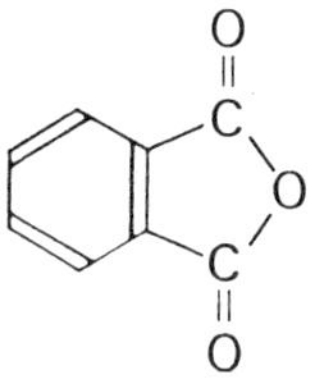

---

35. Acid anhydrides can also be formed by reaction of an acid chloride with the sodium salt of the acid. Write the equation for the preparation of propanoic anhydride from propanoyl chloride and sodium propanoate . . . . (1).

---

(1) $CH_3CH_2COCl + CH_3CH_2COONa \longrightarrow CH_3CH_2\overset{\overset{\displaystyle O}{\|}}{C}-O-\overset{\overset{\displaystyle O}{\|}}{C}CH_2CH_3 + NaCl$

---

36. Acid anhydrides react with water in the presence of acid or excess base to regenerate the carboxylic acids from which they are formed. The hydrolysis of acetic anhydride proceeds according to the equation . . . . (1).

---

(1) $CH_3\overset{\overset{\displaystyle O}{\|}}{C}-O-\overset{\overset{\displaystyle O}{\|}}{C}CH_3 + H_2O \longrightarrow 2CH_3COOH$

(4) In the absence of peroxides the addition of hydrogen bromide obeys the Markovnikov rule. The mode of addition is reversed in the presence of peroxides.

Give the reagents that can be used for the conversion of 1-chlorohexane into 1-fluorohexane . . . . (1), and 2-bromobutane into 2-iodobutane . . . . (2). Write an equation for a possible mechanism for the conversion of methyl chloride into methanol . . . . (3).

---

(1) HgF_2 (2) NaI in acetone

(3)

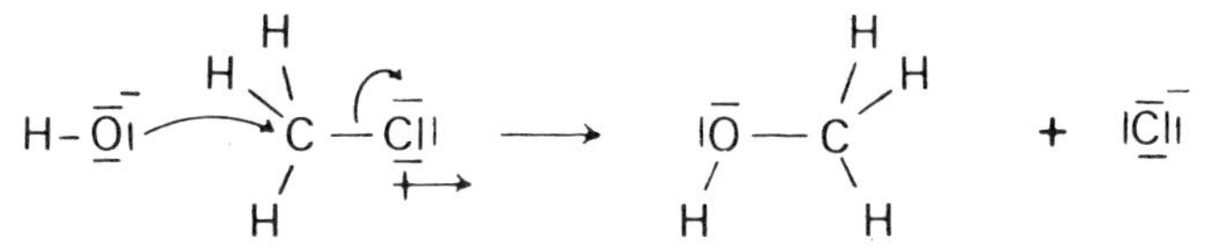

Starting from propan-1-ol and any inorganic reagent necessary, outline preparations of the following compounds using several steps if necessary.

(1) di-n-propyl ether
(2) propene
(3) propane-2-thiol

---

(1) $CH_3CH_2CH_2OH \xrightarrow{HCl} CH_3CH_2CH_2Cl$

$CH_3CH_2CH_2OH + Na \longrightarrow CH_3CH_2CH_2ONa + \frac{1}{2}H_2$

$CH_3CH_2CH_2ONa + CH_3CH_2CH_2Cl \longrightarrow (CH_3CH_2CH_2)_2O + NaCl$

(2) $CH_3CH_2CH_2Cl \xrightarrow[\text{ethanol}]{\text{KOH in}} CH_3CH{=}CH_2$

(3) $CH_3CH{=}CH_2 \xrightarrow{HCl} CH_3CH(Cl)CH_3$

$CH_3CH(Cl)CH_3 + NaSH \longrightarrow CH_3CH(SH)CH_3$

Suggest syntheses of the following compounds starting from 1-bromobutane

(1) pentanenitrile
(2) dec-5-yne
(3) $CH_3COCH(CH_2CH_2CH_2CH_3)COOC_2H_5$

---

29. All the carboxylic acid derivatives can be hydrolysed to give the acid from which they were formed. The hydrolysis of an acid chloride is catalysed by acid and is a vigorous reaction.

$$CH_3COCl + H_2O \xrightarrow{\text{acid}} \ldots\ldots\ldots\ldots \; (1)$$

Acid chlorides also react vigorously with base.

$$RCOCl + 2NaOH \longrightarrow \ldots\ldots\ldots\ldots \; (2)$$

On acidification the free acid is obtained.

---

(1) $CH_3COCl + H_2O \xrightarrow{\text{acid}} CH_3COOH + HCl$

(2) $RCOCl + 2NaOH \longrightarrow RCOO^- + Cl^- + 2Na^+ + H_2O$

---

30. The reaction of an acid chloride with ammonia gives rise to an amide, $RCONH_2$. As an example, acetyl chloride reacts with ammonia according to the equation . . . . (1).

---

(1) $CH_3COCl + 2NH_3 \longrightarrow CH_3CONH_2 + NH_4Cl$

---

31. Esters may be obtained by the reaction of acid chlorides with alcohols, according to the general equation

$$R{-}C(=O){-}Cl + R'OH \longrightarrow R{-}C(=O){-}OR' + HCl$$

Rather than perform the reaction under conditions where hydrogen chloride is evolved, it is usual to add a suitable base, e.g. a tertiary amine, in order to neutralise the liberated acid.
Thus methyl benzoate can be prepared by the reaction of . . . . (1) and . . . . (2) according to the equation . . . . (3).

---

(1) benzoyl chloride (2) methanol

(3) $C_6H_5COCl + CH_3OH \xrightarrow{Et_3N} C_6H_5COOCH_3 + Et_3NHCl$

---

32. Acid chlorides are reduced by $LiAlH_4$ to primary alcohols. The reduction stops at the aldehyde stage when hydrogenation is carried out with a special catalyst or when lithium hydridotri-t-butoxyaluminate, $LiAlH(O\text{-}t\text{-}Bu)_3$ or $LiAlH(O\text{-}t\text{-}C_4H_9)_3$, is used. Propanoyl chloride is reduced by $LiAlH_4$ to . . . . (1) and by $LiAlH(O\text{-}t\text{-}Bu)_3$ to . . . . (2).

---

(1) $CH_3CH_2CH_2CH_2Br + NaCN \longrightarrow CH_3CH_2CH_2CH_2CN + NaBr$

(2) $HC{\equiv}CH + NaNH_2 \rightleftharpoons HC{\equiv}C^-Na^+ + NH_3$

$HC{\equiv}C^-Na^+ + CH_3CH_2CH_2CH_2Br \longrightarrow CH_3(CH_2)_3C{\equiv}CH$

$CH_3(CH_2)_3C{\equiv}C^-Na^+ + CH_3(CH_2)_3Br \longrightarrow CH_3(CH_2)_3C{\equiv}C(CH_2)_3CH_3$

(3) $CH_3COCH_2COOC_2H_5 + NaOC_2H_5 \rightleftharpoons CH_3CO\bar{C}HCOOC_2H_5 + Na^+ + HOC_2H_5$

$$CH_3CO\bar{C}HCOOC_2H_5\ Na^+ + CH_3(CH_2)_3Br \longrightarrow \begin{array}{l} CH_3CO\underset{|}{C}HCOC_2H_5 \\ \qquad\ \ CH_2CH_2CH_2CH_3 \end{array}$$

The reaction of trimethylamine with benzyl chloride gives benzyltrimethyl-ammonium chloride

$$(CH_3)_3N + C_6H_5CH_2Cl \longrightarrow C_6H_5CH_2N^+(CH_3)_3\ Cl^-$$

Using this as an analogy, and remembering that phophorus, like nitrogen, is in Group 5 of the periodic table, predict the consequence of heating methyl iodide with triphenylphosphine ($(C_6H_5)_3P$) . . . . (1).

$(C_6H_5)_3P + CH_3I \longrightarrow (C_6H_5)_3P^+CH_3\ I^-$

methyltriphenylphosphonium iodide

---

27. Carboxylic acids give rise to the following important derivatives, where R can be either an alkyl or aryl group.

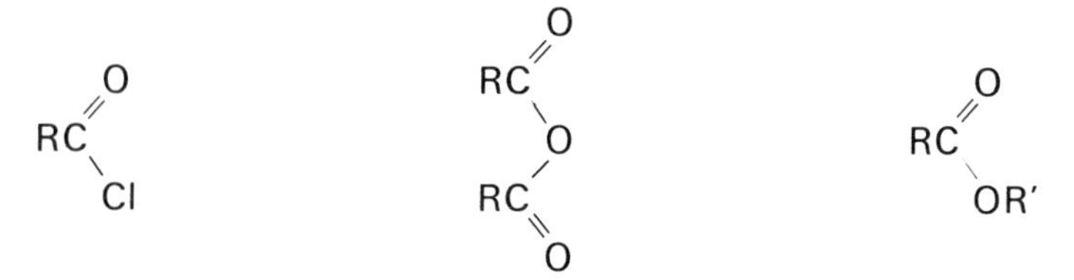

acid chlorides *acid anhydrides* *esters*

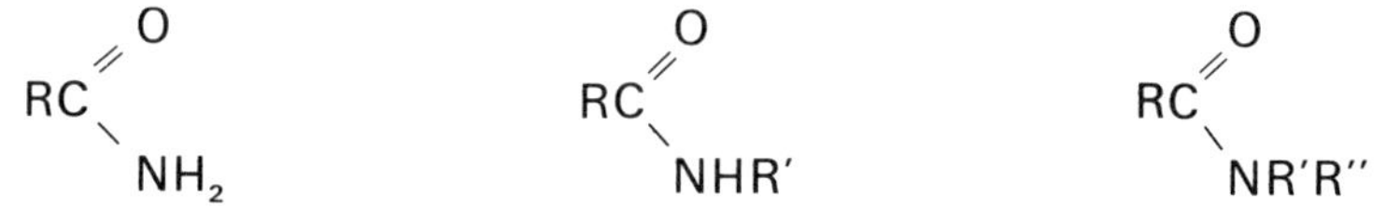

primary amides *secondary amides* *tertiary amides*

Name the following compounds:

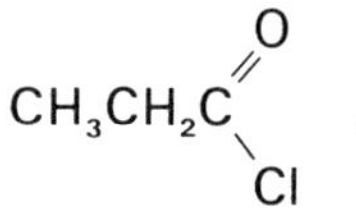 .... (1) 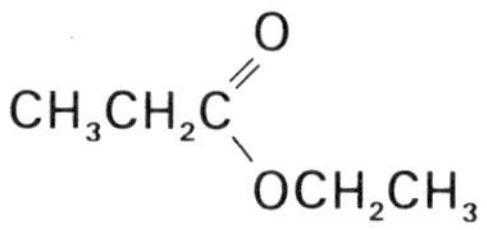 .... (2)

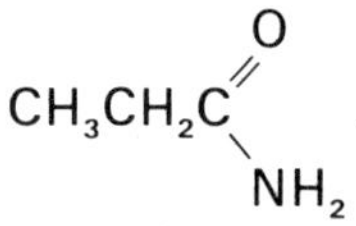 .... (3) 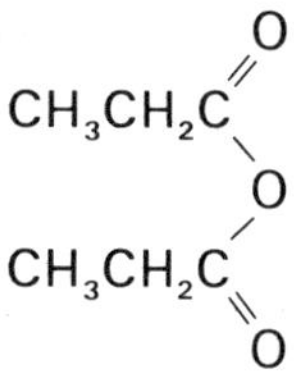 .... (4)

---

(1) propanoyl chloride or propionyl chloride
(2) ethyl propanoate or ethyl propionate
(3) propanamide or propionamide
(4) propanoic anhydride or propionic anhydride

---

28. *Acid Chlorides.* Acid chlorides are the most reactive of the common carboxylic acid derivatives, reacting vigorously with nucleophiles. They are prepared from the corresponding carboxylic acids by reaction with PCl_5 (phosphorus pentachloride) or $SOCl_2$ (thionyl chloride). Frequently thionyl chloride (b.p. 79°) is more convenient to use because the by-products, sulphur dioxide and hydrogen chloride, are both gases. Phosphoryl chloride ($POCl_3$), which is produced in the reaction of a carboxylic acid with PCl_5, has b.p. 107°. Give equations for the preparation of benzoyl chloride using thionyl chloride .... (1) and phosphorus pentachloride .... (2).

---

(1) $C_6H_5COOH + SOCl_2 \longrightarrow C_6H_5COCl + SO_2 + HCl$
(2) $C_6H_5COOH + PCl_5 \longrightarrow C_6H_5COCl + POCl_3 + HCl$

Program 5: ALIPHATIC AMINES

CONTENTS

23. *Decarboxylation*. Carboxylic acids when heated strongly with soda lime (NaOH + CaO) lose carbon dioxide.

$$RCOOH \longrightarrow RH + CO_2 \text{ (absorbed by the soda lime)}$$

Under these conditions benzoic acid decarboxylates to give sodium carbonate and . . . . (1).

---

(1) benzene

---

24. The ease with which a carboxylic acid can be decarboxylated is much influenced by the presence of other functional groups in the molecule. 2,4,6-Trinitrobenzoic acid is decarboxylated on being heated in aqueous acid and the products are . . . . (1).

---

(1) 1,3,5-trinitrobenzene 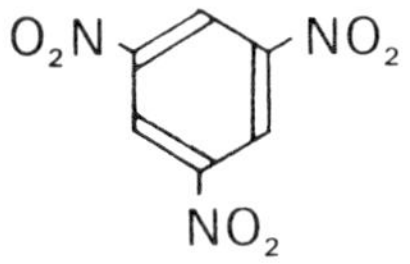 and CO_2

---

25. One mole of carbon dioxide is readily lost from carboxylic acids which have two carboxyl groups separated by one carbon atom. Thus $HOOCCH_2COOH$ on being heated gives . . . . (1).
Similarly, decarboxylation occurs readily when a carbonyl group is separated from the carboxyl group by one carbon atom.
CH_3COCH_2COOH on being heated gives . . . . (2).

---

(1) $HOOCCH_2COOH \longrightarrow CH_3COOH + CO_2$
(2) $CH_3COCH_2COOH \longrightarrow CH_3COCH_3 + CO_2$
For further details see Program 12.

---

26. *Reduction*. A carboxylic acid may be reduced to the corresponding primary alcohol on reduction with lithium tetrahydridoaluminate ($LiAlH_4$).

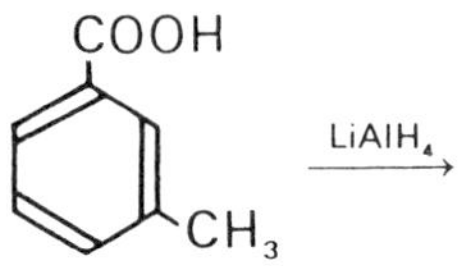

. . . . (1) Give structure and name.

---

(1) 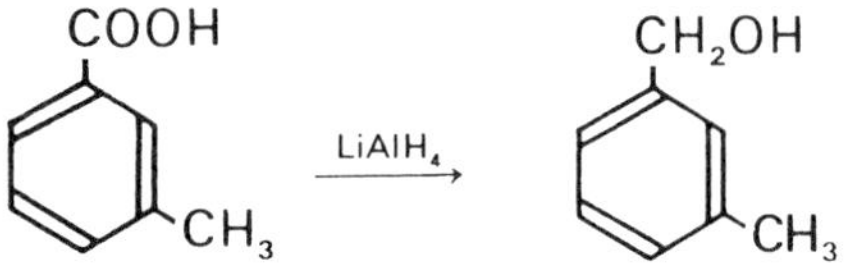

m-methylbenzyl alcohol

ALIPHATIC AMINES

Program 5

NOMENCLATURE AND PROPERTIES

1. The amines are organic (hydrocarbon) derivatives of NH_3. *Primary* alkyl amines RNH_2 have one alkyl group, *secondary* alkyl amines have . . . . (1) such alkyl groups, while *tertiary* amines have three such groups. The general structure for a secondary amine can be written . . . . (2) and for a tertiary amine . . . . (3).

---

(1) two

(2) 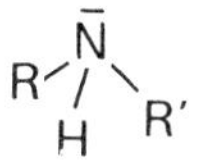

(3) 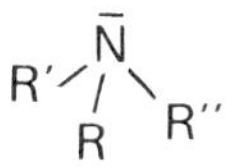

---

2. Amines are best named by specifying, in alphabetical order, the alkyl groups attached to nitrogen. Thus CH_3NH_2 is methylamine;

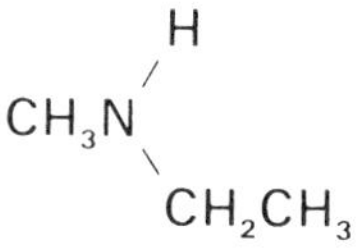

is . . . . (1),

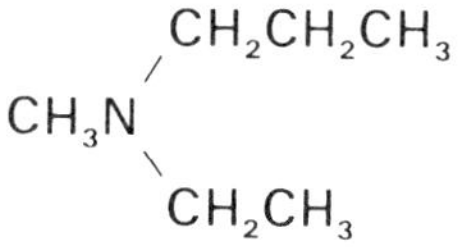

is . . . . (2).

The structure of triethylamine is . . . . (3).

---

(1) ethylmethylamine

(2) ethylmethylpropylamine or 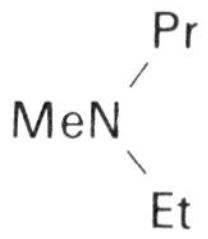

(3) 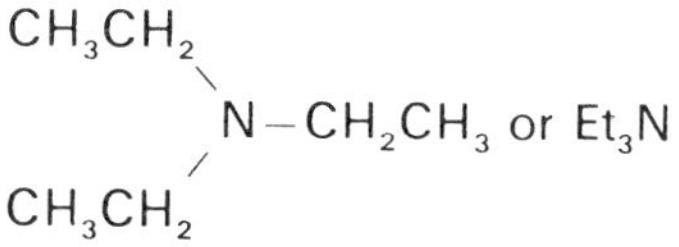

---

3. As a prefix, CH_3NH- is named methylamino, so that $CH_3NHCH_2CO_2H$, named as a derivative of ethanoic acid, CH_3CO_2H, is called . . . . (1). In general the substituent $RNH-$ is alkylamino and R_2N- is . . . . (2).

---

(1) methylaminoethanoic acid (2) dialkylamino

carboxylic acid, RCOOH. The amide is formed when a solution of a nitrile in cold concentrated sulphuric acid is poured into water.

$$RCN + H_2O \xrightarrow{H_2SO_4} RCONH_2$$

If the amide is heated in 6M hydrochloric acid, the carboxylic acid is produced according to the equation

$$RCONH_2 + H_3O^+ \longrightarrow \text{. . . . .} \quad (1)$$

---

(1) $RCONH_2 + H_3O^+ \longrightarrow RCOOH + NH_4^+$

TEST FRAME

Carboxylic acids can be prepared by the following methods:

1. Oxidation of primary alcohols and aldehydes
2. Oxidation of alkyl chains attached to benzene rings
3. Reaction of a Grignard reagent with carbon dioxide
4. Hydrolysis of nitriles
5. The preparation of carboxylic acids from ethyl 3-oxobutanoate (aceto-acetate) and from diethyl propanedioate (malonate) is discussed in Program 12.

Write equations for the synthesis of benzoic acid by methods 1 to 4.

---

(1) $C_6H_5CH_2OH \xrightarrow{\text{acid } KMnO_4} C_6H_5COOH$
benzyl alcohol

and

$C_6H_5CHO \xrightarrow{K_2Cr_2O_7} C_6H_5COOH$
benzaldehyde

(2) $C_6H_5R \xrightarrow[\text{heat}]{\text{alkaline } KMnO_4} C_6H_5COOH$ where R is any alkyl group.

(3) $C_6H_5MgBr + CO_2 \longrightarrow C_6H_5COO^- + Mg^{2+} + Br^-$
$C_6H_5COO^- + H_3O^+ \longrightarrow C_6H_5COOH + H_2O$

(4) $C_6H_5CN + 2H_2O \xrightarrow[\text{or base}]{\text{acid}} C_6H_5COOH + NH_3$

REACTIONS OF CARBOXYLIC ACIDS

22. *Salt formation*. The formation of salts has already been referred to in connection with the acidity of the carboxylic acids. Salts can be formed, not only with the common inorganic bases but also with amines. Write the structure of calcium acetate . . . . (1) and trimethylammonium butyrate . . . . (2).

---

(1) $(CH_3COO^-)_2Ca^{2+}$ (2) $CH_3CH_2CH_2COO^- \ (CH_3)_3NH^+$

Salt formation and nucleophilic character

4. Ammonia has basic properties and forms salts with acids,

e.g., $$NH_3 + H_3O^+ \rightleftharpoons NH_4^+ + H_2O.$$

Its electronic structure is $H:\overset{H}{\underset{..}{N}}:H$ often shown as H_3N: or $H_3N|$, and its basic nature can be understood in terms of the .... (1) of electrons which readily form a fourth bond to an electron-deficient (acidic) atom or group.

---

(1) lone pair

---

5. Primary amines, RNH_2, secondary amines, R_2NH, and tertiary amines, R_3N, can all be regarded as simple derivatives of ammonia, so that it is not surprising that amines possess .... (1) properties.

---

(1) basic

---

6. Amines, being basic, form ionised .... (1) with acids.

$$C_2H_5NH_2 + H_3O^+ + Cl^- \longrightarrow C_2H_5NH_3^+\ Cl^- + H_2O$$

ethylammonium chloride

$$(CH_3)_3N + H_3O^+ + ClO_4^- \longrightarrow \text{............ (2)} + H_2O$$

---

(1) salts (2) $(CH_3)_3NH^+ClO_4^-$ trimethylammonium perchlorate

---

7. By analogy with the behaviour of ammonium chloride, treatment of dimethylpropylammonium chloride with sodium hydroxide would be expected to yield .... (1).

---

(1) dimethylpropylamine or $(CH_3)_2NCH_2CH_2CH_3$ or Me_2NPr

---

8. A nucleophilic reagent is one that either has a full negative charge, e.g., a hydroxide ion, HO^-, or a neutral molecule having a region of .... (1) electron density, most often a lone pair of .... (2).

---

(1) high (2) electrons

---

9. Ammonia and amines are nucleophilic substances, and many of their chemical reactions are a consequence of the presence of a lone pair of electrons on the .... (1) atom. The formation of salts of amines is one such reaction.

If we consider the reaction $HO^- + O{=}C{=}O \longrightarrow HO{-}C(=O)O^-$ and think of R^- as a nucleophile, analogous to OH^-, then the reaction of a Grignard reagent with carbon dioxide can be written as:

$$R^-Mg^{2+}X^- + O{=}C{=}O \longrightarrow \text{. . . . . (1)}$$

---

(1) $R^-Mg^{2+}X^- + O{=}C{=}O \longrightarrow R{-}C(=O)O^- + Mg^{2+} + X^-$

---

18. The reaction above gives rise to the . . . . (1) of the acid and in order to obtain the free acid it is necessary to . . . . (2).

---

(1) anion (2) add strong acid

---

19. Give the equations for the reactions involved in the conversion of bromobenzene into phenylmagnesium bromide and then into benzoic acid . . . . (1).

---

(1) $C_6H_5Br + Mg \xrightarrow{\text{dry ether}} C_6H_5MgBr$

$$C_6H_5MgBr + CO_2 \longrightarrow C_6H_5COO^- + Mg^{2+} + Br^-$$

$$C_6H_5COO^- + H_3O^+ \longrightarrow C_6H_5COOH + H_2O$$

---

20. Which Grignard reagent would be required for the preparation of benzoic acid . . . . (1) and of cyclohexanecarboxylic acid . . . . (2)?

---

(1) C_6H_5MgX where X is Cl, Br or I

(2) 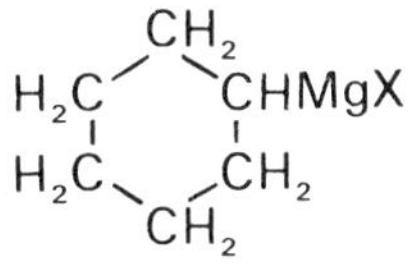

---

21. Carboxylic acids are also obtained by the hydrolysis of nitriles which contain the functional group $-C{\equiv}N$. The hydrolysis of a nitrile in the presence of acid or base is represented by the equation $RCN + 2H_2O \longrightarrow RCOOH + NH_3$. The preparation of nitriles is discussed later in this Program. A nitrile, RCN, can be hydrolysed either to the corresponding amide, $RCONH_2$, or to the

(1) nitrogen

REACTION OF AMINES WITH ALKYL HALIDES

10. We can now consider the reaction of amines with alkyl halides. An alkyl halide such as ethyl bromide, CH_3CH_2Br, has a permanent dipole moment. This is symbolised by $CH_3CH_2-\overset{+\!\!\longrightarrow}{Br}$, as explained in Program 4. The electric vector acts along the line of the carbon-bromine bond, with its positive end near to the bromine nucleus. An amine has a similar lone-pair moment, shown $R_3\overset{+\!\!\longrightarrow}{N}$, and in any collision between an alkyl halide and an amine, the electrostatically most favourable situation will be for the amine to approach the carbon atom which holds the bromine atom *along the line of the carbon-bromine bond* and *from the side opposite to the bromine atom*. For such a collision the dipole orientation can be shown . . . . (1).

---

(1) $+\!\!\longrightarrow$ $+\!\!\longrightarrow$ or $\oplus$——$\ominus$ $\oplus$——$\ominus$

---

11. It is only possible for the new nitrogen-carbon bond to become established if at the same time the carbon-bromine bond is broken. This allows retention of eight electrons around the carbon atom (octet rule). For reaction of a primary amine with ethyl bromide we can therefore write

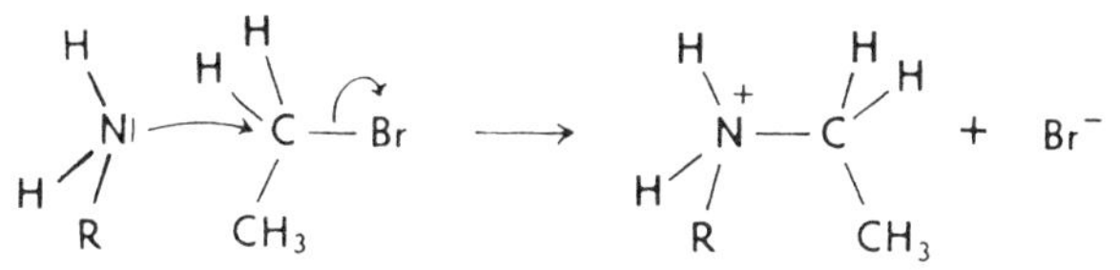

The product is the protonated form (conjugate acid) of a . . . . (1) amine, and the arrows symbolise movements of pairs of electrons.

---

(1) secondary

---

12. Reaction of a secondary amine, R_2NH, with an alkyl halide proceeds in similar fashion, to give the conjugate acid of a . . . . (1) amine.

---

(1) tertiary

---

13. Finally, a tertiary amine R_3N can react with an alkyl halide R′X to give a salt of structure . . . . (1), and such a substance is known as a "quaternary ammonium salt".

---

(1) $R_3NR'^+X^-$

butylbenzene . . . . (1), 4-chloro-1,3-diethylbenzene . . . . (2), *p*-methylbenzaldehyde . . . . (3) and *p*-nitrotoluene . . . . (4).

---

(1) benzoic acid

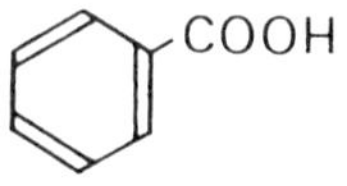

(2) 4-chlorobenzene-1,3-dicarboxylic acid

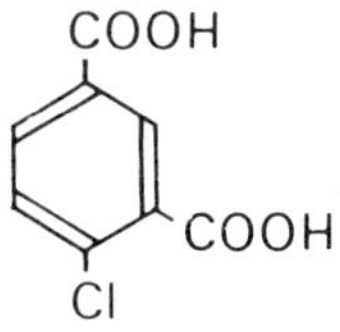

(3) benzene-1,4-dicarboxylic acid, terephthalic acid

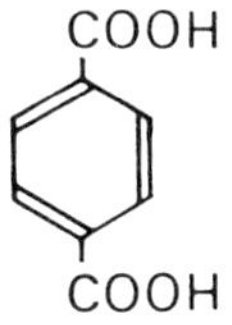

(4) *p*-nitrobenzoic acid

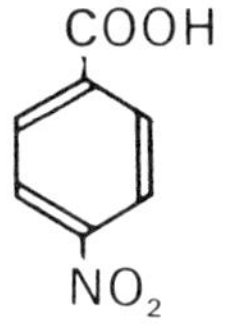

---

15. The production of terephthalic acid from p-methylbenzaldehyde involved both the oxidation of an alkyl group and the oxidation of an aldehyde. Aldehydes are oxidisable by very mild oxidising agents whereas the oxidation of an alkyl side chain requires vigorous oxidation. The product of mild oxidation of *p*-methylbenzaldehyde is . . . . (1).

---

(1) *p*-methylbenzoic acid or *p*-toluic acid

---

16. Sodium benzoate can be prepared by alkaline hydrolysis of phenyltrichloromethane (benzotrichloride, $C_6H_5CCl_3$).

$$C_6H_5CCl_3 + 4NaOH \longrightarrow \ldots\ldots\ldots\ldots \quad (1)$$

Acidification then yields benzoic acid. The trichloro compound is obtained by oxidation of toluene with chlorine in the presence of light.

$$C_6H_5CH_3 + 3Cl_2 \longrightarrow C_6H_5CCl_3 + 3HCl$$

---

(1) $C_6H_5CCl_3 + 4NaOH \longrightarrow C_6H_5COONa + 3NaCl + 2H_2O$

---

17. The oxidation of primary alcohols and aldehydes is a method of preparation of both aliphatic and aromatic carboxylic acids. The oxidation of alkyl side chains is a method for preparing aromatic acids. Both aliphatic and aromatic acids are prepared by the reaction of Grignard reagents with carbon dioxide. Grignard reagents are generally represented as RMgX where X is Cl, Br or I. The Grignard reagent functions chemically as if it were $R^-Mg^{2+}X^-$.

14. The reaction of primary and secondary amines with alkyl halides is rather more complex than was indicated above. We saw for example that a primary amine gives as first product a . . . . (1) amine hydrohalide salt. But as soon as some of this salt appears, it will engage in an acid-base equilibrium with unchanged primary amine, with formation of some primary amine hydrohalide salt and liberation of some free secondary amine. This latter substance can then compete for alkyl halide, and react to give some . . . . (2) amine salt.

---

(1) secondary (2) tertiary

---

15. The tertiary amine salt can in turn become converted, at least in part, into a . . . . (1). Hence in the reaction of primary (and secondary) amines with alkyl halides, a mixture of different products is obtained.

---

(1) quaternary ammonium salt

REACTION OF AMINES WITH ACID DERIVATIVES—FORMATION OF AMIDES

16. Primary and secondary amines, which contain the functional groups . . . . (1) and . . . . (2) respectively, react as nucleophiles towards a variety of carboxylic acid derivatives, in particular acid chlorides, acid anhydrides and esters. Reaction of ethanoyl chloride (acetyl chloride) with excess ammonia at room temperature gives ethanamide (acetamide) according to the equation:

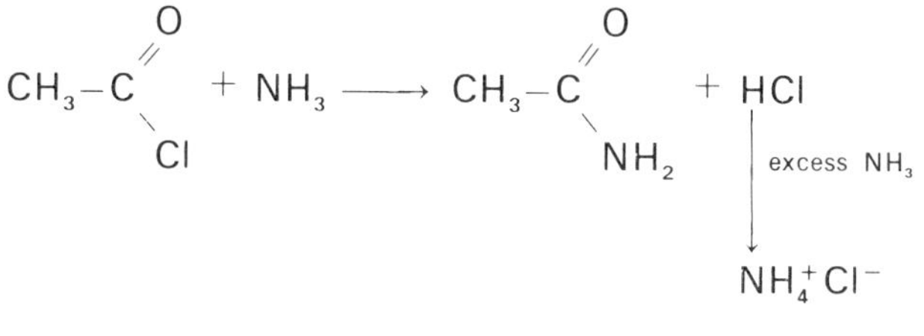

In the same way, ethylamine reacts vigorously with acetyl chloride to yield an amide of structure . . . . (3), while dimethylamine reacts with butanoyl chloride to yield an amide of structure . . . . (4).

---

(1) $-NH_2$ (2) $-\overset{|}{N}H$ (3) $CH_3C(=O)NHC_2H_5$ (4) $CH_3CH_2CH_2C(=O)N(CH_3)_2$

(1) addition of strong mineral acid　(2) $C_6H_5COO^-$ benzoate ion

(3) aqueous

PREPARATION OF CARBOXYLIC ACIDS

11. Carboxylic acids can be prepared by the oxidation of primary alcohols and aldehydes using reagents such as acid permanganate or dichromate, i.e.

$$RCH_2OH \xrightarrow{oxidation} RCOOH; \qquad RCHO \xrightarrow{oxidation} RCOOH$$

Acetic acid can be prepared by the oxidation of both . . . . (1) and . . . . (2). Oxidation of 2-methylbutanal gives . . . . (3).

---

(1) ethanol, CH_3CH_2OH　(2) ethanal, CH_3CHO

(3) 2-methylbutanoic acid, $CH_3CH_2CH(CH_3)COOH$

---

12. Carboxylic acids, with the exception of formic acid, are resistant to further oxidation except under very vigorous conditions. Formic acid, HCOOH, contains the arrangement H–C=O (C also bonded to a further group) which is also found in . . . . (1). Formic acid is oxidised readily to give . . . . (2) and . . . . (3).

---

(1) aldehydes　(2) CO_2　(3) H_2O

---

13. The oxidation of alkanes cannot normally be limited to produce aliphatic acids. However, alkyl groups attached to benzene rings can be oxidised by alkaline permanganate to carboxyl groups, while the benzene ring remains intact. Oxidation of toluene (methylbenzene) gives . . . . (1) and oxidation of o-dimethylbenzene gives . . . . (2).

---

(1) 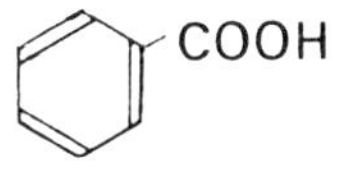

benzoic acid

(2) 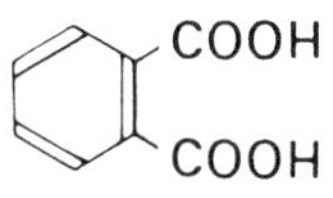

benzene-1,2-dicarboxylic acid or phthalic acid

---

14. An alkyl side chain of any length, attached to a benzene ring, is oxidised to a carboxyl group and the remainder of the alkyl chain is lost. Give the products of vigorous oxidation of:

17. Ethanoic anhydride (acetic anhydride), $CH_3\overset{\displaystyle O}{\overset{\|}{C}}-O-\overset{\displaystyle O}{\overset{\|}{C}}CH_3$ likewise reacts easily with primary or secondary amines. The amide obtained with, for example, a secondary amine, is identical with that which would be formed from acetyl chloride, so that the equation can be written

$$CH_3\overset{\displaystyle O}{\overset{\|}{C}}-O-\overset{\displaystyle O}{\overset{\|}{C}}CH_3 + 2RR'NH \longrightarrow \text{............} \quad (1)$$

$$(1)\ CH_3\overset{\displaystyle O}{\overset{\|}{C}}-O-\overset{\displaystyle O}{\overset{\|}{C}}CH_3 + 2RR'NH \longrightarrow CH_3\overset{\displaystyle O}{\overset{\|}{C}}NRR' + CH_3COO^- \quad RR'NH_2^+$$

---

18. Amines also yield amides on being heated with esters:

$$\text{e.g.,} \qquad RC\begin{matrix}\nearrow\!\!\!/\ O \\ \searrow OC_2H_5\end{matrix} + R'NH_2 \longrightarrow RC\begin{matrix}\nearrow\!\!\!/\ O \\ \searrow NHR'\end{matrix} + HOC_2H_5$$

Note that in the conversion of an amine, $-\overset{|}{N}-H$, into an amide, $-\overset{|}{N}-\overset{\displaystyle O}{\overset{\|}{C}}-R$, a nitrogen .... (1) bond is replaced by a nitrogen-carbon bond. Hence it will be clear that tertiary amines .... (do/do not) (2) give these reactions, because in the structure R_3N there is no .... (3) bond.

---

(1) hydrogen (2) do not (3) nitrogen-hydrogen

---

19. Sulphonic acids, $R-SO_2OH$, can be converted into sulphonyl chlorides, $R-SO_2Cl$, which are analogous to carboxylic acid chlorides, $R-CO-Cl$. Hence it is not surprising that RSO_2Cl will react with $R'NH_2$ to yield a *sulphonamide* of structure .... (1), while an amine R'_2NH will yield a sulphonamide of structure .... (2). An amine R'_3N .... (will/will not) (3) yield this type of product.

---

(1) RSO_2NHR' (2) $RSO_2NR'_2$ (3) will not

REACTION OF ALKYLAMINES WITH NITROUS ACID

20. When ammonium nitrite is heated, it decomposes into water and nitrogen:

$$NH_3 + HO-N=O \rightleftharpoons NH_4^+NO_2^- \xrightarrow{\text{heat}} N_2 + 2H_2O$$

A similar decomposition occurs when a primary aliphatic amine is treated with nitrous acid in aqueous solution, so that evolution of .... (1) gas in this test is evidence for the presence of a .... (2) amine.

(1) benzoate anion $C_6H_5COO^-$ (2) benzoic acid C_6H_5COOH

---

6. On the other hand, the carboxylic acids are stronger acids than carbonic acid, $(HO)_2CO$. On addition of a carboxylic acid (RCOOH) to a saturated solution of sodium hydrogencarbonate, Na^+HOCOO^-, the acid donates a proton to . . . . (1) according to the equation . . . . (2).

---

(1) hydrogencarbonate ion 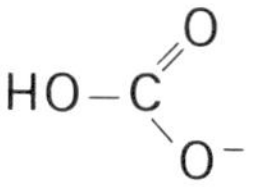

(2) $RCOOH + Na^+HOCOO^- \rightleftharpoons RCOO^-Na^+ + (HO)_2CO$, where $(HO)_2CO \rightarrow CO_2 + H_2O$

---

7. A simple test for carboxylic acids is the evolution of . . . . (1) when the acid is added to a saturated solution of sodium hydrogencarbonate.

---

(1) carbon dioxide

---

8. The relative proportions of free acid and anion in a solution of a carboxylic acid can be varied by adjusting the pH of the solution. The addition of a base will increase the proportion of the . . . . (1) and the addition of strong acid will cause more of the . . . . (2) to be formed.

---

(1) carboxylate ion or anion (2) free acid or carboxylic acid

---

9. A typical carboxylic acid has a dissociation constant, K_a, of about 10^{-4} ($pK_a = -\log K_a = 4$). In aqueous solution at pH 4, this acid will be present as 50% of the free acid, RCOOH, and 50% of . . . . (1).

---

(1) the anion $RCOO^-$

---

10. The ability to control the form in which the carboxylic acid occurs, by adjustment of pH, is of considerable practical value. Whereas the sodium salt of an acid is usually soluble in water and insoluble in non-polar solvents, the free acid is often insoluble in water and soluble in non-polar solvents. Benzoic acid can be precipitated from an aqueous solution of sodium benzoate by . . . . (1). If a solution of benzoic acid in toluene is shaken with aqueous sodium hydroxide, two layers will separate on standing —a lower aqueous layer and a toluene layer. The benzoic acid will be found in the form of . . . . (2) in the . . . . (3) layer.

The equation is often written

$$RNH_2 + HONO \longrightarrow ROH + H_2O + N_2$$

but the reaction is in fact complex and the yield of alcohol, ROH, is often low.

---

(1) nitrogen   (2) primary

---

21. Secondary amines react quite differently with aqueous nitrous acid and give rise to nitrosamines, $R_2NH + HONO \longrightarrow R_2N-N=O + H_2O$. The nitrosamine from ethylmethylamine has structure . . . . (1) while that from methylaminoethanoic acid (cf. Frame 3) has structure . . . . (2).

$$\underset{(1)}{CH_3CH_2-\overset{\displaystyle CH_3}{\overset{|}{N}}-N=O} \qquad \underset{(2)}{CH_3-\overset{\displaystyle N=O}{\overset{|}{N}}-CH_2CO_2H}$$

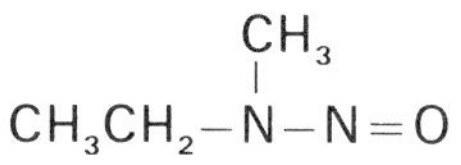

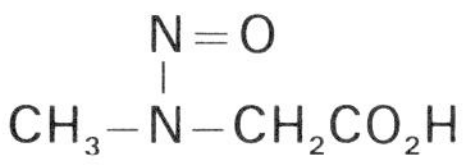

TEST FRAME

Primary, secondary, and tertiary amines are basic (nucleophilic) substances. They form . . . . (1) with acids and react as nucleophiles towards alkyl . . . . (2). Primary and secondary amines and also ammonia react with acid chlorides, acid anhydrides, or esters to form . . . . (3). Sulphonyl chlorides similarly react with these bases to form . . . . (4). The equation for the reaction of ethylmethylamine with propanoyl chloride is . . . . (5).

---

(1) salts   (2) halides   (3) amides   (4) sulphonamides

(5) $CH_3CH_2COCl + 2HN\overset{\displaystyle CH_3}{\underset{\displaystyle C_2H_5}{<}} \longrightarrow CH_3CH_2C\overset{\displaystyle O}{\underset{\displaystyle N}{<}}\overset{\displaystyle CH_3}{\underset{\displaystyle C_2H_5}{<}} + \overset{\displaystyle CH_3}{\underset{\displaystyle C_2H_5}{>}}NH_2^+Cl^-$ 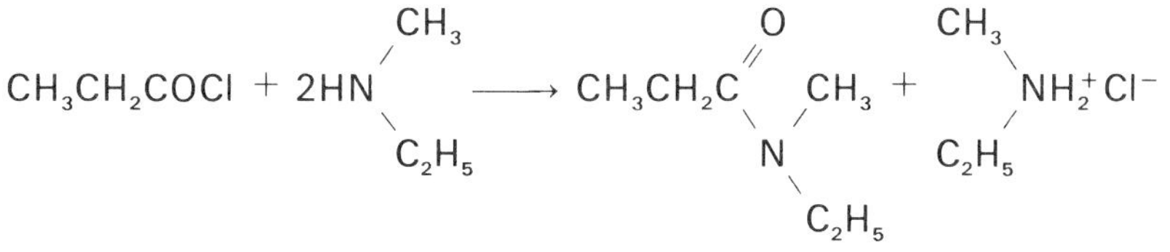

SYNTHESIS OF ALIPHATIC AMINES
Introduction

22. We can now turn to methods for the synthesis of aliphatic amines. Since amines are hydrocarbon derivatives of . . . . (1) it is not surprising that a method exists for preparing amines from this substance.

---

(1) ammonia

From alkyl halides

23. Ammonia reacts as a nucleophile towards alkyl halides in the same way as do amines themselves, as discussed in Frames 10-15. Thus when ethyl

CARBOXYLIC ACIDS

Program 8

ACIDITY OF CARBOXYLIC ACIDS

1. The functional group of the carboxylic acids is $-C(=O)OH$, which is called the .... (1) group. Acetic acid (ethanoic acid) .... (2) is an example of an aliphatic acid and benzoic acid .... (3) is an aromatic carboxylic acid.

---

(1) carboxyl   (2) CH_3COOH (3) 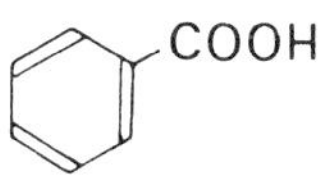

---

2. Carboxylic acids can transfer a .... (1) to a base, thereby forming a carboxylate anion .... (2).

---

(1) proton   (2) $R-C(=O)O^-$

---

3. The carboxylate ion is often called the *conjugate base* of the acid since it is able to .... (1) and re-form the acid. That is, the carboxylate ion functions as a .... (2).

---

(1) combine with a proton   (2) base

---

4. When a carboxylic acid is dissolved in water, the equilibrium between the carboxylic acid and the carboxylate ion can be represented by the equation:

$$RCOOH + H_2O \rightleftharpoons \text{............} \quad (1).$$

---

(1) $RCOOH + H_2O \rightleftharpoons RCOO^- + H_3O^+$

---

5. Carboxylic acids are weaker acids than the strong mineral acids, such as hydrochloric acid. If an aqueous solution of sodium benzoate is titrated with hydrochloric acid, the .... (1) anion which is originally present, is converted into .... (2).

---

bromide is heated with ammonia the nitrogen lone-pair forms a new bond to the carbon atom with simultaneous displacement of a .... (1) ion, so that the reaction can be written .... (2).

---

(1) bromide (2)

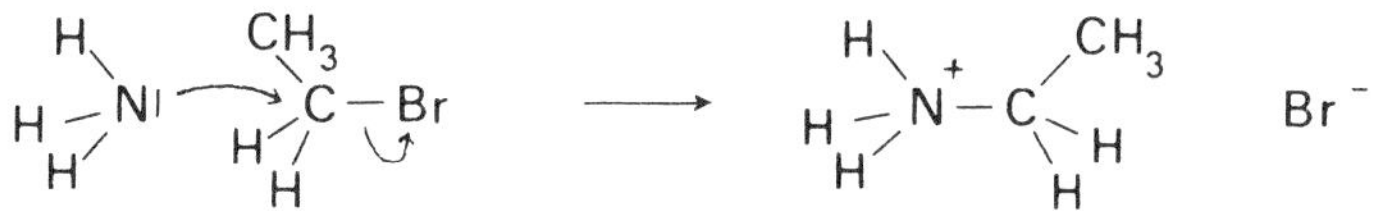

---

24. We say that the bromine atom has been *substituted* by ammonia, so that this is a nucleophilic .... (1) reaction, for which we use the symbol S_N.

---

(1) substitution

---

25. The product of this particular .... (1) reaction, $CH_3CH_2NH_3^+Br^-$, is named ethylammonium bromide.

---

(1) S_N or nucleophilic substitution

---

26. Just as ammonium bromide on treatment with alkali gives free ammonia, so ethylammonium bromide on treatment with alkali gives ethylamine. Hence, by allowing ammonia to react with ethyl bromide and treating the product with alkali we have synthesised a .... (1) aliphatic amine.

---

(1) primary

---

27. This reaction of ammonia with ethyl bromide is in fact more complex than we have indicated. We have already seen (Frame 14) that the reaction of primary or secondary amines with an alkyl halide yields a mixture of products. In the same way, in this reaction, the ammonia can liberate some free ethylamine from its salt, and this can then itself react as a .... (1) towards more of the .... (2), leading to the formation of diethylammonium bromide, of structure .... (3).

---

(1) nucleophile (2) ethyl bromide (3) $(CH_3CH_2)_2NH_2^+Br^-$

---

28. Further reaction can lead to the formation of $(CH_3CH_2)_3NH^+Br^-$, triethylammonium bromide, and of $(CH_3CH_2)_4N^+Br^-$, named .... (1).

Program 8: CARBOXYLIC ACIDS

CONTENTS

It is not always easy to separate the mixture of products one from the other and this is a very common situation in practical organic chemistry.

---

(1) tetraethylammonium bromide

---

29. *Gabriel method.* There are several very much better methods for preparing amines than by reaction of ammonia with alkyl halides. In the Gabriel method, an alkyl halide is allowed to react with the easily-available compound potassium phthalimide,

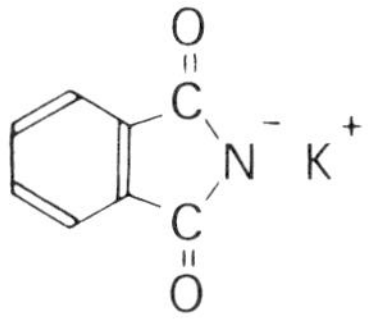

prepared in turn from phthalimide,

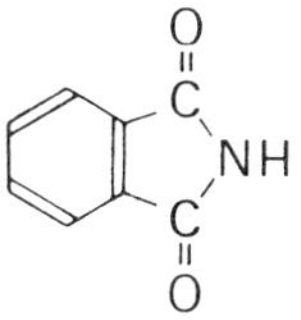

The phthalimide ion has a full . . . . (1) charge, and acts as a nucleophile towards the alkyl halide.

---

(1) negative

---

30. Nucleophilic substitution of 1-chloropropane by the phthalimide ion gives as product the substance

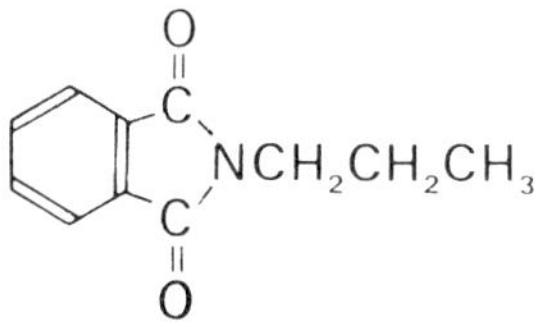

in which a new carbon- . . . . (1) bond is evident.

---

(1) nitrogen

---

31. Hydrolysis of this phthalimide derivative by alkali gives propylamine and a salt of benzene-1,2-dicarboxylic acid (phthalic acid).

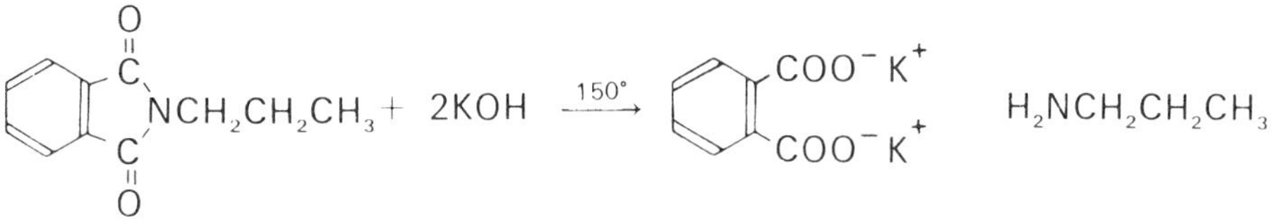

In this hydrolysis, the CH_2–N bond remains intact, while the two . . . . (1) bonds are broken.

---

(1) amide

(2) $HOCH_2(CH_2)_3CHO$ + CH_3OH $\overset{HCl}{\underset{}{\rightleftharpoons}}$ 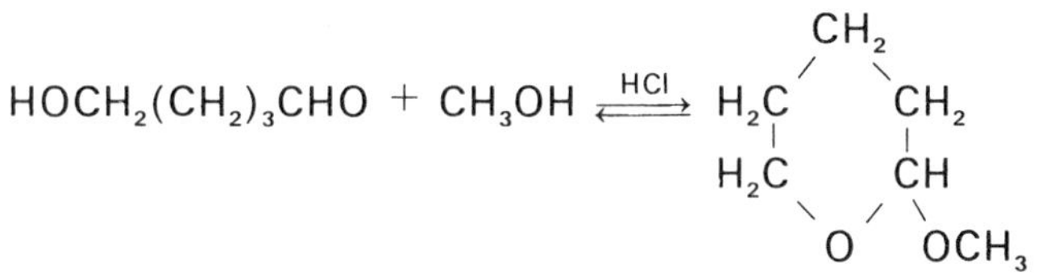

$$\begin{array}{c} CH_2 \\ H_2C \quad\quad CH_2 \\ H_2C \quad\quad CH \\ O \quad OCH_3 \end{array}$$

The hydroxy group in the 5-position participates in formation of the acetal.

(3) C_6H_5CHO + $HOCH_2CH_2OH$ $\overset{HCl}{\longrightarrow}$ $C_6H_5CH \overset{O-CH_2}{\underset{O-CH_2}{|}}$ + H_2O 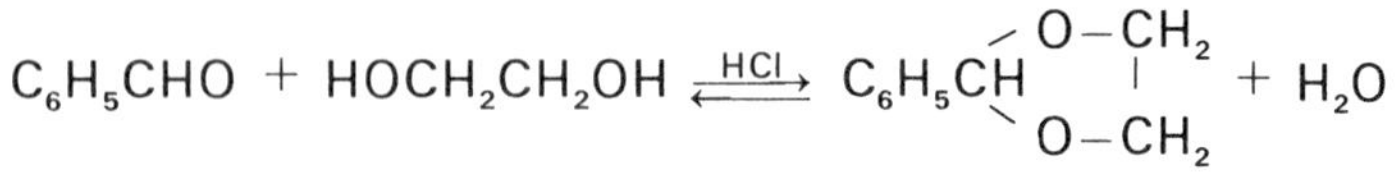

32. The Gabriel procedure gives exclusively a .... (1) amine as product with no possibility of forming any secondary or tertiary amine.

---

(1) primary

From amides by reduction

33. Another convenient synthesis of primary amines is found in the reductive process

$$RCONH_2 \xrightarrow{LiAlH_4} RCH_2NH_2$$

That is, reduction of a primary .... (1) by lithium aluminium hydride yields a primary amine.

---

(1) amide

---

34. The analogous reduction of a secondary amide RCONHR′ yields a .... (1) of general structure .... (2).

---

(1) secondary amine (2) RCH_2NHR'

---

35. As a specific example, the reduction by $LiAlH_4$ of N-methylpropanamide (the N-methylamide of propanoic acid) of structure .... (1) yields a secondary amine of structure .... (2).

---

(1) $CH_3CH_2CONHCH_3$ (2) $CH_3CH_2CH_2NHCH_3$

---

36. Reduction of a tertiary amide can yield a tertiary amine; e.g. $RCONMe_2 \longrightarrow$ .... (1). However, the yields in this process are not always good and other products can be formed.

---

(1) $RCH_2N(CH_3)_2$

From nitriles by reduction

37. Reduction of the cyano (nitrile) functional group, $-C{\equiv}N$, is a good method for introducing the $-CH_2NH_2$ function. Thus $CH_3C{\equiv}N \longrightarrow CH_3CH_2NH_2$. Virtually any cyano compound can be reduced in this way, and it is clearly a method for the synthesis of .... (1) amines in which the amino group is attached of necessity to a .... (2) group. Reducing agents such as Na/C_2H_5OH, Zn/H_2SO_4, $LiAlH_4$ or H_2/metal catalyst can be employed.

---

(1) $CH_3CH_2CHO + 2CH_3OH \xrightarrow[\text{catalyst}]{\text{acid}} CH_3CH_2CH(OCH_3)_2 + H_2O$

---

61. Ketals, $RC(OR'')_2R'$, are the corresponding derivatives of ketones, but they cannot be formed readily by reaction of the ketone, RCOR', with the alcohol, R''OH, in the presence of an acid catalyst. Indirect methods have to be used for the preparation of such compounds. However, ketal formation does occur readily if a five-membered ring is formed. Suggest a structure for the ketal formed by the reaction of acetone with ethane-1,2-diol in the presence of an anhydrous mineral acid . . . . (1).

---

(1) 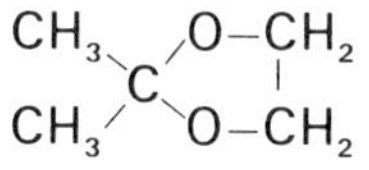

TEST FRAMES

Write the structures of the following compounds:
Benzophenone oxime . . . . (1); 2-butenal semicarbazone . . . . (2); benzaldehyde phenylhydrazone . . . . (3); formaldehyde 2,4-dinitrophenylhydrazone . . . . (4).

---

(1) $C_6H_5C(=NOH)C_6H_5$

(2) $CH_3CH{=}CHCH{=}NNHCONH_2$

(3) $C_6H_5C(H){=}NNHC_6H_5$

(4) 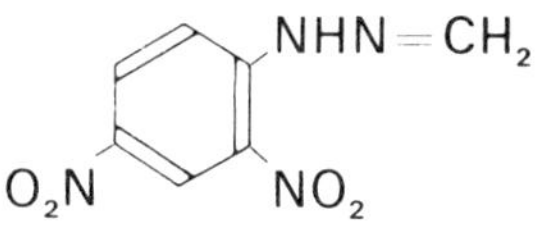

Write equations for the following reactions. Dry hydrogen chloride is passed into the following mixtures: pentanal dissolved in excess methanol . . . . (1); 5-hydroxypentanal dissolved in methanol . . . . (2) (Hint: the product is a cyclic structure of molecular formula $C_6H_{12}O_2$); benzaldehyde and 1,2-dihydroxyethane . . . . (3).

---

(1) $CH_3(CH_2)_3CHO + 2CH_3OH \xrightleftharpoons{HCl} CH_3(CH_2)_3CH(OCH_3)_2 + H_2O$

(1) primary (2) CH_2 or methylene

From amides by oxidation

38. The so-called Hofmann reaction of primary amides of carboxylic acids, $RCONH_2$, is also a good method for preparing primary amines. The amide is added to a solution of bromine (1 molar proportion) and KOH (at least 4 molar proportions) in water, and the solution is heated rapidly to 70-80°. The overall oxidative reaction is

$$RCONH_2 + Br_2 + 4OH^- \longrightarrow RNH_2 + CO_3^{2-} + 2Br^- + 2H_2O$$

The reaction involves intermediate formation of an N-bromoamide, RCONHBr, which decomposes in alkali, with rearrangement of the R group from carbon to nitrogen, to give a further intermediate of structure RN=C=O. This undergoes hydrolysis to give the amine and carbonate. Hofmann reaction of benzamide, $C_6H_5CONH_2$, yields .... (1).

---

(1) aniline, $C_6H_5NH_2$

TEST FRAME

Some methods for preparing aliphatic amines are as follows:
Reaction of ammonia with an .... (1), reaction of an alkyl halide according to the Gabriel procedure with .... (2), reduction of a primary, secondary, or tertiary amide with .... (3), reduction of the .... (4), or .... (5) functional groups with formation of the $-CH_2NH_2$ group. Application of the .... (6) degradation of primary amides, $RCONH_2$, whereby treatment with .... (7) followed by heating to 70-80° yields an amine of formula .... (8).

---

(1) alkyl halide (2) potassium phthalimide
(3) $LiAlH_4$ (4) and (5) $-C{\equiv}N$, $-CONH_2$
(6) Hofmann (7) bromine in alkali
(8) RNH_2

Hydration of aldehydes and ketones and the formation of acetals and ketals

58. In an aqueous solution of formaldehyde, water adds to the carbonyl group according to the equilibrium

$$\begin{array}{c} H \\ \diagdown \\ C=O \\ \diagup \\ H \end{array} + H_2O \rightleftharpoons \begin{array}{c} H \\ | \\ C \diagdown OH \\ \diagup \diagdown \\ H \quad OH \end{array}$$

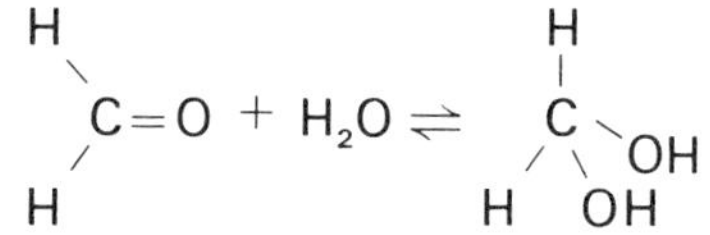

In the case of other aldehydes, the position of the corresponding equilibrium lies further to the left, that is they are less hydrated, and only certain highly reactive ketones hydrate to an appreciable extent. Write the equilibrium for the hydration of propanal (propionaldehyde) . . . . (1).

---

(1) $CH_3CH_2CHO + H_2O \rightleftharpoons \begin{array}{c} H \\ | \\ CH_3CH_2C-OH \\ | \\ OH \end{array}$ 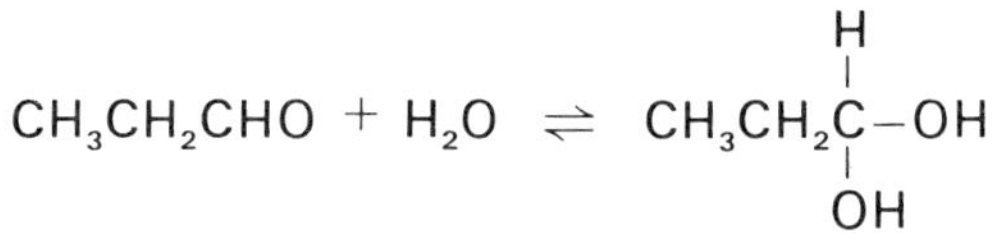

---

59. Aldehydes, when dissolved in anhydrous alcohols in the presence of an acid catalyst, undergo a similar reaction. For example, acetaldehyde, when dissolved in ethanol containing a small amount of anhydrous HCl, reacts according to the equilibrium

$$\begin{array}{c} H \\ \diagdown \\ C=O \\ \diagup \\ CH_3 \end{array} + HOC_2H_5 \underset{acid}{\rightleftharpoons} \begin{array}{c} H \\ | \\ C \diagdown OC_2H_5 \\ \diagup \diagdown \\ CH_3 \quad OH \end{array}$$

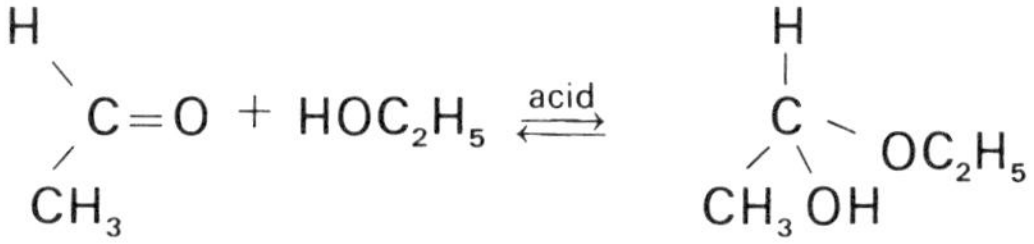

The product is known as a hemiacetal. Write the structure of the hemiacetal formed between propanal and methanol . . . . (1).

---

(1) $\begin{array}{c} H \\ | \\ CH_3CH_2C-OCH_3 \\ | \\ OH \end{array}$ 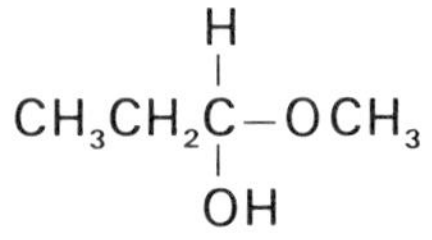

---

60. Acetaldehyde reacts with an excess of ethanol in the presence of an anhydrous acid catalyst to give 1,1-diethoxyethane (acetaldehyde diethyl acetal), which is commonly known as acetal.

$$CH_3CHO + 2C_2H_5OH \underset{catalyst}{\overset{acid}{\rightleftharpoons}} CH_3CH(OC_2H_5)_2 + H_2O$$

The reaction is reversible and the acetal can be hydrolysed readily to the aldehyde. Write the equation for the reaction of propanal with excess methanol in the presence of some dry hydrogen chloride . . . . (1).

Program 6: ALCOHOLS

CONTENTS

Several other substituted hydrazines are commonly used to prepare derivatives of aldehydes and ketones. These include 2,4-dinitrophenylhydrazone, of structure . . . . (2), and "semicarbazide", of structure $H_2NNHCONH_2$. By analogy with the structure of acetone phenylhydrazone, the structure of the semicarbazone of 3-methylbutanal is . . . . (3).

---

(1) water

(2) 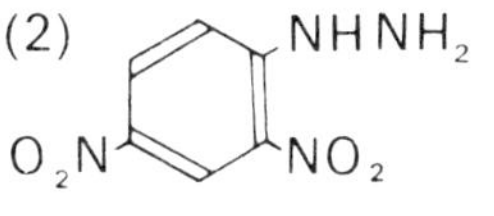

(3) $(CH_3)_2CHCH_2CH \quad NNHCONH_2$

---

55. Aldehydes and ketones also react with hydroxylamine, H_2NOH, under mildly acidic conditions to give analogous products known as oximes. Thus benzaldehyde reacts with hydroxylamine to give benzaldoxime according to the equation . . . . (1).

$$\text{(1)}\ C_6H_5\overset{\overset{\displaystyle H}{|}}{C}{=}O + H_2NOH \longrightarrow C_6H_5\overset{\overset{\displaystyle H}{|}}{C}{=}NOH + H_2O$$

---

56. Butanal reacts with hydroxylamine in the presence of weak acid to give butanal oxime. This again is an addition-elimination reaction (cf. Frame 54) and the equation can therefore be shown . . . . (1).

$$\text{(1)}\ CH_3CH_2CH_2CHO + H_2NOH \longrightarrow \left[CH_3CH_2CH_2CH\begin{matrix}\diagup OH\\ \diagdown NHOH\end{matrix}\right] \longrightarrow$$

$$CH_3CH_2CH_2\overset{\overset{\displaystyle H}{|}}{C}{=}NOH + H_2O$$

---

57. The oximes of aldehydes are useful intermediates for the synthesis of nitriles.

$$R\overset{\overset{\displaystyle H}{|}}{C}{=}NOH \xrightarrow{-H_2O} RC{\equiv}N$$

This dehydration can be effected by hot acetic anhydride (cf. Program 8, Frame 57), the equation being

$$R\overset{\overset{\displaystyle H}{|}}{C}{=}NOH + (CH_3CO)_2O \longrightarrow \text{. . . . . (1)}$$

$$\text{(1)}\ RCH{=}NOH + (CH_3CO)_2O \longrightarrow RC{\equiv}N + 2CH_3COOH$$

ALCOHOLS

Program 6

FUNCTIONAL GROUP: NOMENCLATURE

1. Alcohols are characterised by the presence of a hydroxyl (–OH) functional group attached to a saturated carbon atom. Methanol (methyl alcohol) and ethanol (ethyl alcohol) are the simplest alcohols, containing respectively one and two carbon atoms per molecule. Their structures are . . . . (1) and . . . . (2) respectively.

---

(1) 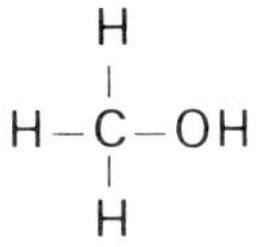 or CH_3OH or MeOH

(2) 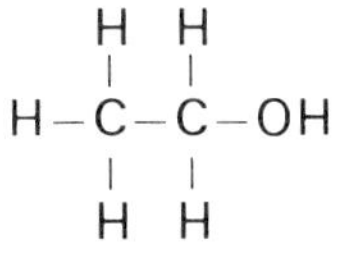 or CH_3CH_2OH or C_2H_5OH or EtOH

---

2. Hexadecan-1-ol (cetyl alcohol), $CH_3(CH_2)_{15}OH$, is obtained by chemical degradation of sperm whale oil and is used in lipsticks and also for retarding the evaporation of water from large reservoirs in Australia. Glycerol ("glycerine" or propane-1,2,3-triol), a by-product in the manufacture of soap from animal fats, has the structure . . . . (1).

---

(1) 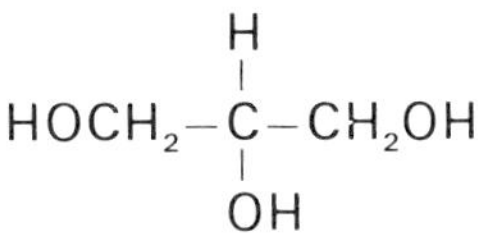 or $HOCH_2CH(OH)CH_2OH$

---

3. These substances have in common a . . . . (1) group or groups attached to a . . . . (2) carbon atom or atoms.

---

(1) hydroxyl (2) saturated

---

4. Note that compounds having *two* hydroxyl groups attached to the *same* carbon atom are usually unstable and decompose readily to aldehydes RCHO, or ketones R_2CO according to the equations:

$$CH_3CH_2CH_2\overset{\displaystyle H}{\underset{\displaystyle OH}{C}}SO_3Na + HCl \longrightarrow CH_3CH_2CH_2CHO + SO_2 + H_2O + NaCl$$

The procedure constitutes a chemical separation of an aldehyde and alcohol.

Addition reactions followed by elimination—formation of hydrazones and oximes.

51. Hydrazine has the structure H_2NNH_2 so that phenylhydrazine has the structure . . . . (1).

---

(1) 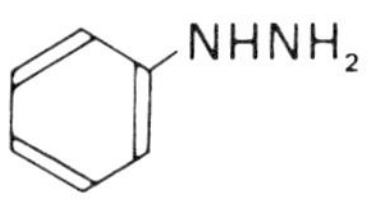

or $C_6H_5NHNH_2$

---

52. Aldehydes and ketones react with phenylhydrazine in the presence of weak acid to form phenylhydrazones. Thus acetone reacts with phenylhydrazine in the presence of an acid catalyst to form a crystalline derivative named acetone phenylhydrazone, $C_9H_{12}N_2$, which contains a carbon-nitrogen double bond at carbon atom 2 of the original acetone molecule. The equation of the reaction in terms of molecular formulae is

C_3H_6O + ... $\xrightarrow{\text{acid}}$ . . . . + ... (1)

---

(1) $C_3H_6O + C_6H_8N_2 \xrightarrow{\text{acid}} C_9H_{12}N_2 + H_2O$

---

53. The equation of the reaction in terms of structures and names is . . . . (1).

---

(1) 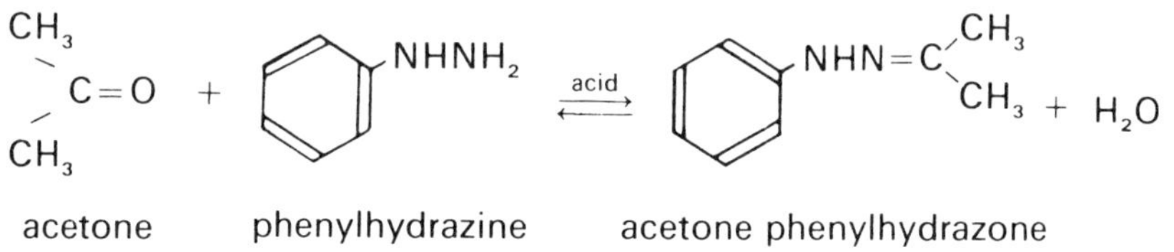

---

54. The above reaction can be pictured as *addition* of phenylhydrazine followed by *elimination* of . . . . (1).

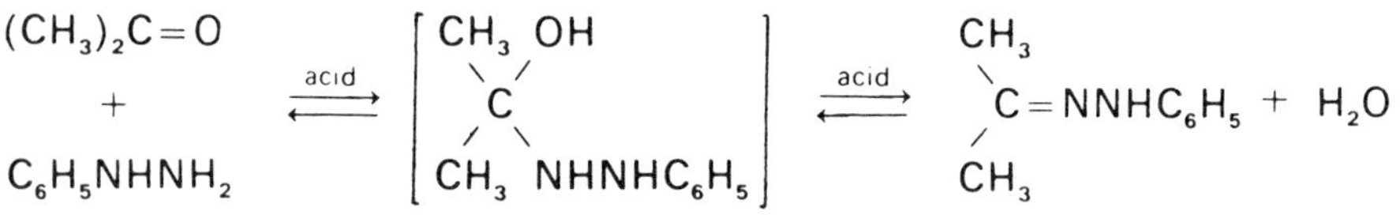

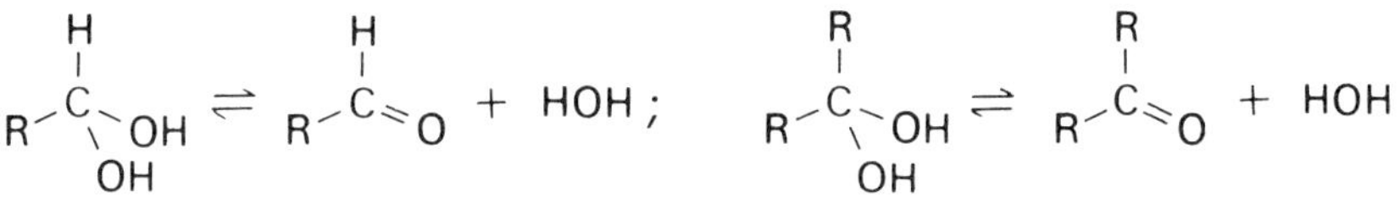

where R represents any hydrocarbon or substituted hydrocarbon group. For this reason, such compounds are not regarded as alcohols but rather as hydrated . . . . (1) or . . . . (2) (cf. Program 7).

---

(1) aldehydes (2) ketones

---

5. Three types of simple alcohol are possible: *primary* alcohols RCH_2OH, where *one* organic group (R) is attached to the saturated carbon atom holding the hydroxyl group: *secondary* alcohols $RR'CHOH$ where . . . . (1) R-groups, which can be the same or different, are attached to this carbon atom; and *tertiary* alcohols of structure . . . . (2) where . . . . (3) groups other than hydrogen are attached to the carbon atom holding the hydroxyl group.

---

(1) two (2) $R-C(R')(R'')-OH$ (3) three

---

6. The compound ethane-1,2-diol (ethylene glycol) is used as an antifreeze in car radiators and as a raw material in the manufacture of "Terylene". It has the structure $HOCH_2CH_2OH$ and thus contains two . . . . (1) hydroxyl groups. Glycerol, a trihydroxy alcohol (see Frame 2), has two . . . . (2) and one . . . . (3). Sugars and carbohydrates are polyhydroxy compounds, that is, each molecule contains . . . . (4) hydroxyl groups, both primary and secondary.

---

(1) primary (2) primary hydroxyl groups

(3) secondary hydroxyl group (4) many or several

REACTIONS OF ALCOHOLS

7. A molecule of a simple alcohol such as $CH_3CH_2CH_2OH$ possesses carbon-carbon and carbon-hydrogen bonds, one carbon- . . . . (1) bond and one oxygen- . . . . (2) bond. In the various reactions of alcohols we shall discuss, one or more of these types of bond may be broken.

---

50. The reduction of aldehydes and ketones with $LiAlH_4$ and $NaBH_4$ (Frames 41, 43) can be considered as addition of a hydride ion. For example the reduction of benzaldehyde with these reagents may involve the initial reactions

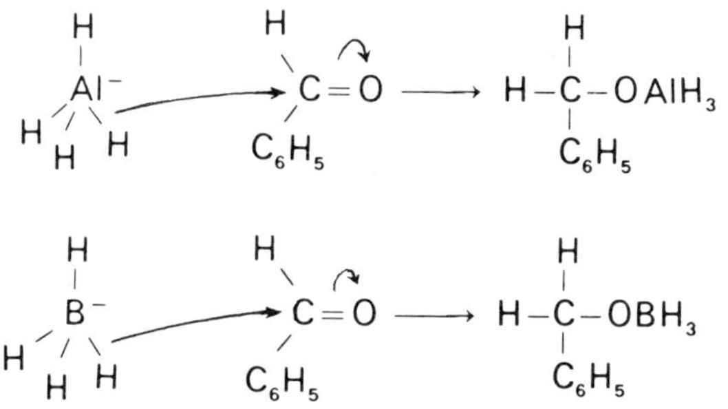

The complexes resulting from further reaction of these intermediates with benzaldehyde, when treated with dilute acid yield . . . . (1).

---

(1) benzyl alcohol

TEST FRAME

Give equations for the following reactions:

Methyl phenyl ketone (acetophenone) is added to a solution of the sodium salt of but-1-yne in liquid ammonia and, after removal of the ammonia, dilute hydrochloric acid is added . . . . (1).

A sample of butanal containing a small amount of butan-1-ol is shaken with a saturated aqueous solution of sodium hydrogensulphite and the precipitate collected. The precipitate is dissolved in dilute hydrochloric acid . . . . . .
. . . . (2).

---

(1) $CH_3CH_2C \equiv C^-Na^+ + C_6H_5COCH_3 \longrightarrow CH_3CH_2C \equiv \overset{\overset{\displaystyle CH_3}{|}}{\underset{\underset{\displaystyle C_6H_5}{|}}{C}}C-O^-Na^+$

$CH_3CH_2C \equiv \overset{\overset{\displaystyle CH_3}{|}}{\underset{\underset{\displaystyle C_6H_5}{|}}{C}}C-O^-Na^+ + HCl \longrightarrow CH_3CH_2C \equiv \overset{\overset{\displaystyle CH_3}{|}}{\underset{\underset{\displaystyle C_6H_5}{|}}{C}}C-OH + NaCl$

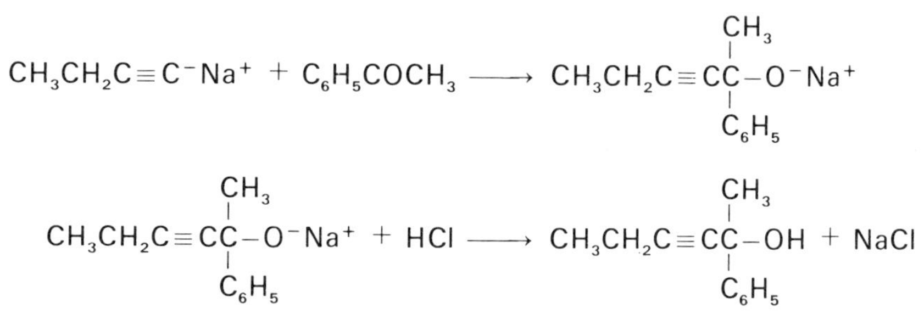

(2) $CH_3CH_2CH_2CHO$

and $+ NaHSO_3 \rightleftharpoons$

$CH_3CH_2CH_2CH_2OH$

$CH_3CH_2CH_2\overset{\overset{\displaystyle}{}}{\underset{\underset{\displaystyle OH}{|}}{C}}HSO_3Na$ (precipitate)

$CH_3CH_2CH_2CH_2OH$ unchanged

Collection of the precipitate gives the pure "bisulphite addition compound".

(1) oxygen (2) hydrogen

Oxygen-hydrogen fission—formation of alkoxides, ethers and esters

8. We shall consider first the reaction of alcohols with electropositive metals whereby the oxygen-hydrogen bond is broken. Ethanol reacts readily with metallic sodium to give hydrogen gas and sodium ethoxide, C_2H_5ONa, according to the equation . . . . (1). This reaction may be compared with that of sodium and water to give hydrogen gas and sodium hydroxide according to the equation . . . . (2).

---

(1) $2Na + 2C_2H_5OH \longrightarrow H_2 + 2C_2H_5O^-Na^+$

(2) $2Na + 2HOH \longrightarrow H_2 + 2HO^-Na^+$

---

9. Sodium ethoxide is CH_3CH_2ONa. In general, a metal derivative of an alcohol is called a "metal alkoxide". Sodium prop-1-oxide has the structure . . . . (1), potassium 2-methylprop-2-oxide (potassium t-butoxide) has structure . . . . (2) and aluminium prop-2-oxide (isopropoxide) has structure . . . . (3).

---

(1) $CH_3CH_2CH_2ONa$ (2) 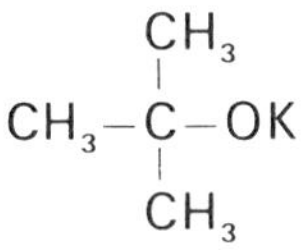 or $(CH_3)_3COK$

(3) 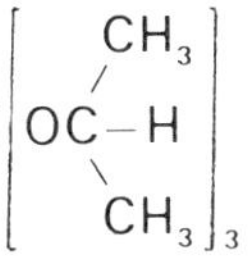 or $Al(OCHMe_2)_3$

---

10. In the reaction of an alcohol with an electropositive metal, the alcohol is behaving like a very weak acid. The metal alkoxide can thus be regarded as a salt. We therefore expect (and find) that treatment of a metal alkoxide with acid regenerates the . . . . (1). Thus the reaction of magnesium methoxide with hydrogen chloride proceeds according to the equation . . . . (2). Alkoxides cannot be prepared by reaction of alcohols with sodium hydroxide since the water formed is a much stronger acid than the alcohol, so that the equilibrium that is set up lies well to the left.

e.g., $CH_3CH_2OH + OH^- \rightleftharpoons CH_3CH_2O^- + H_2O$

The solution contains only a small concentration of ethoxide ions.

---

(1) alcohol (2) $Mg(OCH_3)_2 + 2HCl \longrightarrow MgCl_2 + 2CH_3OH$

48. Sodium hydrogensulphite (sodium bisulphite) adds to aldehydes and a few reactive ketones, but not to ketones in general, to form the so-called "bisulphite addition compounds".

$$RCHO + NaHSO_3 \rightleftharpoons R\underset{|\atop OH}{\overset{H\atop |}{C}}SO_3Na$$

The reaction is reversible and the addition compound can be decomposed easily with dilute acid e.g.

$$R\underset{|\atop OH}{\overset{H\atop |}{C}}SO_3Na + H_3O^+ + Cl^- \longrightarrow RCHO + SO_2 + 2H_2O + NaCl$$

In the hydrogensulphite addition compound the carbon atom is attached to the sulphur atom. Draw the structure of the product resulting from reaction of hydrogensulphite ion with acetaldehyde . . . . (1).

---

(1) $CH_3-\underset{|\atop HO}{\overset{H\atop |}{C}}-\underset{\|\atop O}{\overset{O\atop \|}{S}}-O^-$ or other mesomeric structures such as $CH_3-\underset{|\atop HO}{\overset{H\atop |}{C}}-\underset{|\atop O^-}{\overset{O^-\atop |}{S^{2+}}}-O^-$

---

49. Grignard reagents react readily with aldehydes and ketones to form intermediate magnesium alkoxides which on treatment with dilute acid liberate the corresponding alcohols.

$$RMgX + R'CHO \longrightarrow RR'CHOMgX \longrightarrow RR'CHOH + MgX_2$$

(the group R in RMgX can be either alkyl or aryl).

Write equations for the reaction of phenyl magnesium chloride with acetone . . . . (1) and for the reaction of the intermediate with hydrochloric acid . . . . (2). (cf. Program 6, Frames 33-40).

---

(1) $C_6H_5MgCl + \begin{matrix} CH_3 \\ \quad\searrow \\ \quad\quad C=O \\ \quad\nearrow \\ CH_3 \end{matrix} \longrightarrow C_6H_5-\underset{|\atop CH_3}{\overset{CH_3\atop |}{C}}-OMgCl$

(2) $C_6H_5-\underset{|\atop CH_3}{\overset{CH_3\atop |}{C}}-OMgCl + HCl \longrightarrow C_6H_5\underset{|\atop CH_3}{\overset{CH_3\atop |}{C}}OH + MgCl_2$

11. By suitable chemical manipulation, alcohols can be converted into ethers. An ether contains two hydrocarbon groups linked by an oxygen atom, and a generalised structure can be shown .... (1).

---

(1) R–O–R′ or $-\overset{|}{\underset{|}{C}}-O-\overset{|}{\underset{|}{C}}-$

---

12. As an example of the conversion of an alcohol into an alkoxide and thence into an ether we can discuss the reaction of sodium methoxide with iodobutane (n-butyl iodide) to give butyl methyl ether of structure .... (1).

---

(1) $CH_3OCH_2CH_2CH_2CH_3$

---

13. Sodium metal is dissolved in methanol, giving sodium methoxide ($NaOCH_3$ or $Na^+CH_3O^-$). The methoxide ion can displace the .... (1) atom from iodobutane according to the equation -

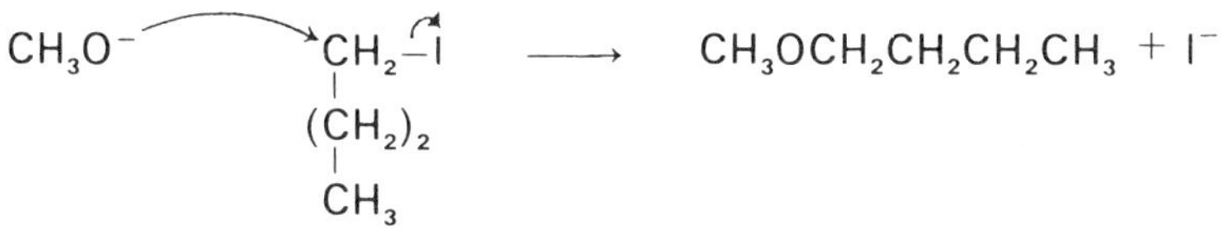

The products are butyl methyl ether and .... (2) ion.

---

(1) iodine (2) iodide

---

14. In this reaction we observe *substitution* of the iodine atom by the incoming methoxide group on carbon, so we call this a .... (1) reaction.

---

(1) substitution

---

15. The methoxide ion belongs to the general class of nucleophiles, so the above reaction can be further described as a .... (1) substitution reaction.

---

(1) nucleophilic

---

16. We have discussed the preparation of butyl methyl ether as a reaction of methoxide ion with iodobutane. It is equally valid to discuss this as a reaction of iodobutane, the reagent being methoxide ion (cf. Program 4, Frames 41, 42).

45. Write equations for the reaction of hydrogen cyanide with benzaldehyde, ... (1) and cyclopentanone, ... (2).

---

(1) $C_6H_5CHO + HCN \longrightarrow C_6H_5CH(OH)CN$

(2) 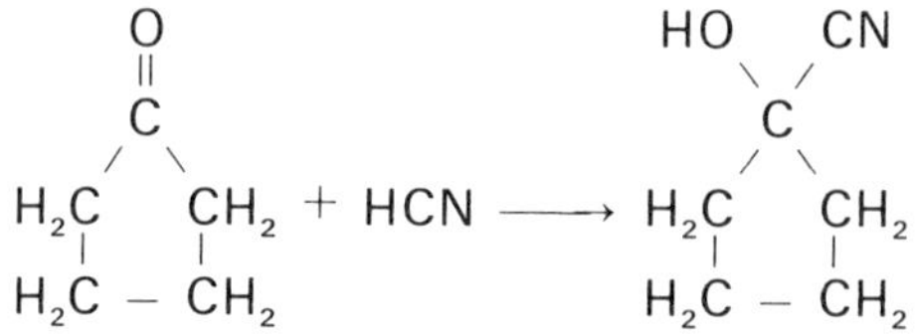

---

46. Sodium acetylide can be prepared by passing acetylene into a solution of sodium amide (sodamide) in liquid ammonia

$$H-C\equiv C-H + NH_2^- + Na^+ \rightleftharpoons H-C\equiv C|^- + Na^+ + NH_3$$

The acetylide ion is a powerful nucleophile and readily attacks aldehydes and ketones. For instance, it will react with acetophenone according to the equation:

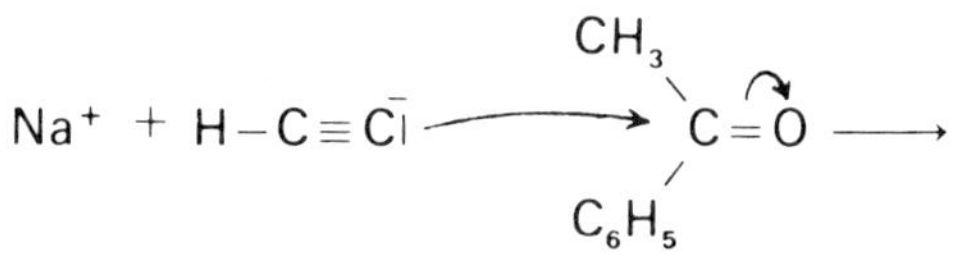

. . . . (1).

---

(1) 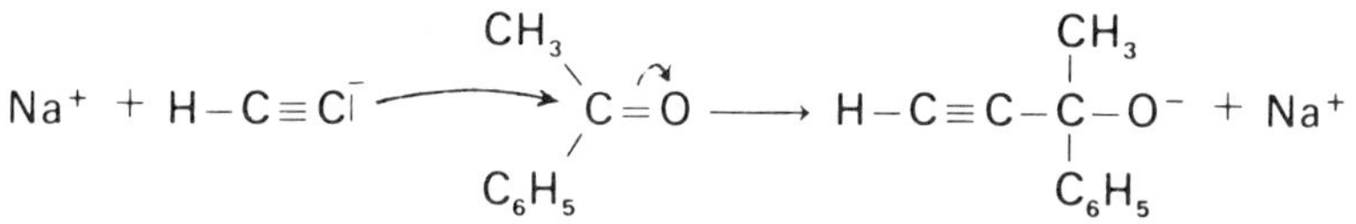

---

47. The reaction is completed by allowing the liquid ammonia to evaporate and adding dilute acid.

$$H-C\equiv C-C(CH_3)(C_6H_5)-O^- + H_3O^+ \longrightarrow \text{. . . . .} \quad (1)$$

---

(1) $H-C\equiv C-C(CH_3)(C_6H_5)-O^- + H_3O^+ \longrightarrow H-C\equiv C-C(CH_3)(C_6H_5)-OH$ 2-phenylbut-3-yn-2-ol

Addition of benzyl bromide to a solution of potassium prop-2-oxide in propan-2-ol yields potassium bromide and . . . . (1).

---

(1) benzyl isopropyl ether, $C_6H_5CH_2OCH(CH_3)_2$

---

17. The conversion of an alcohol into an ester is another example of a reaction where the oxygen-hydrogen bond of the alcohol is broken. Alcohols, R′OH, react with acid chlorides, RCOCl, to give esters of general structure . . . . (1). The reaction works best with primary and secondary alcohols. 2-Methylpropan-1-ol reacts with acetyl chloride according to the equation

$$(CH_3)_2CHCH_2OH + CH_3\overset{O}{\overset{\|}{C}}\,Cl \longrightarrow CH_3\overset{O}{\overset{\|}{C}}OCH_2CH(CH_3)_2 + HCl$$

The name of this ester is . . . . (2).

---

(1) $R\overset{O}{\overset{\|}{C}}OR'$ (2) 2-methylpropyl ethanoate or isobutyl acetate

---

18. Acid anhydrides can also be used to prepare esters from alcohols. The general equation for such a reaction is

$$R\overset{O}{\overset{\|}{C}}-O-\overset{O}{\overset{\|}{C}}R + R'OH \longrightarrow RCOOR' + RCOOH.$$

Reaction of *p*-methoxybenzyl alcohol with acetic anhydride gives the ester named . . . . (1).

---

(1) *p*-methoxybenzyl acetate

TEST FRAME

Complete the following equations:

$$(CH_3)_2CHOH + K \longrightarrow \;. . . . .\; (1)$$
$$Mg(OCH_3)_2 + 2HBr \longrightarrow \;. . . . .\; (2)$$
$$CH_3I + (CH_3)_2CHOK \longrightarrow \;. . . . .\; (3)$$
$$(CH_3CO)_2O + CH_3CH_2CH_2OH \longrightarrow \;. . . . .\; (4)$$

---

(1) $(CH_3)_2CHOH + K \longrightarrow (CH_3)_2CHOK + \tfrac{1}{2}H_2$

(2) $Mg(OCH_3)_2 + 2HBr \longrightarrow MgBr_2 + 2CH_3OH$

(3) $CH_3I + (CH_3)_2CHOK \longrightarrow (CH_3)_2CHOCH_3 + KI$

(4) $(CH_3CO)_2O + CH_3CH_2CH_2OH \longrightarrow CH_3COOCH_2CH_2CH_3 + CH_3COOH$

Complete the following equations:

$$C_6H_5CH_2COCH_3 \xrightarrow{KMnO_4,\ H_3O^+} \ldots\ldots\ldots\ldots (2)$$

$$CH_3CH_2CH(OH)CH_3 \xrightarrow{I_2\ +\ NaOH} \ldots\ldots\ldots\ldots (3)$$

$$CH_3COCH_2COCH_3 \xrightarrow[(2)\ acid]{(1)\ LiAlH_4} \ldots\ldots\ldots\ldots (4)$$

$$CH_3CH{=}CHCHO \xrightarrow{H_2/Pt} \ldots\ldots\ldots\ldots (5).$$

---

(1) Red copper(I) oxide is precipitated

(2) $C_6H_5CH_2COCH_3 \xrightarrow{KMnO_4,\ H_3O^+} C_6H_5COOH,\ C_6H_5CH_2COOH$, and CH_3COOH ($HCOOH$ is oxidised to CO_2)

(3) $CH_3CH_2CH(OH)CH_3 \xrightarrow{I_2\ +\ NaOH} CH_3CH_2COOH + CHI_3$

(4) $CH_3COCH_2COCH_3 \xrightarrow[(2)\ acid]{(1)\ LiAlH_4} CH_3CH(OH)CH_2CH(OH)CH_3$

(5) $CH_3CH{=}CHCHO \xrightarrow{H_2/Pt} CH_3CH_2CH_2CH_2OH$

ADDITION REACTIONS

44. We have already met one type of addition reaction in Frames 25-31. The addition of hydrogen cyanide to an aldehyde or ketone takes place readily, yielding products known as cyanohydrins.

$$\underset{\overset{\|}{O}}{RCH} + HCN \longrightarrow R\overset{H}{\underset{OH}{C}}CN \qquad \underset{\overset{\|}{O}}{RCR'} + HCN \longrightarrow R\overset{R'}{\underset{OH}{C}}CN$$

The addition reaction takes place by an ionic mechanism. The cyanide ion from HCN acts as a nucleophile towards the carbon of the carbonyl group which behaves as an .... (1). As the new carbon-carbon bond is formed, a pair of electrons moves from the carbonyl double bond to the oxygen atom which then carries the .... (2) charge.

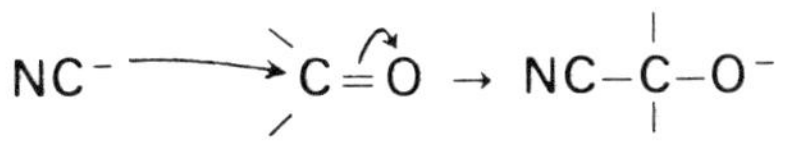

The reaction is completed by a proton transfer, e.g.,

$$NC{-}\overset{|}{\underset{|}{C}}{-}O^- + H_3O^+ \rightleftharpoons NC{-}\overset{|}{\underset{|}{C}}{-}OH + H_2O$$

---

(1) electrophile (2) negative

Carbon-oxygen fission—formation of alkyl halides and alkenes

19. Many alcohols react with halogen acids, HX, with substitution of the hydroxyl group by the halide X. For example, when pentan-1-ol is treated with anhydrous HBr, it is converted into 1-bromopentane with breaking of the .... (1) bond, and formation of a .... (2) bond.

---

(1) carbon-oxygen (2) carbon-bromine

---

20. Other reagents are also available for replacing a hydroxyl group by a halogen atom, e.g. a chlorine atom can be introduced with reagents such as PCl_5, PCl_3, $POCl_3$, or thionyl chloride, $SOCl_2$. A bromine atom is commonly introduced by using PBr_3, and an iodine atom by the use of a mixture of iodine and red phosphorus. Complete the equation

$$(CH_3)_2CHOH + SOCl_2 \longrightarrow \text{............} \quad (1)$$

---

(1) $(CH_3)_2CHOH + SOCl_2 \longrightarrow (CH_3)_2CHCl + SO_2 + HCl$

---

21. The *acid-catalysed dehydration* of alcohols, $-\underset{\underset{H}{|}}{\overset{|}{C}}-\underset{|}{\overset{|}{C}}-OH \longrightarrow \;\rangle C=C\langle\; + HOH$

is a reaction in which both a carbon-oxygen and a carbon- .... (1) bond are broken. For example, if a mixture of 2-methylpropan-1-ol (isobutanol) and phosphoric acid is heated to around 150°, the lower-boiling alkene of structure .... (2) distils out.

---

(1) hydrogen (2) $(CH_3)_2C{=}CH_2$

---

22. Since a molecule of water is *eliminated* from the alcohol to form an alkene, we call this an .... (1) reaction.

---

(1) elimination

TEST FRAME

Complete the following equations:

$$(CH_3)_2CHOH \xrightarrow[NaBr]{H_2SO_4} \text{............} \quad (1)$$

$$C_6H_5CH_2OH \xrightarrow{PBr_3} \text{............} \quad (2)$$

$$RCH_2CH_2OH \xrightarrow[heat]{H_3PO_4} \text{............} \quad (3).$$

---

(1) $CH_3\underset{\underset{O}{\|}}{C}CI_3 + \bar{O}H \longrightarrow CH_3COO^- + CHI_3$

REDUCTION OF ALDEHYDES AND KETONES

40. Aldehydes are reduced catalytically by hydrogen in the presence of finely divided platinum or palladium to yield primary alcohols according to the general equation . . . . $\xrightarrow{H_2/Pt\ or\ Pd}$ . . . . (1).

---

(1) $RCHO \xrightarrow{H_2/Pt\ or\ Pd} RCH_2OH$

---

41. Reduction of an aldehyde with either lithium tetrahydridoaluminate, $LiAlH_4$, or sodium tetrahydridoborate, $NaBH_4$, leads to the formation of a metal complex which on acidification yields the primary alcohol (see Frame 50 and also Program 6, Frames 42, 43).

$$RCHO \xrightarrow[(2)\ acid]{(1)\ LiAlH_4\ or\ NaBH_4} \text{. . . . . (1).}$$

---

(1) $RCHO \xrightarrow[(2)\ acid]{(1)\ LiAlH_4\ or\ NaBH_4} RCH_2OH$

---

42. Ketones are reduced by hydrogen over platinum or palladium to form . . . . (1) according to the equation . . . . (2).

---

(1) secondary alcohols

(2) $R\underset{\underset{O}{\|}}{C}R' \xrightarrow{H_2/Pt\ or\ Pd} R\underset{\underset{OH}{|}}{C}HR'$

---

43. Reduction of ketones with $LiAlH_4$ or $NaBH_4$ also leads to the formation of a metal complex which on acidification yields the secondary alcohol according to the equation . . . . (1).

---

(1) $R\underset{\underset{O}{\|}}{C}R' \xrightarrow[(2)\ acid]{(1)\ LiAlH_4\ or\ NaBH_4} R\underset{\underset{OH}{|}}{C}HR'$

TEST FRAME

What observation is made when butanal is boiled with Fehling's reagent? . . . . (1).

(1) $(CH_3)_2CHOH \xrightarrow[NaBr]{H_2SO_4} (CH_3)_2CHBr$

(2) $C_6H_5CH_2OH \xrightarrow{PBr_3} C_6H_5CH_2Br$

(3) $RCH_2CH_2OH \xrightarrow[heat]{H_3PO_4} RCH{=}CH_2$

Fission of carbon-hydrogen and oxygen-hydrogen bonds—oxidation of alcohols

23. Secondary alcohols, RR′CHOH, on treatment with various oxidising agents such as CrO_3, give rise to ketones RR′CO by the loss of two . . . . (1) atoms. For this process to work, the two hydrogens must be on *adjacent atoms*, one oxygen, one carbon.

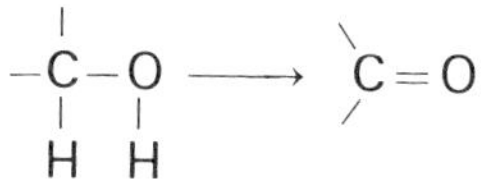

---

(1) hydrogen

---

24. Such an oxidative process is therefore not applicable to . . . . (1) alcohols which do not have a hydrogen atom on the carbon carrying the hydroxyl group.

---

(1) tertiary

---

25. Tertiary alcohols can be degraded by strong oxidants in acid solution. When this occurs the first step is an acid-catalysed dehydration of the alcohol to an . . . . (1) which then reacts with the oxidant with fission of the carbon-carbon . . . . (2) bond (cf. Program 3).

---

(1) alkene or olefin (2) double

---

26. The oxidation process is applicable to primary alcohols, RCH_2OH, but the final product is nearly always the corresponding carboxylic acid, RCOOH. The intermediate product corresponding to the removal of two hydrogen atoms is the . . . . (1) of structure . . . . (2), but this easily undergoes further oxidation to the acid.

---

(1) aldehyde (2) 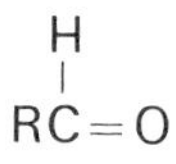

$CH_3\overset{\overset{O}{\|}}{C}R$ are oxidised by iodine in alkaline solution to give iodoform, CHI_3, and the conjugate base of the carboxylic acid RCOOH. In the case of the alcohols the first step is oxidation to the corresponding ketone CH_3COR (or acetaldehyde when R = H). The reaction of interest is

$$CH_3\underset{\underset{O}{\|}}{C}R + 3I_2 + 4OH^- \longrightarrow RCOO^- + CHI_3 + 3I^- + 3H_2O$$

Iodine in alkaline solution is a .... (weak/strong) (1) oxidising agent.

---

(1) weak

---

38. This ready oxidation of certain ketones by a weak oxidising agent in alkaline solution stands in contrast to their resistance to oxidation in acidic media. Ketones which have an α-hydrogen atom are oxidised in alkaline solution by way of the enolate anion since it is easier to remove an electron from an anion than from a neutral species. (Oxidation = removal of electrons). The iodoform reaction on a methyl ketone proceeds by a number of steps. The three hydrogen atoms of the methyl group are successively replaced by iodine atoms by the following process. Complete the second and third equations by analogy with the first.

$$CH_3\underset{\underset{O}{\|}}{C}CH_3 + \bar{O}H \rightleftharpoons H_2O + CH_3\underset{\underset{O}{\|}}{C}\bar{C}H_2 \quad I{-}I \longrightarrow CH_3\underset{\underset{O}{\|}}{C}CH_2I + I^-$$

$$CH_3\underset{\underset{O}{\|}}{C}CH_2I + \bar{O}H \rightleftharpoons H_2O + \text{............ (1)} \longrightarrow CH_3\underset{\underset{O}{\|}}{C}CHI_2 + I^-$$

$$CH_3\underset{\underset{O}{\|}}{C}CHI_2 + \bar{O}H \rightleftharpoons H_2O + \text{............ (2)} \longrightarrow CH_3\underset{\underset{O}{\|}}{C}CI_3 + I^-$$

---

(1) $CH_3\underset{\underset{O}{\|}}{C}\bar{C}HI \quad I{-}I \longrightarrow CH_3\underset{\underset{O}{\|}}{C}CHI_2 + I^-$

(2) $CH_3\underset{\underset{O}{\|}}{C}\bar{C}I_2 \quad I{-}I \longrightarrow CH_3\underset{\underset{O}{\|}}{C}CI_3 + I^-$

---

39. The triiodoacetone, $CH_3\underset{\underset{O}{\|}}{C}CI_3$, then reacts with a hydroxide ion to form the acetate ion and iodoform according to the equation .... (1).

---

27. Catalytic dehydrogenation of primary and secondary alcohols over hot copper affords aldehydes or ketones respectively, together with hydrogen gas. The equations can be written

$RCH_2OH \xrightarrow[250\text{-}300^\circ]{Cu}$ . . . . (1) and $R_2CHOH \xrightarrow[250\text{-}300^\circ]{Cu}$ . . . . (2).

---

(1) $RCH_2OH \xrightarrow[250\text{-}300^\circ]{Cu} RCHO + H_2$

(2) $R_2CHOH \xrightarrow[250\text{-}300^\circ]{Cu} R_2CO + H_2$

TEST FRAME

An alcohol, C_7H_8O, on oxidation with CrO_3 in acetic acid solution, gave a carboxylic acid $C_7H_6O_2$; on being passed over hot copper the alcohol gave a substance C_7H_6O. Write a possible structure for each compound, and give its name . . . . (1).

An alcohol, $C_4H_{10}O$, is resistant to mild oxidising agents, but is degraded by $K_2Cr_2O_7$ in hot, strongly acid solution. Write its structure and name . . . . (2).

---

(1) $C_6H_5CH_2OH$, benzyl alcohol
C_6H_5COOH, benzoic acid
C_6H_5CHO, benzaldehyde

(2) $(CH_3)_3COH$, 2-methylpropan-2-ol or t-butanol

SYNTHESIS OF ALCOHOLS

Methods of commercial importance

28. Methanol, CH_3OH, is made on a commercial scale by the catalytic reduction of carbon monoxide with hydrogen according to the equation . . . . (1). Ethanol, CH_3CH_2OH, is made by the fermentation of sugars and also by the hydration of ethylene, as discussed below.

---

(1) $CO + 2H_2 \longrightarrow CH_3OH$

---

29. Another two-carbon alcohol is ethane-1, 2-diol (ethylene glycol) of structure . . . . (1). This alcohol, in which the hydroxyl groups are *vicinal*, that is, ... (2), is prepared on a large scale by acid-catalysed hydrolysis of ethylene oxide, $CH_2{-}CH_2$.
\ /
O

---

adjacent (α-) carbon atoms is broken and the resulting fragments are both oxidised to carboxylic acids:

$$RCH_2\vdots C(=O)-CH_2R' \longrightarrow RCOOH \text{ and } HOOCCH_2R'$$

$$RCH_2-C(=O)\vdots CH_2R' \longrightarrow RCH_2COOH \text{ and } HOOCR'$$

Oxidation of pentan-3-one, a symmetrical ketone, yields only two products according to the equation

$$\ldots\ldots\ldots\ldots \xrightarrow[\text{heat}]{KMnO_4,\ H_3O^+} \ldots\ldots\ldots\ldots \text{ and } \ldots\ldots\ldots\ldots \quad (1)$$

---

(1) $$CH_3CH_2C(=O)CH_2CH_3 \xrightarrow[\text{heat}]{KMnO_4,\ H_3O^+} CH_3COOH \text{ and } HOOCCH_2CH_3$$

---

35. Oxidation of hexan-3-one, an unsymmetrical ketone, with hot acid permanganate yields three different acids, namely . . . . (1)

---

(1) $$CH_3CH_2C(=O)CH_2CH_2CH_3 \xrightarrow[\text{heat}]{KMnO_4,\ H_3O^+} CH_3COOH,\ HOOCCH_2CH_2CH_3, \text{ and } CH_3CH_2COOH.$$

---

36. Oxidation of cyclohexanone with concentrated nitric acid yields a dicarboxylic acid, hexandioic acid $C_6H_{10}O_4$, according to the equation . . . . (1).

---

(1) 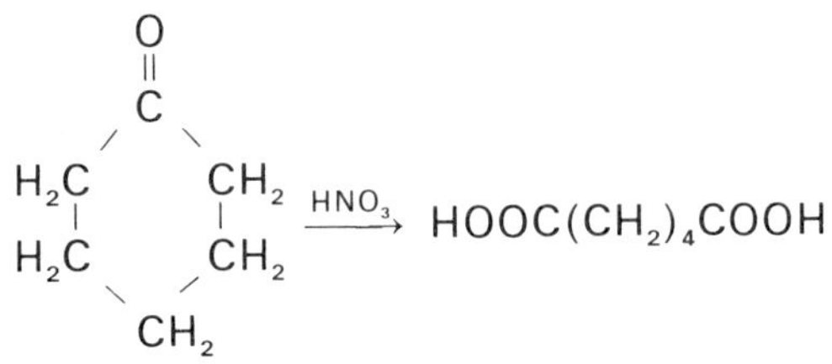

Note that the same acid is obtained irrespective of which CH_2-CO bond is cleaved.

The iodoform reaction

37. Alcohols having the structure $CH_3CH(OH)R$ and ketones having the structure

(1) $HOCH_2CH_2OH$ (2) on adjacent atoms

---

30. Hydration of the carbon-carbon double bond of an alkene (olefin) is another important method for preparing an alcohol. In this type of reaction which is catalysed by strong acid, the elements of water (H, OH) are *added* across the .... (1) bond,

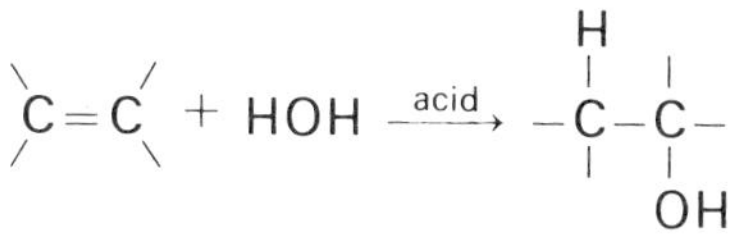

and it is therefore called an .... (2) reaction. This reaction is essentially the reverse of the acid-catalysed dehydration of alcohols discussed in Frames 21-22.

---

(1) double   (2) addition

---

31. Ethanol is made commercially on a large scale by hydration of .... (1) in hot aqueous sulphuric acid under pressure. This addition reaction of alkenes is also of importance in laboratory syntheses.

---

(1) ethene (ethylene), $CH_2{=}CH_2$

---

32. Certain alcohols are prepared industrially by the controlled catalytic oxidation of alkanes with oxygen, which can be shown schematically as

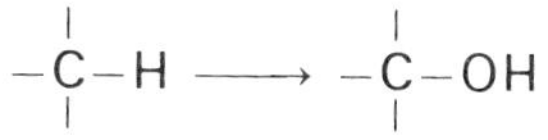

It is not easy to control this type of reaction on a laboratory scale, since fission of .... (1) bonds competes with the fission of carbon-hydrogen bonds, and the ultimate products can very easily be carbon dioxide and water. (cf. use of hydrocarbons as motor fuels).

---

(1) carbon-carbon

Laboratory methods.

The Grignard method. Conversion of an alkyl or aryl halide, RX, into a primary, secondary, or tertiary alcohol.

33. The Grignard method allows the conversion of a halide RX into an alcohol R-C-OH. It can be seen from these formulae that the method results in an .... (1) in the number of carbon atoms in the product as compared to the number in the starting halide. From a given halide RX it is possible

TEST FRAMES

List the following compounds under the headings "enolisable" and "non-enolisable": Benzaldehyde . . . . (1); 3,3-dimethylpentan-2-one . . . . (2); methyl phenyl ketone . . . . (3); diphenyl ketone . . . . (4); 2,2-dimethylpropanal . . . . (5); cyclopentanone . . . . (6).

Enolisable: (2) $CH_3CH_2C(CH_3)_2COCH_3$, (3) $C_6H_5COCH_3$

(6) 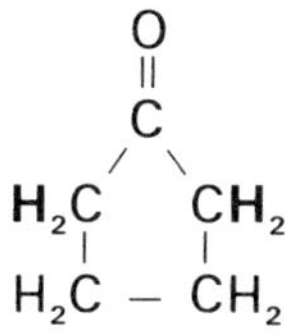

Non-enolisable: (1) C_6H_5CHO (4) $C_6H_5COC_6H_5$ (5) $(CH_3)_3CCHO$

Write the structure and give the name of the product obtained when propanal (propionaldehyde) is treated with dilute aqueous sodium hydroxide under conditions which produce aldol addition . . . . (1).

---

(1) $CH_3CH_2CH(OH)CH(CH_3)CHO$ 3-hydroxy-2-methylpentanal

OXIDATION OF ALDEHYDES AND KETONES

33. A characteristic property of aldehydes which is not shared by ketones is their ease of oxidation. Aldehydes give rise to a silver mirror when they are treated with ammoniacal silver nitrate solution (Tollen's reagent) whereas ketones do not. Similarly aldehydes give a red precipitate of copper(I) oxide when they are boiled with an ammoniacal solution of a copper tartrate complex (Fehling's reagent) or a copper citrate complex (Benedict's reagent) but ketones do not. These observations show that aldehydes are . . . . (1) agents and are themselves easily . . . . (2).

---

(1) reducing (2) oxidised

---

34. Ketones are more difficult to oxidise. When oxidised with a powerful oxidising agent the bond between the carbonyl group and one or other of the

to prepare primary, secondary, and tertiary alcohols, and these different cases will be discussed below.

---

(1) increase

---

34. Grignard reagents are prepared from alkyl halides (cf. Program 4) or aryl halides (cf. Program 9) by reaction with metallic magnesium in anhydrous ether. The resulting Grignard reagent, obtained as a solution in ether, reacts rapidly with any added acid with formation of the parent alkane or arene, so that care must be taken to exclude atmospheric moisture. In the case of bromoethane (ethyl bromide), the equation for the formation of the reagent is $CH_3CH_2Br + Mg \longrightarrow CH_3CH_2MgBr$. The organometallic Grignard reagent is polarised in the sense $-\overset{|}{\underset{|}{C}}{}^{\delta-}-\overset{\delta+}{Mg}Br$ so that the organic group attached to magnesium can function as a . . . . (1).

---

(1) nucleophile

---

35. By contrast, the carbon-oxygen double bond present in an aldehyde, RCHO, ketone, R_2CO, or ester, RCOOR′, is polarised in the opposite sense to that of a Grignard reagent, so that the polarity is . . . . (1).

---

(1) $\gt\overset{\delta+}{C}=\overset{\delta-}{O}$

---

36. The Grignard reagent adds readily to the $\gt C=O$ bond and on the basis of simple electrostatics we expect (and find) that reaction leads to the organic (hydrocarbon) part of the organometallic reagent becoming attached to the . . . . (1) atom of the $\gt C=O$ group with formation of a new . . . . (2) bond.

---

(1) carbon (2) carbon-carbon

---

37. The general equation for the above addition reaction can be written

$$\gt C=O + RMgX \longrightarrow R-\overset{|}{\underset{|}{C}}-OMgX.$$

If we compare the product $-\overset{|}{\underset{|}{C}}-OMgX$ with compounds $-\overset{|}{\underset{|}{C}}-ONa$ we see that the former, like the latter, is an . . . . (1) and on treatment with dilute acid it should (and does) yield an . . . . (2).

$$R-\overset{|}{\underset{|}{C}}-OMgX + HX \longrightarrow R-\overset{|}{\underset{|}{C}}-OH + MgX_2$$

---

(1) alkoxide (2) alcohol

(1) $CH_3\underset{OH}{CH}CH_2CHO \xrightarrow{acid} CH_3CH{=}CHCHO + H_2O$

(2) conjugated

---

31. Acetone undergoes the same type of condensation when it is passed through solid barium hydroxide to form diacetone alcohol (4-hydroxy-4-methyl-pentan-2-one) in low yield according to the equation

$$2CH_3COCH_3 \xrightarrow{Ba(OH)_2} CH_3\underset{OH}{\overset{CH_3}{C}}CH_2COCH_3$$

Write a sequence of equations for the formation of this product as follows:
Formation of the acetone enolate ion . . . . (1)
Attack by the enolate ion on a second molecule of acetone . . . . (2)
Transfer of a proton from water to the alkoxide ion to complete the formation of diacetone alcohol . . . . (3).

---

(1) $CH_3\underset{O}{\underset{\|}{C}}CH_3 + OH^- \rightleftharpoons CH_3\underset{O}{\underset{\|}{C}}\bar{C}H_2 + H_2O$

(2) $CH_3\underset{O}{\underset{\|}{C}}\bar{C}H_2 \curvearrowright (CH_3)_2C{=}O \rightleftharpoons CH_3\underset{O}{\underset{\|}{C}}CH_2\underset{CH_3}{\overset{CH_3}{C}}{-}O^-$

(3) $CH_3\underset{O}{\underset{\|}{C}}CH_2\underset{CH_3}{\overset{CH_3}{C}}{-}O^- + H_2O \rightleftharpoons CH_3\underset{O}{\underset{\|}{C}}CH_2\underset{CH_3}{\overset{CH_3}{C}}{-}OH + OH^-$

---

32. Diacetone alcohol is dehydrated to mesityl oxide, 4-methylpent-3-en-2-one, on being heated with acid, according to the equation

$$CH_3\underset{OH}{\overset{CH_3}{C}}CH_2COCH_3 \xrightarrow{acid} \text{. . . . .} \quad (1)$$

---

(1) $CH_3\underset{OH}{\overset{CH_3}{C}}CH_2COCH_3 \xrightarrow{acid} CH_3\overset{CH_3}{C}{=}CHCOCH_3 + H_2O$

38. The reaction of RMgBr with formaldehyde, HCHO, yields a bromomagnesium alkoxide of structure . . . . (1), which with acid yields a . . . . (2) alcohol.

---

(1) RCH_2OMgBr or $RCH_2O^-Mg^{2+}Br^-$ (2) primary

---

39. With aldehydes other than formaldehyde, the final product of the Grignard process is a secondary alcohol, the appropriate equation being

RMgX + R′CHO ⟶ . . . . (1)

---

(1) RMgX + R′CHO ⟶ R′CH(R)(OMgX) $\xrightarrow{\text{acid}}$ R′CH(R)(OH)

---

40. When the Grignard reagent phenylmagnesium bromide, C_6H_5MgBr which is prepared from bromobenzene and magnesium in anhydrous ether, is allowed to react with methyl propyl ketone and the intermediate bromomagnesium alkoxide is treated with acid, the alcohol formed has the structure . . . . (1). In general, reaction of a Grignard reagent with a ketone is a useful method for the synthesis of . . . . (2) alcohols.

---

(1) $CH_3C(OH)(C_6H_5)CH_2CH_2CH_3$ (2) tertiary

---

41. An alcohol is also obtained by reaction of a Grignard reagent with an ester, which likewise contains a carbonyl or >C=O group. Reaction of RMgBr with an ester such as R′C(=O)OEt yields finally a tertiary alcohol containing two identical R groups derived from the Grignard reagent. The first formed intermediate, R′–C(OMgBr)(R)–OEt which has two oxygen functions attached to the same saturated carbon atom, decomposes to BrMgOEt and (R)(R′)C=O.

reacts with the carbonyl group of a second molecule of acetaldehyde according to the equation

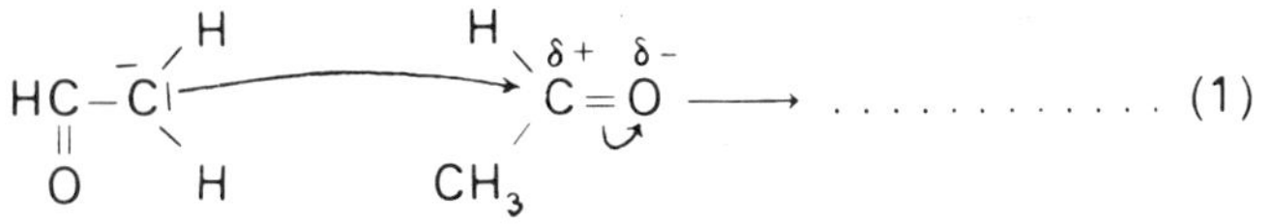

$$
\underset{\underset{O}{||}}{HC}-\overset{\overset{H}{\diagup}}{\underset{\diagdown}{C}}_{H} \quad \longrightarrow \quad \overset{H}{\underset{CH_3}{\diagdown}}\overset{\delta+\ \ \delta-}{C=O} \quad \longrightarrow \quad \ldots\ldots\ldots\ldots \text{ (1)}
$$

---

(1) $\underset{\underset{O}{||}}{HC}-\overset{\overset{H}{\diagup}}{\underset{\diagdown}{C}}_{H} \quad \overset{H}{\underset{CH_3}{\diagdown}}\overset{\delta+\ \ \delta-}{C=O} \quad \rightleftharpoons \quad \underset{\underset{O}{||}}{HCCH_2}\overset{\overset{H}{|}}{\underset{\underset{CH_3}{|}}{C}}-O^-$ 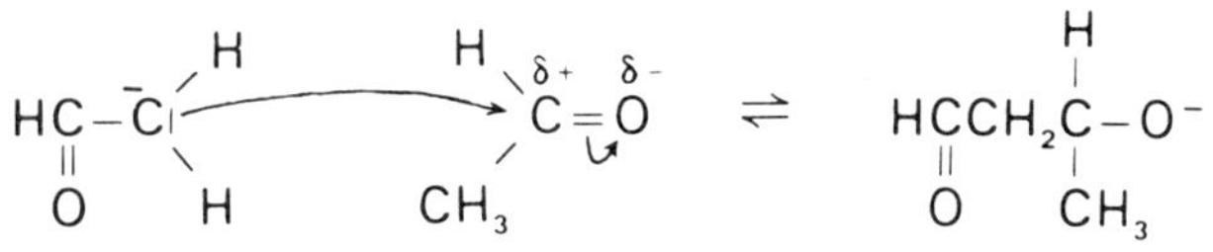

---

28. Finally, in the mildly alkaline solution, the resulting alkoxide ion removes a proton from water according to the equilibrium

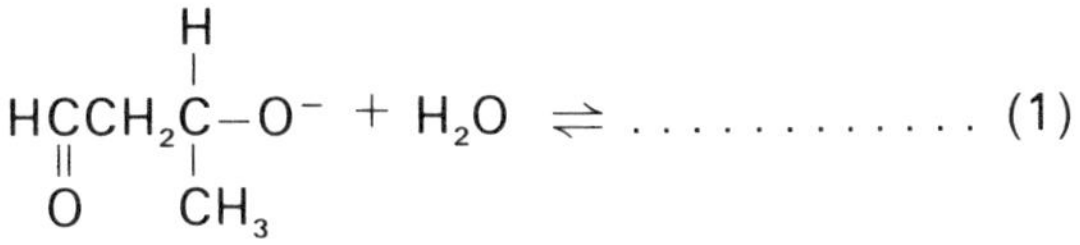

$$
\underset{\underset{O}{||}}{HCCH_2}\overset{\overset{H}{|}}{\underset{\underset{CH_3}{|}}{C}}-O^- + H_2O \;\rightleftharpoons\; \ldots\ldots\ldots\ldots \text{ (1)}
$$

---

(1) $\underset{\underset{O}{||}}{HCCH_2}\overset{\overset{H}{|}}{\underset{\underset{CH_3}{|}}{C}}-O^- + H_2O \;\rightleftharpoons\; \underset{\underset{O}{||}}{HCCH_2}\overset{\overset{H}{|}}{\underset{\underset{CH_3}{|}}{C}}-OH + OH^-$ 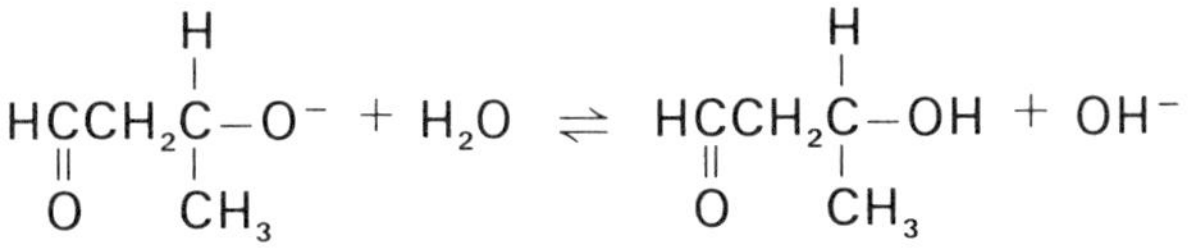

---

29. The aldol addition can be summarised in the form

$CH_3CHO + NaOH \rightleftharpoons {}^-CH_2CHO + Na^+ + H_2O$

$CH_3CHO + {}^-CH_2CHO + Na^+ \rightleftharpoons \underset{\underset{O^-}{|}}{CH_3CH}CH_2CHO + Na^+$

$\underset{\underset{O^-}{|}}{CH_3CH}CH_2CHO + Na^+ + H_2O \rightleftharpoons \underset{\underset{OH}{|}}{CH_3CH}CH_2CHO + NaOH$

Note that the reactions are .... (reversible/non-reversible) (1)
and the sodium hydroxide .... (is/is not) (2) consumed.
Note also that the product of the reaction is itself an aldehyde with enolisable
α-hydrogen atoms and further aldol additions can take place.

---

(1) reversible   (2) is not

---

30. If 3-hydroxybutanal (3-hydroxybutyraldehyde) is heated with acid it is dehydrated to but-2-enal (crotonaldehyde) according to the equation .... (1). In but-2-enal the C=C and C=O bonds are .... (conjugated/unconjugated) (2).

The latter .... (1) then reacts rapidly with a second mole of RMgBr and after acid work-up, the product is the tertiary alcohol of structure .... (2).

---

(1) ketone

(2) $$\begin{array}{ccccc} & & R & & \\ & & | & & \\ R' & - & C & - & OH \\ & & | & & \\ & & R & & \end{array}$$

TEST FRAME

The Grignard method for the synthesis of alcohols can be summarised as follows:

The halide RX is converted into an organomagnesium compound RMgX which on reaction with CH_2O followed by acidification yields RCH_2OH,
on similar reaction with R′CHO yields RCH(OH)R′,
on similar reaction with a ketone R′COR″ yields RR′R″COH,
and on similar reaction with $R'CO_2CH_3$ (or other ester) yields $$\begin{array}{ccccc} & & R & & \\ & & | & & \\ R' & - & C & - & OH. \\ & & | & & \\ & & R & & \end{array}$$

That is, the Grignard method allows us to derive a .... (1) from formaldehyde, a .... (2) from higher aldehydes, and tertiary alcohols from .... (3) or from .... (4) of carboxylic acids.

---

(1) primary alcohol (2) secondary alcohol (3) ketones (4) esters

---

Methods for preparing primary and secondary alcohols by the reduction of more highly oxidised compounds

42. In Frames 23-27 we saw that alcohols can be oxidised to products such as aldehydes, ketones, and acids. The reduction of an aldehyde or ketone, or an ester of a carboxylic acid, is often a convenient method for preparing an alcohol. Suitable reduction of an aldehyde RCHO yields a primary alcohol, RCH_2OH, while reduction of a ketone RR′C=O necessarily yields a .... (1) alcohol of structure .... (2).

---

(1) secondary

(2) $$\begin{array}{ccccc} & & R' & & \\ & & | & & \\ R & - & C & - & OH \\ & & | & & \\ & & H & & \end{array}$$ or RR′CHOH

---

43. These reductions are conveniently effected with lithium tetrahydridoaluminate ($LiAlH_4$) in dry ether solution, or with sodium tetrahydridoborate ($NaBH_4$) in

24. Simple aldehydes and ketones also enolise. For example the tautomerism of acetone would be shown by the equilibrium

$$\underset{\text{keto}}{CH_3C(=O)CH_3} \rightleftharpoons \underset{\text{enol}}{CH_2{=}C(OH)CH_3}$$

However the equilibrium lies far on the keto side where the hydrogen atom is attached to the carbon atom and very little of the enolic form is present. Removal, by a base, of the proton from the oxygen atom of the enolic form results in the formation of the same mesomeric anion as that obtained by removal of a proton from the carbon atom of the keto form. The anion is consequently called the . . . . (1) anion.

---

(1) enolate

The effect of alkali on enolisable aldehydes and ketones

25. *Aldol addition*. Two molecules of acetaldehyde react together in the presence of dilute aqueous sodium hydroxide to give a low yield of 3-hydroxybutanal (acetaldol), along with a complex mixture of other products. This type of reaction is known as the aldol addition or aldol condensation.

$$2CH_3CHO \xrightarrow{\text{dilute NaOH}} \ldots\ldots\ldots\ldots \quad (1)$$

---

(1) $2CH_3CHO \xrightarrow{\text{dilute NaOH}} CH_3CH(OH)CH_2CHO$

---

26. The first step in the aldol addition is the formation of the enolate anion according to the equation $CH_3CHO + NaOH \rightleftharpoons$ . . . . (1).

---

(1) $CH_3CHO + NaOH \rightleftharpoons Na^+ + {}^-CH_2CHO + H_2O$

Note that in the next frame the enolate ion is turned around and written:

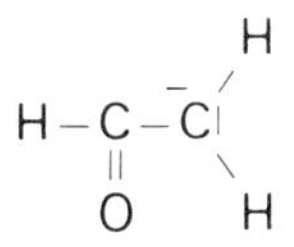

---

27. The acetaldehyde carbanion is strongly nucleophilic towards the carbon atom of a carbonyl group which has the property of a weak electrophile (cf. Frame 3). Since only part of the acetaldehyde is converted into the enolate ion there is a high concentration of acetaldehyde present. The enolate ion

water or aqueous methanol. Reduction of the compounds $(CH_3)_2CHCHO$ and $(CH_3)_2CHCOCH_3$ yields alcohols of structure .... (1) and .... (2) respectively

---

(1) $(CH_3)_2CHCH_2OH$ (2) $(CH_3)_2CH\overset{\overset{\large OH}{|}}{\underset{\underset{\large H}{|}}{C}}CH_3$ or $(CH_3)_2CHCH(OH)CH_3$

---

44. Esters, which have the structure $R\overset{\overset{O}{\|}}{C}OR'$, can also be reduced by $LiAlH_4$ according to the overall equation:

$$R\overset{\overset{O}{\|}}{C}OR' + 4(H) \longrightarrow R-\overset{\overset{OH}{|}}{\underset{\underset{H}{|}}{C}}-H + HOR'$$

That is, cleavage of the ester molecule occurs, and two different alcohols are produced, RCH_2OH and $R'OH$. One of these products, $R'OH$, is the alcohol, commonly methanol or ethanol, that was used in the first place to prepare the ester. The other product, RCH_2OH, is the one desired, being the .... (1) alcohol corresponding to the acid .... (2).

---

(1) primary (2) RCOOH

---

45. Hence $LiAlH_4$ reduction of the ester $EtCH(CH_3)CO_2CH_3$ yields methanol and the primary alcohol of structure .... (1).

---

(1) $EtCH(CH_3)CH_2OH$

Preparation of alcohols by hydrolysis of alkyl halides

46. Primary alkyl halides have the structure RCH_2X where X is a halogen atom. Hydrolysis of these compounds in aqueous alkali, $RCH_2X + OH^- \longrightarrow RCH_2OH + X^-$, is clearly a method for the synthesis of primary alcohols. The reaction is one of .... (1) substitution of the halogen atom by .... (2) ion. A carbon-halogen bond is broken and a .... (3) bond is formed.

---

(1) nucleophilic (2) hydroxide (3) carbon-oxygen

Keto-enol tautomerism

21. The diketone acetylacetone is normally a mixture of two interconvertible forms, the keto and the enol.

$$\underset{\text{keto}}{CH_3\underset{\underset{O}{\|}}{C}CH_2\underset{\underset{O}{\|}}{C}CH_3} \qquad \underset{\text{enol}}{CH_3\underset{\underset{OH}{|}}{C}=CH\underset{\underset{O}{\|}}{C}CH_3}$$

The term "enol" is used to show that the particular structure has both a double bond (en) and a hydroxyl group (ol). The presence of these interconvertible forms can be shown by chemical and physical means and under carefully controlled conditions the two distinct compounds can sometimes be separated. Chemical species which have the same molecular formulae but which differ in the way in which the component atoms are joined together are called structural isomers (cf. Program 11, Frames 2-8). Structural isomers which can co-exist in rapid equilibrium are called *tautomers*, and the phenomenon is known as *tautomerism*. The keto and enol forms of acetylacetone are therefore tautomers and the keto-enol equilibrium is an example of tautomerism. The keto and enol tautomers differ structurally in the position of attachment of a .... (1) atom.

---

(1) hydrogen

---

22. Removal, by a base, of a proton from the α-carbon atom of the keto form of acetylacetone yields the *same mesomeric anion* as removal of a proton from the oxygen atom of the enol form. The resulting anion is consequently known as an *enolate anion*. Acidification of a solution containing the sodium salt of acetylacetone results in the formation of .... (1).

---

(1) a mixture of the keto and enol forms of acetylacetone

---

23. The phenomenon of keto-enol tautomerism is also shown by ethyl 3-oxobutanoate (ethyl acetoacetate), CH_3COCH_2COOEt. The presence of the two forms can again be shown by chemical and physical means and they can also be separated. The position of equilibrium, i.e. the proportion of the enolic form present in the tautomeric mixture, differs from compound to compound. In ethyl acetoacetate, the ketone enolises giving rise to the tautomeric equilibrium .... (1).

---

(1) $$\underset{\text{keto}}{CH_3\underset{\underset{O}{\|}}{C}CH_2COOEt} \rightleftharpoons \underset{\text{enol}}{CH_3\underset{\underset{OH}{|}}{C}=CHCOOEt}$$

47. The analogous hydrolysis of a secondary alkyl halide R_2CHX, yields a secondary alcohol, R_2CHOH. The equation for this nucleophilic substitution reaction can be written

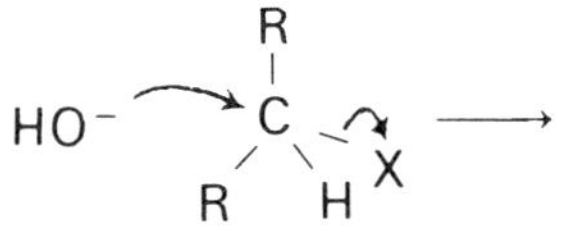

. . . . (1)

---

(1) 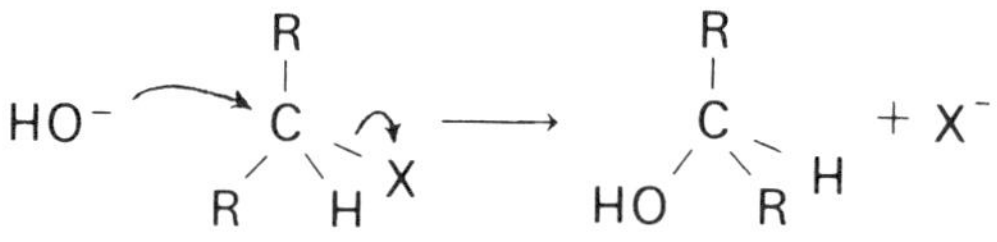

---

48. The hydrolysis of tertiary alkyl halides, R_3CX, is not always a good method for preparing the corresponding tertiary alcohol. A competing reaction is the elimination of HX from adjacent carbon atoms, with the formation of an alkene.

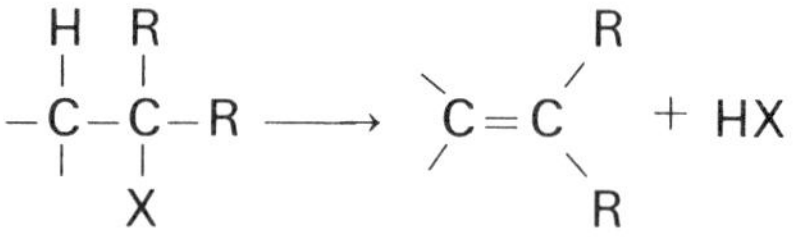

Hydrolysis of 2-chloro-2-methylpropane (t-butyl chloride) gives a mixture of the tertiary alcohol . . . . (1) and the alkene . . . . (2).

---

(1) $(CH_3)_3COH$
2-methylpropan-2-ol
or t-butanol

(2) $(CH_3)_2C{=}CH_2$
2-methylpropene
or isobutylene

Methods for preparing alcohols from alkenes

49. A general method for the preparation of 1,2-diols is found in the hydroxylation of alkenes with aqueous alkaline potassium permanganate, whereby an –OH group is added to each of the carbon atoms at each end of the double bond. Thus the alkene $CH_3CH{=}CHCH_2CH_3$ yields a diol of structure . . . . (1). Since alcohols are easily oxidised by permanganate to other products, it is important not to use an . . . . (2) of the reagent.

---

(1) $CH_3CH{-}CHCH_2CH_3$ (with OH on each of the two CH carbons)
OH OH

(2) excess

---

50. 1,2-Diols may also be prepared by hydrolysis of epoxides, prepared in turn from alkenes and peracids (cf. Program 3, Frame 18).

(1) 

(2) 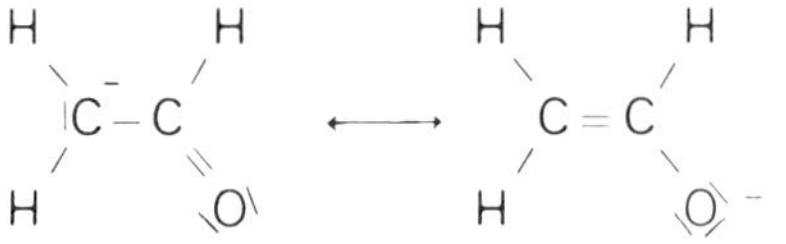

(3) 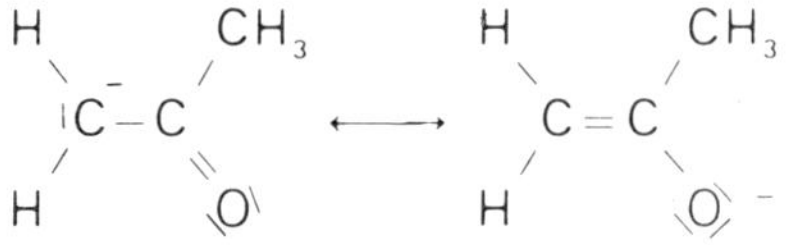

The acidity of α-hydrogen atoms in β-diketones

19. Examples of compounds containing two carbonyl groups which are in the β-position with respect to one another $-COCH_2CO-$ (cf. Frame 16) are β-dialdehyde .... (1), β-keto aldehyde .... (2), β-diketone .... (3), β-keto ester .... (4), and β-diester .... (5). Show the simplest structures possible using methyl ketones and ethyl esters.

---

(1) $HC(=O)CH_2C(=O)H$ (2) CH_3COCH_2CHO (3) $CH_3COCH_2COCH_3$

(4) CH_3COCH_2COOEt (5) $EtOOCCH_2COOEt$

---

20. Acetylacetone or pentane-2,4-dione, $CH_3COCH_2COCH_3$, is a β-diketone. The hydrogen atoms of the $-CH_2-$ group between the carbonyl groups are called the α-hydrogen atoms. Because of the proximity of the two carbonyl groups the acidity of the α-hydrogen atoms in acetylacetone is greatly increased compared with the acidity of those in a simple ketone. The theory of mesomerism can be used to explain this increase in acidity since the negative charge of the carbanion or enolate ion can be delocalised over a greater number of atoms than in the case of a simple ketone.

$$CH_3COCH_2COCH_3 + NaOC_2H_5 \rightleftharpoons CH_3CO\bar{C}HCOCH_3Na^+ + C_2H_5OH$$

Draw mesomeric structures for the acetylacetone enolate ion .... (1).

---

(1) 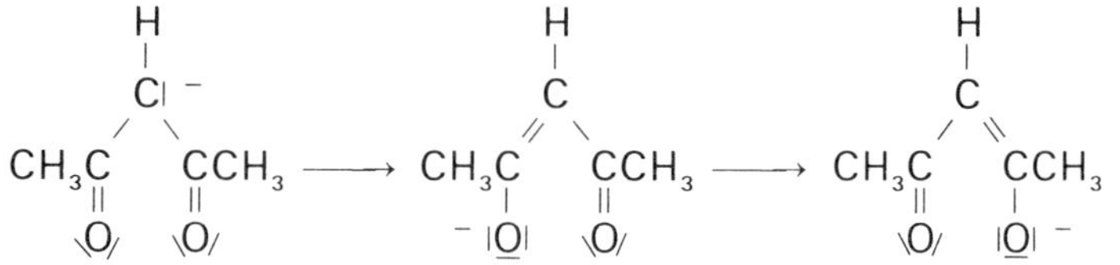

Note that the remaining α-hydrogen atom could be replaced by an alkyl group without affecting the mesomeric structures.

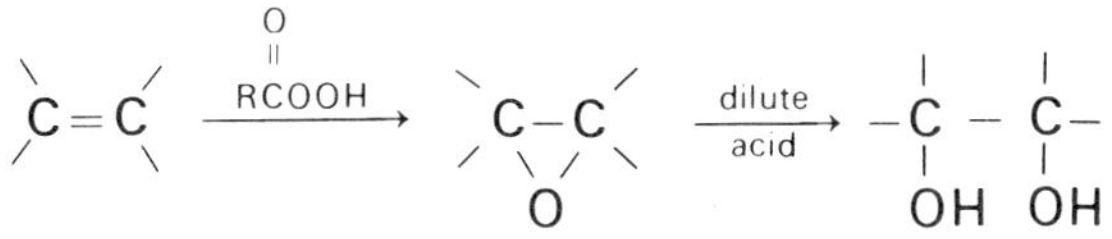

Write equations for the conversion of prop-2-en-1-ol (allyl alcohol) into propane-1,2,3-triol (glycerol) . . . . (1).

---

(1) 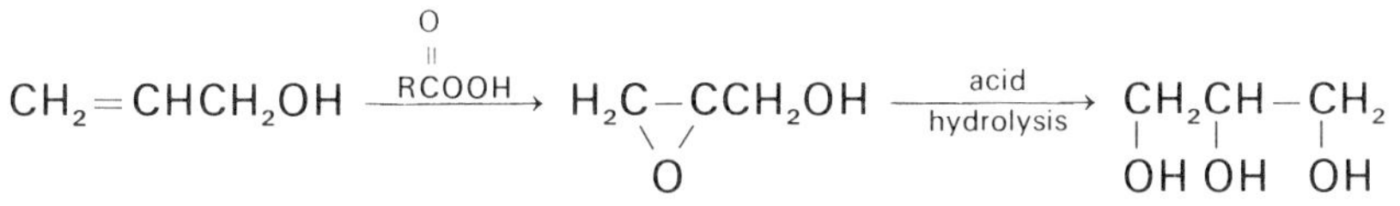

---

51. It was shown in Frames 30 and 31 that alkenes can be hydrated to give alcohols:

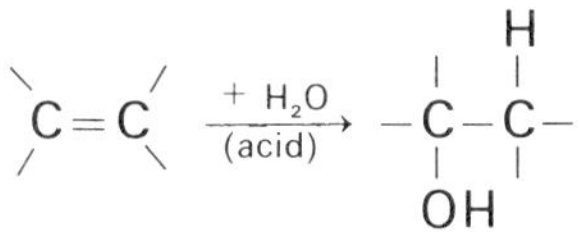

In the case of a symmetrical alkene, e.g., ethene, only one alcohol can be formed on hydration. With an unsymmetrical alkene such as $CH_3CH_2CH=CH_2$ it might appear that hydration could yield two isomeric products, namely the alcohols of structure . . . . (1).

---

(1) $CH_3CH_2CH(OH)CH_3$ and $CH_3CH_2CH_2CH_2OH$

---

52. In practice only the secondary alcohol, $CH_3CH_2CH(OH)CH_3$, is obtained by hydration of but-1-ene. In such cases, the empirical "Markovnikov rule" is useful in deciding which isomeric alcohol will be formed. The general rule states that "the negative part of the adding molecule becomes joined to that carbon of the double bond which holds the least number of hydrogen atoms". This is equivalent to saying, for an addendum HX (such as HOH), that the hydrogen goes to the carbon already holding the greater number of hydrogens (cf. Programs 3 and 4). Hence acid-catalysed hydration of the alkenes $(CH_3)_2C=CH_2$ and $CH_3CH=C(CH_3)_2$ yields respectively the alcohols of structure . . . . (1) and . . . . (2).

---

(1) $(CH_3)_2\underset{\displaystyle OH}{\underset{|}{C}}CH_3$ or $(CH_3)_3COH$ (2) $CH_3CH_2\underset{\displaystyle OH}{\underset{|}{C}}(CH_3)_2$

---

53. For alkenes such as $CH_3CH=CHCH_2CH_3$, $(CH_3)_2C=C(CH_2CH_3)_2$ etc., where the number of hydrogens attached to each of the unsaturated carbon atoms

THE WEAK ACIDITY OF HYDROGEN ATOMS ATTACHED TO THE α-CARBON ATOMS IN ALDEHYDES AND KETONES

Enolate anions

16. The carbon atoms of an alkyl chain attached to a carbonyl group are often designated by the Greek letters alpha (α), beta (β), gamma (γ) etc.

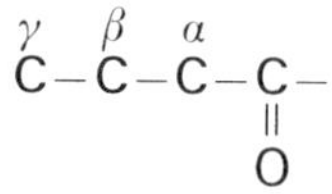

If the hydrogen atoms attached to the α-carbon atoms are defined as being α-hydrogen atoms, then the number of α-hydrogen atoms in acetaldehyde, and acetone is respectively .... (1), and .... (2).

---

(1) three (2) six

---

17. The α-hydrogen atoms in aldehydes and ketones are weakly acidic and a strong base, such as the ethoxide anion, can remove such a proton, leaving a negative charge on the α-carbon atom. Carbon atoms carrying a negative charge are known as carbanions. In the case of acetone the following equilibrium would exist:

$$CH_3COCH_3 + Na^+C_2H_5O^- \rightleftharpoons \text{............ (1)}$$

---

(1) $CH_3COCH_3 + Na^+C_2H_5O^- \rightleftharpoons {}^-CH_2COCH_3\,Na^+ + C_2H_5OH$

---

18. This weak acidity is due to the fact that when the proton is removed, the charge on the resulting carbanion is not localised but is spread over a number of atoms. When a molecule or ion can be represented by structures having the *same arrangement of atomic nuclei* but *different distributions of valence electrons*, the actual electronic structure is a weighted average of these different distributions. The delocalisation of the charge in the carbanion can be indicated by writing such *mesomeric structures*. An α-hydrogen atom has the relationship to the carbonyl group indicated in the partial structure:

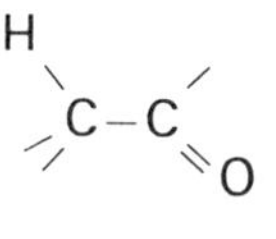

Write mesomeric structures for the anions resulting from loss of a proton from the above partial structure .... (1), from ethanal (acetaldehyde) propan-2-one .... (2), and from (acetone) .... (3). It is helpful to remember the structure of the carboxylate ion.

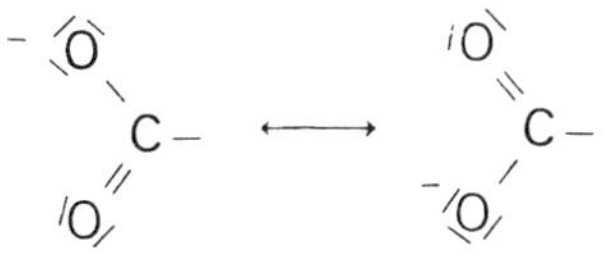

---

is either the same or zero, the situation is more evenly balanced, and a
. . . . (1) of isomeric alcohols is obtained.

---

(1) mixture

PERIODATE CLEAVAGE OF 1,2-DIOLS AND RELATED COMPOUNDS
54. The bond between two carbon atoms, each of which carries a hydroxyl
substituent, can be split readily by oxidation. With sodium periodate ($NaIO_4$)
in aqueous solution, reaction occurs as shown in the equation below.

$$\begin{matrix} -\overset{|}{C}-OH \\ -\overset{|}{\underset{|}{C}}-OH \end{matrix} + IO_4^- \longrightarrow \begin{matrix} -\overset{|}{C}=O \\ -\underset{|}{C}=O \end{matrix} + IO_3^- + H_2O$$

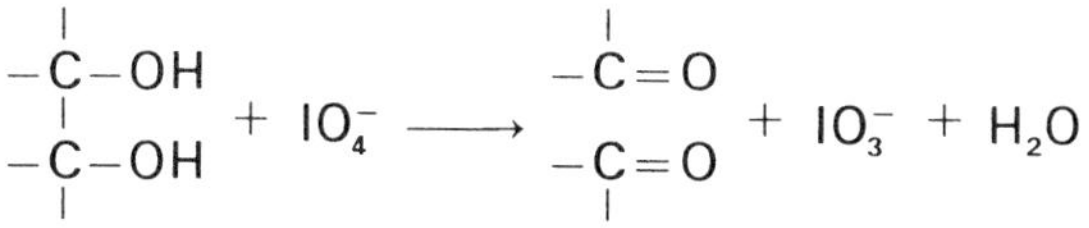

Provided we are dealing with a 1,2-diol, a functional group $-CH_2OH$ is
converted into $HCHO$, a group $-CH(OH)R$ into $RCHO$, and a group
$-C(OH)RR'$ into $RCOR'$. Oxidation of propane-1,2-diol with sodium
periodate proceeds according to the equation . . . . (1).

$$(1)\ CH_3\underset{OH}{\overset{|}{CH}}-\underset{OH}{\overset{|}{CH_2}} + IO_4^- \longrightarrow CH_3\underset{O}{\overset{||}{CH}} + H\underset{O}{\overset{||}{CH}} + IO_3^- + H_2O$$ 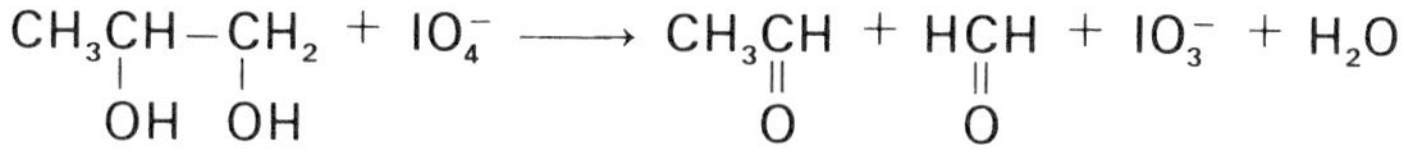

---

55. A similar cleavage occurs with α-hydroxy aldehydes or ketones,

$$\begin{matrix} -\overset{|}{C}-OH \\ \underset{|}{C}=O \end{matrix} + IO_4^- \longrightarrow \begin{matrix} -\overset{|}{C}=O \\ OH \\ \underset{|}{C}=O \end{matrix} + IO_3^-$$

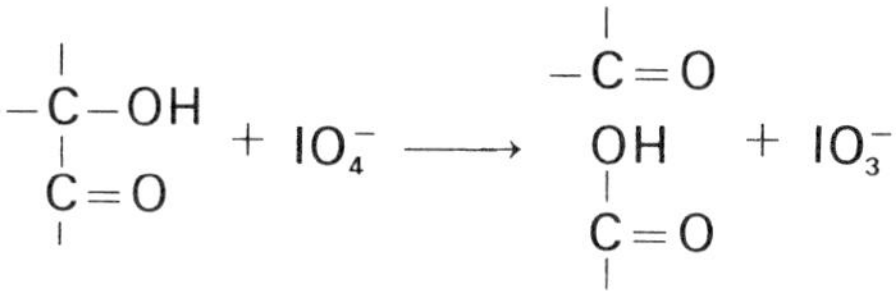

the products being an aldehyde or ketone, and a carboxylic acid. Write
down the products of periodate oxidation of $CH_3CH(OH)CHO$ . . . . (1),
$C_6H_5COCH_2OH$ . . . . (2), $CH_3COCH(OH)CH_3$ . . . . (3).

---

(1) CH_3CHO, $HCOOH$ (2) C_6H_5COOH, $HCHO$

(3) CH_3COOH, CH_3CHO

---

56. Oxidation of a 1,2,3-triol with periodate proceeds according to the overall
equation

$$\begin{matrix} -\overset{|}{C}-OH \\ -\overset{|}{C}-OH \\ -\underset{|}{C}-OH \end{matrix} + 2IO_4^- \longrightarrow \begin{matrix} -\overset{|}{C}=O \\ -COOH \\ -\underset{|}{C}=O \end{matrix} + 2IO_3^- + H_2O$$

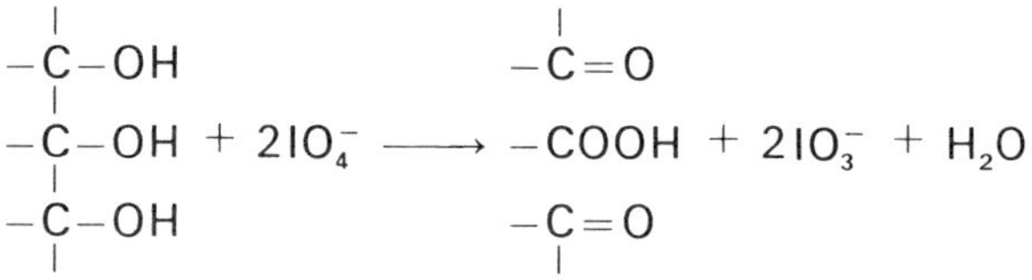

14. The oxidising agent can be acidified permanganate or chromic acid, e.g.,

$$3RCH(OH)R' + Cr_2O_7^{2-} + 8H^+ \longrightarrow \ldots\ldots\ldots\ldots (1).$$

---

(1) $3RCH(OH)R' + Cr_2O_7^{2-} + 8H^+ \longrightarrow 3RCOR' + 2Cr^{3+} + 7H_2O$

An abbreviated form of this equation would be shown

$$RCH(OH)R' \xrightarrow{H_2Cr_2O_7} RCOR'$$

---

15. Substituted methyl ketones can be prepared from ethyl 3-oxobutanoate (see Program 12). Methyl ketones can also be prepared by reacting acid chlorides with dimethyl cadmium according to the equation

$$2RCOCl + Cd(CH_3)_2 \longrightarrow \ldots\ldots\ldots\ldots (1).$$

---

(1) $2RCOCl + Cd(CH_3)_2 \longrightarrow 2RCOCH_3 + CdCl_2$

TEST FRAMES

3-Methylbutanal can be prepared by dehydrogenation of . . . . (name) (1) or by reduction of . . . . (name) (2). The equations for these processes showing structural formulae, are . . . . (3) and . . . . (4) respectively.

---

(1) 3-methylbutan-1-ol (2) 3-methylbutanoyl chloride

(3) $CH_3\overset{\overset{CH_3}{|}}{C}HCH_2CH_2OH \xrightarrow[250\text{-}300°]{Cu} CH_3\overset{\overset{CH_3}{|}}{C}HCH_2\overset{\overset{O}{\|}}{C}H + H_2$

(4) $CH_3\overset{\overset{CH_3}{|}}{C}HCH_2\overset{\overset{O}{\|}}{C}Cl \xrightarrow{LiAlH(OtBu)_3} CH_3\overset{\overset{CH_3}{|}}{C}HCH_2\overset{\overset{O}{\|}}{C}H$

Hexan-3-one (ethyl propyl ketone) can be prepared by dehydrogenation of . . . . (name) (1) over . . . . (2), or by oxidation with . . . . (3). Abbreviated equations for these processes, showing structural formulae are . . . . (4) and . . . . (5) respectively.

---

(1) hexan-3-ol (2) copper at 250-300°

(3) acid permanganate or chromic acid

(4) $CH_3CH_2\underset{\underset{OH}{|}}{C}HCH_2CH_2CH_3 \xrightarrow[250\text{-}300°]{Cu} CH_3CH_2COCH_2CH_2CH_3$

(5) $CH_3CH_2\underset{\underset{OH}{|}}{C}HCH_2CH_2CH_3 \xrightarrow[\text{or } H_2Cr_2O_7]{KMnO_4,\ H_3O^+} CH_3CH_2COCH_2CH_2CH_3$

Reaction occurs in two stages, with cleavage first of one of the carbon-carbon bonds, followed by cleavage of the second:

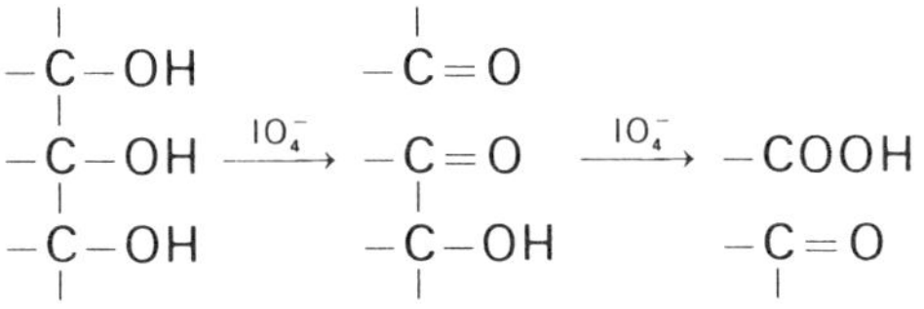

In the case of glycerol (propane-1,2,3-triol), the intermediate oxidation products will be . . . . (1) and the final products . . . . (2).

---

(1) $HOCH_2CHO + HCHO$ (2) $HCHO + HCOOH$

---

57. Periodate cleavage of 1,2-diols and related compounds is a quantitative reaction, and the cleavage reaction has been widely employed in the study of carbohydrates and polysaccharides which contain these functional groups. The simplest carbohydrate is glyceraldehyde $CH_2(OH)CH(OH)CHO$, and oxidation of this compound with two moles of periodate follows the equation

$$CH_2(OH)CH(OH)CHO + 2IO_4^- \longrightarrow \text{. . . . . (1)}$$

---

(1) $CH_2(OH)CH(OH)CHO + 2IO_4^- \longrightarrow HCHO + HCOOH + HCOOH + 2IO_3^-$

TEST FRAME

Complete the following equations:

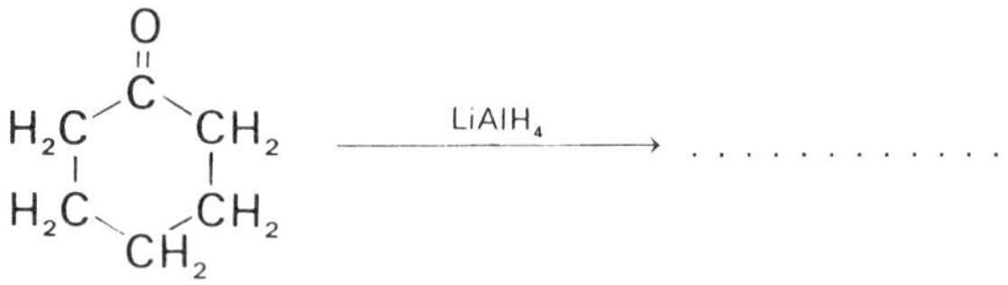

(1)

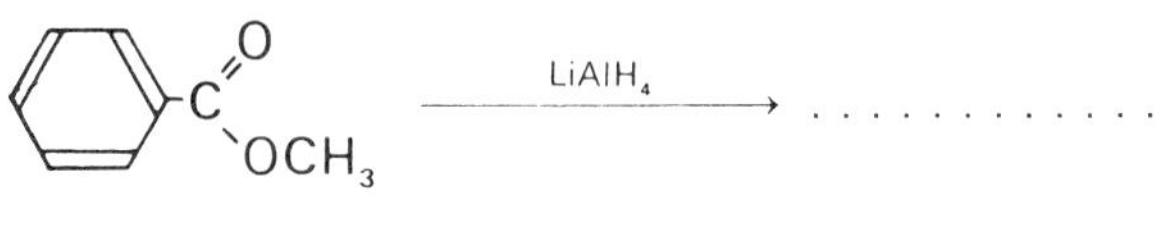

(2)

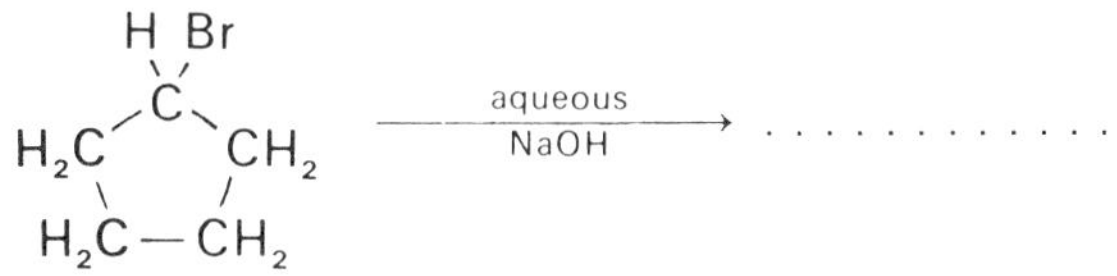

(3)

$(CH_3)_2C{=}CHCH_2CH_3 \xrightarrow[\text{hydration}]{\text{acid-catalysed}}$ . . . . (4)

---

9. The strength of lithium tetrahydridoaluminate as a reducing agent can be modified by first reacting it with 2-methylpropan-2-ol (t-butanol) which has the structure .... (1).

---

(1) $(CH_3)_3COH$

---

10. The selective reducing agent is prepared according to the equation:

$$LiAlH_4 + 3(CH_3)_3COH \longrightarrow LiAlH[OC(CH_3)_3]_3 + 3H_2.$$

The formula of the reagent can also be written $LiAlH(O\text{-}tBu)_3$. The general equation for the preparation of an aldehyde from an acid chloride is then

$$\ldots\ldots\ldots\ldots \xrightarrow{LiAlH[OC(CH_3)_3]_3} \ldots\ldots\ldots\ldots \quad (1)$$

---

(1) $RCOCl \xrightarrow{LiAlH[OC(CH_3)_3]_3} RCHO$

Ketones

11. The difference between the formula of a secondary alcohol RCH(OH)R′ and a ketone RCOR′ is .... (1). A ketone can be formed from a secondary alcohol if a molecule of .... (2) can be eliminated. Such a reaction is known as .... (3).

---

(1) 2H (2) hydrogen (3) dehydrogenation

---

12. Secondary alcohols can be dehydrogenated over hot copper according to the equation $\ldots\ldots\ldots\ldots \xrightarrow[250\text{-}300^\circ]{Cu} \ldots\ldots\ldots\ldots + \ldots\ldots\ldots\ldots$ (1).

---

(1) $RCH(OH)R' \xrightarrow[250\text{-}300^\circ]{Cu} RCOR' + H_2$

---

13. Aldehydes are readily oxidised because they have a .... (1) atom attached to the carbonyl group. In contrast, ketones are fairly stable to oxidation, especially in acid solution, and can therefore be prepared by oxidation of .... (2).

---

(1) hydrogen (2) secondary alcohols

(1) 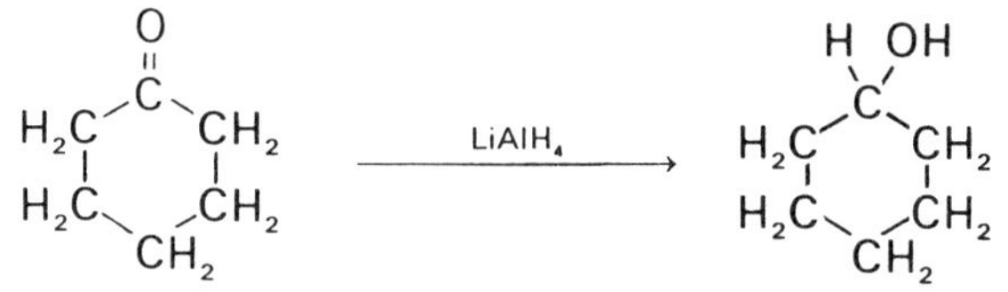

(2) 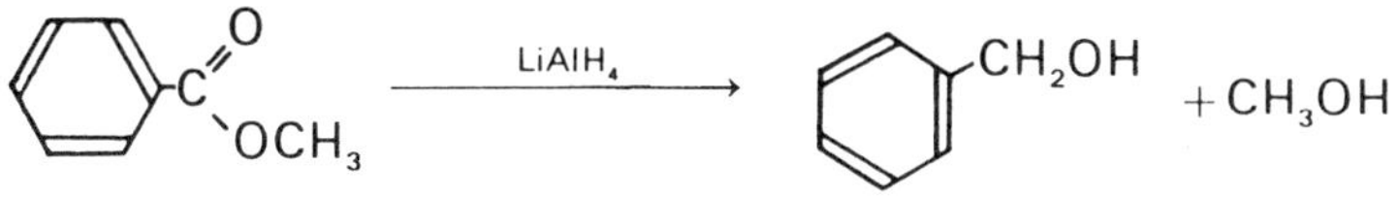

(3) 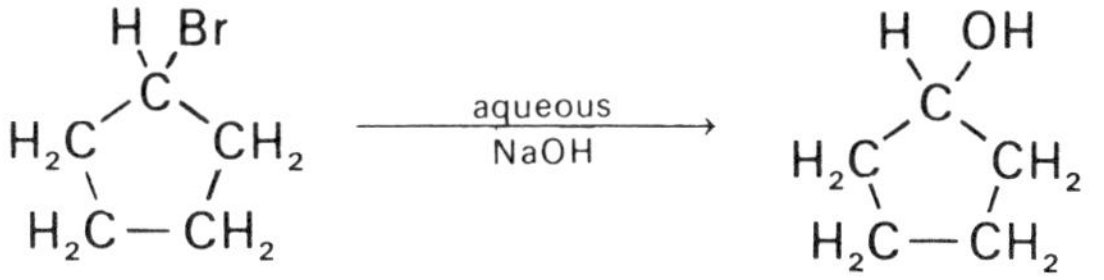 plus a small amount of cyclopentene

(4) $(CH_3)_2C{=}CHCH_2CH_3 \xrightarrow[\text{hydration}]{\text{acid-catalysed}} (CH_3)_2C(OH)CH_2CH_2CH_3$

be formed from a primary alcohol if a molecule of . . . . (2) can be removed.

---

(1) 2H (2) hydrogen

---

5. The elimination of hydrogen from a molecule is known as dehydrogenation. Primary alcohols can be dehydrogenated over copper at 250-300° to give . . . . (1). The equation for the reaction is . . . . (2).

---

(1) aldehydes (2) $RCH_2OH \xrightarrow[250\text{-}300°]{Cu} RCHO + H_2$

---

6. Aldehydes *cannot usually* be prepared directly from primary alcohols by using the common oxidising agents such as potassium permanganate or chromic acid since aldehydes are more easily oxidised than the starting alcohol. Oxidation of a primary alcohol therefore leads to the formation of a . . . . (1). The general equation of the reaction is . . . . (2).

---

(1) carboxylic acid (2) $RCH_2OH \xrightarrow{KMnO_4} RCOOH$

---

7. The substitution of a hydrogen atom by a chlorine atom can be considered as an oxidation e.g. $RH + Cl_2 \longrightarrow RCl + HCl$.
The reverse reaction, the substitution of a chlorine atom by a hydrogen atom is a . . . . (1). An aldehyde RCHO can be prepared from the corresponding acid chloride . . . . (2) by reacting it with a . . . . (3) agent.

---

(1) reduction (2) RCOCl (3) reducing

---

8. The reaction requires a selective reducing agent since reduction of the acid chloride must not proceed beyond the aldehyde stage. Thus lithium tetrahydridoaluminate, of formula . . . . (1), is a powerful reducing agent which readily reduces aldehydes to . . . . (2) and therefore cannot be used directly for the preparation of aldehydes from acid chlorides.

---

(1) $LiAlH_4$ (2) primary alcohols

INDEX

ALDEHYDES AND KETONES

Program 7

1. Aldehydes, RCHO, and ketones, RCOR′, (where R and R′ represent alkyl or aryl groups), both contain a carbonyl group, $\rangle C=O$, and give many reactions in common. Aldehydes differ importantly from ketones in that they have a .... (1) atom attached to the carbonyl carbon.

---

(1) hydrogen

---

2. The octet rule applies to both carbon and oxygen and the valency diagram of the carbonyl group can be written as .... (1).

---

(1) 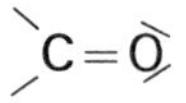 Each line represents an electron pair

---

3. The electrons forming the carbonyl bond are not distributed equally between the carbon and oxygen atoms, but are polarised or shifted towards the oxygen atom causing an electric dipole which is readily detected experimentally. The dipole can be indicated by the symbols $\delta+$ and $\delta-$ which represent small equal and opposite charges. The dipolar nature of the carbonyl group would therefore be indicated by the symbol .... (1) and the structures of acetaldehyde and acetone showing their dipolar nature are .... (2) and .... (3).

---

(1) 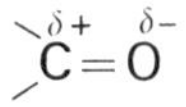

(2) 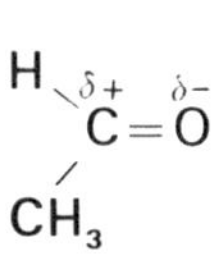

(3) 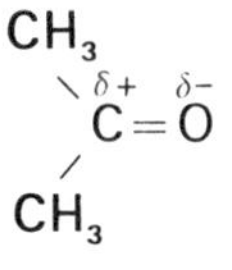

PREPARATION OF ALDEHYDES AND KETONES

Aldehydes

4. Aldehydes lie between primary alcohols and carboxylic acids in oxidation level.

$$RCH_2OH \underset{reduction}{\overset{oxidation}{\rightleftarrows}} RCHO \underset{reduction}{\overset{oxidation}{\rightleftarrows}} RCOOH$$

There are many methods for preparing aldehydes but we shall consider only two. The difference between the formula of a primary alcohol RCH_2OH and the corresponding aldehyde RCHO is .... (1). An aldehyde can

INDEX

Program 7: ALDEHYDES AND KETONES

CONTENTS

INDEX

CONTENTS

Programs 7 to 12 are on the right-hand pages when reading the book from this end and Programs 1 to 6 when reading from the other end.

ORGANIC CHEMISTRY

A FIRST UNIVERSITY COURSE
IN TWELVE PROGRAMS

by

F. W. EASTWOOD
M.Sc., D.Phil., A.R.A.C.I.
Reader in Chemistry, Monash University

J. M. SWAN
D.Sc., Ph.D., F.A.A., F.R.A.C.I.
Professor of Organic Chemistry, Monash University

JEAN B. YOUATT
M.Sc., Ph.D.
Senior Lecturer in Chemistry, Monash University

THIRD EDITION

Programs 7 to 12

CAMBRIDGE
AT THE UNIVERSITY PRESS
1972

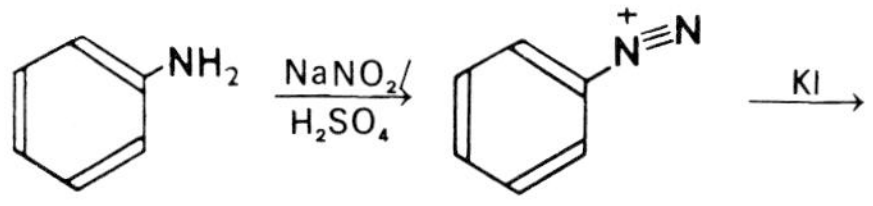	$RCHO + NaHSO_3 \longrightarrow \ldots$
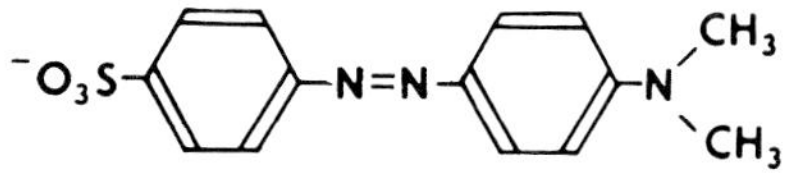 methyl orange method of synthesis	Synthesis of oximes 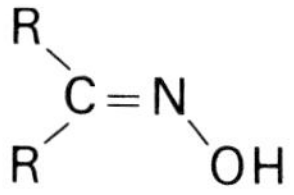
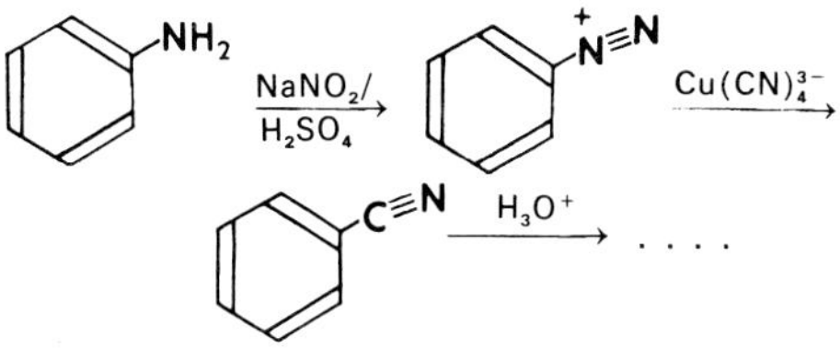	Synthesis of a primary alcohol RCH_2OH by reductive methods
Preparation of esters, $RCOOR'$, amides, $RCONHR'$ and methyl ketones, $RCOCH_3$, from acid chlorides	Synthesis of aldehyde semicarbazones $RCH{=}NNHCONH_2$
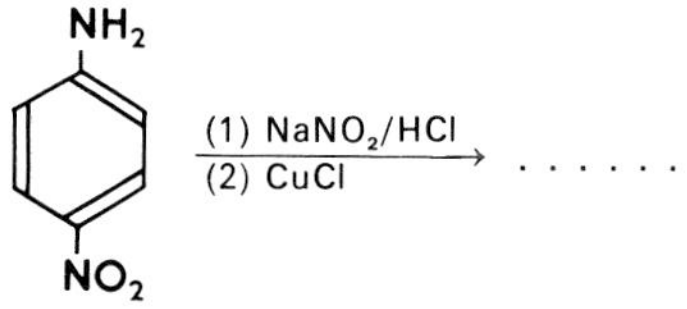	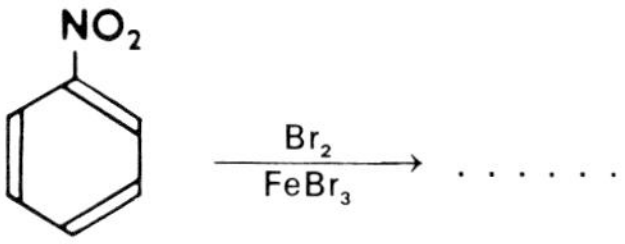
$CH_3\underset{\underset{O}{\|}}{C}Cl + CH_3COO^- \ Na^+ \longrightarrow$	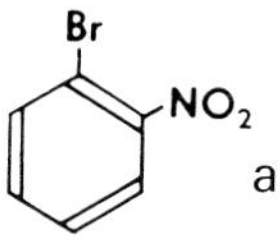 and  method of synthesis
$RCOCl \xrightarrow{?} RCHO$	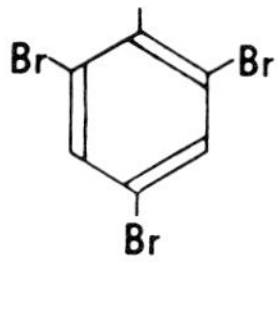 method of synthesis

Please cut cards along the lines. Do not tear out.

Synthesis of $RCH(OH)SO_3Na$

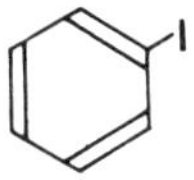

method of synthesis from aniline

$R_2C{=}O + H_2NOH \xrightarrow{\text{acid}}$ ....

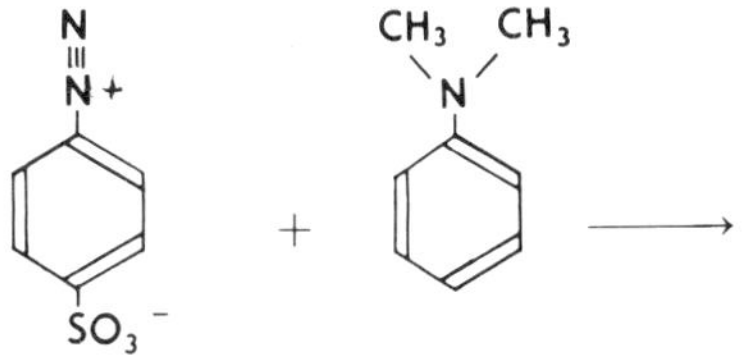

$RCHO \xrightarrow[\text{or } LiAlH_4]{H_2/Pt}$ ......

$RCOOC_2H_5 \xrightarrow{LiAlH_4}$ ....

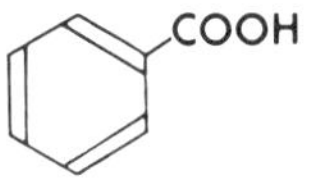

Synthesis in three steps from aniline

$RCHO + H_2NNHCONH_2 \xrightarrow{\text{acid}}$

semicarbazide

$RCOCl \xrightarrow{R'OH}$ ......

$RCOCl \xrightarrow{R'NH_2}$ ......

$RCOCl \xrightarrow{Cd(CH_3)_2}$ ......

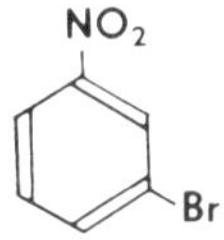

method of synthesis

method of synthesis from p-nitroaniline

$\xrightarrow[H_2SO_4]{HNO_3}$ ....

Synthesis of acetic anhydride CH_3COCCH_3 (with =O on each C of CO–C)

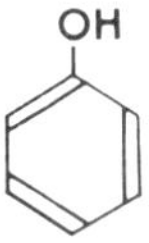

$\xrightarrow{Br_2/H_2O}$ ....

$RCOCl \xrightarrow{LiAlH(O\text{-}tBu)_3}$ ......

Synthesis of pent-1-yne $CH_3CH_2CH_2C{\equiv}CH$ from acetylene	Propan-2-one (acetone) and ethanal (acetaldehyde) are formed by ozonolysis of the alkene . . . .
Synthesis of methyl ketones, RCH_2COCH_3, from ethyl 3-oxobutanoate (acetoacetate)	Synthesis of $CH_3CH{-}CHCH_3$ (with O bridging the two CH carbons, epoxide) from but-2-ene
Synthesis of substituted acetic acids RCH_2COOH from diethyl malonate	Synthesis of $(CH_3)_2CHCH_2Br$ by an anti-Markovnikov addition
$R'COOCH_3 \xrightarrow[\text{(2) acid}]{\text{(1) RMgX}}$ . . . . $R_2CO \xrightarrow[\text{(2) acid}]{\text{(1) R'MgX}}$	Synthesis of $(CH_3)_2CBrCH_3$ by Markovnikov addition
$C_6H_5NH_2 \xrightarrow{NaNO_2/H_2SO_4} C_6H_5N_2^+$ $C_6H_5N_2^+ \xrightarrow{Na_2SO_3}$ (a) ? $C_6H_5N_2^+ \xrightarrow{H_2P(O)(OH)}$ (b) ?	Synthesis of benzyl chloride from toluene
$RCONH_2 \xrightarrow{NaOBr}$	$RX \xrightarrow{\text{NaOH in } H_2O}$ . . . .
$-CH_3$ (or alkyl), $-OCH_3$, $-OH$, $-NH_2$, $-$halogen atom. These groups are ... directing in electrophilic aromatic substitution	$CH_3I + NaCN \longrightarrow$ . . . . . . $CH_3CONH_2 \xrightarrow{P_2O_5}$ . . . . $CH_3CH{=}NOH \xrightarrow{(CH_3CO)_2O}$ . .

Please cut cards along the lines. Do not tear out.

$(CH_3)_2C{=}CHCH_3 \xrightarrow[(2)\ Zn/acetic\ acid]{(1)\ O_3}$ 2-methylbut-2-ene	$HC{\equiv}CH \xrightarrow[(2)\ CH_3CH_2CH_2I]{(1)\ NaNH_2/NH_3}$
Reaction of but-2-ene with peracetic acid, $CH_3\overset{\displaystyle O}{\overset{\|}{C}}OOH$	$CH_3COCH_2COOC_2H_5 \xrightarrow{(1)\ NaOC_2H_5;\ (2)\ RX;\ (3)\ cold\ NaOH;\ (4)\ acid}$
Reaction of $(CH_3)_2C{=}CH_2$ with hydrogen bromide in the *presence* of light or peroxides	$CH_2(COOC_2H_5)_2 \xrightarrow{(1)\ NaOC_2H_5;\ (2)\ RX;\ (3)\ NaOH;\ (4)\ heat\ with\ acid}$
Reaction of $(CH_3)_2C{=}CH_2$ with hydrogen bromide in the *absence* of light or peroxides	Grignard synthesis of the tertiary alcohol $R_2R'COH$ (two different methods)
Reaction of methylbenzene (toluene) with chlorine in the presence of ultraviolet light gives	Synthesis of (a) phenyl-hydrazine and (b) benzene from aniline
$RX \xrightarrow{?} ROH$	$RCONH_2 \xrightarrow{?} RNH_2$
Synthesis of CH_3CN by three different methods	List five or more groups which are *ortho-para* directing when attached to a benzene ring

List five groups which are *meta*- directing when attached to a benzene ring	$R'CHO \xrightarrow[(2)\ acid]{(1)\ RMgX}$ . . . . $R'COR'' \xrightarrow[(2)\ acid]{(1)\ RMgX}$ . . . .
Synthesis of ArR or ArCOR (Friedel-Crafts method)	Synthesis of $(CH_3)_2C(OH)CH_2COCH_3$
Synthesis of RCH_2NH_2 by reductive methods	$ArCH_3 \xrightarrow{KMnO_4}$; $ArCCl_3 \xrightarrow{NaOH}$ $ArMgCl \xrightarrow[(2)\ acid]{(1)\ CO_2}$; $ArCN \xrightarrow{acid}$
$RCOCH_3 \xrightarrow[\text{sodium hydroxide solution}]{\text{Iodine and}}$	Preparation of an acid chloride RCOCl
$C_6H_5CHO + HCN \longrightarrow$	$CH_3CHO + 2C_2H_5OH \xrightarrow{acid}$ . . . .
Two molecules of ethanal react together in the presence of dilute sodium hydroxide (aldol addition)	Preparation of cyclohexene (two methods)
Oxidative cleavage of the glycol RCH(OH)CH(OH)R′ with periodate ion, IO_4^-	Preparation of 2-bromo-pentane from the corresponding alcohol

Please cut cards along the lines. Do not tear out.

Synthesis of alcohols by the Grignard method	$-NO_2$, $-COOH$, $-COOR$, $-COR$, $-CHO$. These groups are . . . . directing in electrophilic aromatic substitution
Hot acetone is passed through barium hydroxide $2CH_3COCH_3 \xrightarrow{Ba(OH)_2}$ . . . . (aldol addition)	$ArH \xrightarrow[AlCl_3]{RCl}$ . . . . $ArH \xrightarrow[AlCl_3]{RCOCl}$ . . . .
Synthesis of aromatic carboxylic acids ArCOOH (four methods)	$RCN \xrightarrow{LiAlH_4}$ . . . . $RCONH_2 \xrightarrow{LiAlH_4}$ . . . . . . $RCH{=}NOH \xrightarrow{LiAlH_4}$ . . . .
$RCOOH \xrightarrow[\text{or } SOCl_2]{PCl_5}$	Iodoform reaction, giving $RCOO^- Na^+ + CHI_3$
Synthesis of 1,1-diethoxy-ethane (acetal) $CH_3CH(OC_2H_5)_2$	Synthesis of $C_6H_5CH(OH)CN$
(1) Cyclohexyl chloride heated with KOH in ethanol $\longrightarrow$? (2) Cyclohexanol heated with H_2SO_4 or H_3PO_4 $\longrightarrow$?	Synthesis of $CH_3CH(OH)CH_2CHO$
$CH_3CH_2CH_2CH(OH)CH_3 \xrightarrow[PBr_3]{HBr \text{ or}}$	Degradation of the glycol RCH(OH)CH(OH)R′ to give RCHO AND R′CHO

"FLIP CARDS" FOR REVISION →

Separate cards by cutting along the lines.

Students are advised to prepare additional "flip cards".

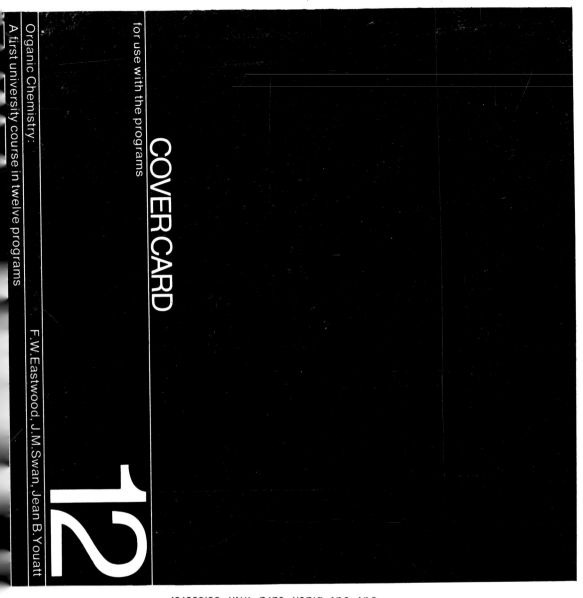

COVER CARD

for use with the programs

12

Organic Chemistry:
A first university course in twelve programs

F.W.Eastwood, J.M.Swan, Jean B.Youatt

Cut out black card with scissors.